Wounded Planet

Wounded Planet

How Declining Biodiversity Endangers Health
and How Bioethics Can Help

Henk A.M.J. ten Have, MD, PhD

Johns Hopkins University Press
Baltimore

Johns Hopkins University Press
2715 North Charles Street
Baltimore, Maryland 21218-4363
www.press.jhu.edu

Library of Congress Cataloging-in-Publication Data

Names: ten Have, H., author.
Title: Wounded planet : how declining biodiversity endangers health
    and how bioethics can help / Henk A.M.J. ten Have.
Description: Baltimore : Johns Hopkins University Press, 2019. | Includes
    bibliographical references and index.
Identifiers: LCCN 2018029956 | ISBN 9781421427454 (hardcover : alk. paper) |
    ISBN 1421427451 (hardcover : alk. paper) | ISBN 9781421427461
    (electronic) | ISBN 142142746X (electronic)
Subjects: | MESH: Bioethical Issues | Biodiversity | Environmental Health—
    ethics | Global Health—ethics
Classification: LCC R724 | NLM WB 60 | DDC 174.2—dc23
LC record available at https://lccn.loc.gov/2018029956

A catalog record for this book is available from the British Library.

Special discounts are available for bulk purchases of this book.
For more information, please contact Special Sales at 410-516-6936 or
specialsales@press.jhu.edu.

For Carien and Krijn, custodians of the future

Man is but a reed, the most feeble thing in nature; but he is a thinking reed. The entire universe need not arm itself to crush him. A vapor, a drop of water suffices to kill him. But, if the universe were to crush him, man would still be more noble than that which killed him, because he knows that he dies and the advantage which the universe has over him; the universe knows nothing of this. All our dignity consists, then, in thought. By it we must elevate ourselves, and not by space and time which we cannot fill. Let us endeavor, then, to think well; this is the principle of morality.

—Blaise Pascal, *Pensées* (1670)

Contents

Preface

The planet has become a patient. This is one of Wolfgang Sachs's central conclusions in his studies on development and globalization. Since their republication in 2015, many have endorsed this observation. Healthcare experts and policymakers are now very much aware that the environment is one of the crucial determinants of health, not only for individual persons but also for the human population of the planet. Without a healthy planet, human beings cannot be healthy. For our most essential needs—clean air, safe water, adequate nutrition, development of medicinal drugs, and protection from infectious diseases—we all depend on biodiversity.

This book elaborates the thesis that human health depends on healthy biodiversity, that current loss of biodiversity is a serious threat to planetary health, and that this predicament requires a different bioethical discourse. Growing awareness that health is dependent on sound environmental conditions has implications for the ethical debate on health and healthcare. A new and broad concept of bioethics is needed that pays serious attention to declining biodiversity. The emerging field of global bioethics provides ethical principles that transcend the limited vision of mainstream bioethics and its focus on individual autonomy. It introduces a normative horizon in which other perspectives become visible, such as justice, equality, vulnerability, human rights, and solidarity.

Wounded Planet grows out of two of my previous books. In *Global Bioethics: An Introduction* (2016), I argue that the concerns of bioethics should not be trained solely on the individual patient but rather also on the social and environmental contexts of health and disease. The work explores the theory and practice of global bioethics. It points out the need to move from an ethics oriented toward the individual person and medical challenges to a social ethics focused on global problems. The second book, *Vulnerability: Challenging Bioethics* (2016), elaborates the social dimension of global bioethics. It argues that the language of vulnerability offers perspectives beyond the traditional autonomy model and can help bioethics to evolve into a global enterprise. The current volume may be regarded as the final component of a trilogy, exploring threats to the environment and arguing that such concerns can no longer be separated from healthcare challenges—and that both of these areas should thus be included in the domain of global bioethics. Together, the three books argue in favor of a conception and practice of bioethics that encompasses medical, social, and environmental dimensions of present-day life and that can address the challenges of the future.

Wounded Planet

1 | Global Bioethics and the Environment

B ioethics developed rapidly as a new discipline in the 1970s. Van Rensselaer Potter (1911–2001), who coined the term in 1971, was surprised four years later by how widely it had been applied to new activities in research, education, and policy. Specific institutes, books, journals, and conferences emerged that all dealt with "bioethical problems." At the same time, Potter was disappointed about the narrow focus of the new bioethics. It concentrated exclusively on the challenges of medical science and technology, such as the lifesaving interventions of dialysis and organ transplantation. It was also primarily concerned with the benefits and harms for individual patients deciding on possible medical treatments. Potter believed that bioethics should have a wider scope. It should address the social and environmental context of challenges. As a cancer researcher, he realized that medical and biological knowledge, though necessary, would be of limited use as long as the social contexts of smoking and the environmental impact of carcinogenic substances were not addressed. An example is vinyl chloride, an industrial petrochemical used to make plastic; exposure to it in air and water is associated with liver cancer. The environment can therefore be an important source of damage to human health and contribute to significant diseases. However, it seems that from the start, bioethics did not take into account these environmental concerns. At the same time, environmental ethics as it developed during the 1970s as a special branch of applied ethics did not connect with bioethics. While Potter saw the notion of bioethics as an umbrella term to cover medical, environmental, and social ethical concerns in one theoretical and practical approach, the subsequent development of the new discipline separated these matters. This chapter examines why this separation occurred. Why was the broad notion of bioethics as initially envisioned by Potter narrowed into the biomedical ethics that would dominate bioethical discourse during five decades?

Examining this question is important because the current context has changed. Nowadays, bioethical analysis and debate can no longer ignore environmental issues. Climate change, toxic waste, air pollution, ozone depletion, extreme weather events, and loss of biodiversity have had significant impacts on health and health-care. Deforestation and destruction of habitat are associated with the emergence of new viral diseases such as Ebola or Zika. Focusing on care, treatment, or vaccination of individual patients can thus not be disconnected from the wider environmental context in the management of epidemics. The main reason that mainstream bioethics cannot disregard the environment today is the global nature of bioethical problems. The ethical challenges are no longer confined to specific countries or healthcare systems; they can be addressed only at a global level, necessitating international cooperation and coordination. In other words, globalization compels bioethics to assume a broader scope and to develop into a global bioethics. Such scope was exactly Potter's intent when he introduced the term *bioethics*. The first part, 'bio,' indicates that all living beings should be taken into account, not only human beings, and regardless of where they are located. The interconnections between health and environment are now the subject of many studies and policies. In bioethics itself there is growing recognition that ethical concerns should extend to the environment. A milestone is the adoption by member states of the United Nations Educational, Scientific and Cultural Organization (UNESCO) in 2005 of the Universal Declaration on Bioethics and Human Rights, confirming that protecting future generations and protection of the environment, the biosphere, and biodiversity are principles of bioethics. The aim of this book is to explore the current rapprochement between biomedical and environmental ethics and to demonstrate how a broader concept of bioethics as global bioethics can be applied in practical settings across the world, combining medical, social, and environmental concerns in a coherent ethical framework of analysis and action.

Bioethics

In a nutshell, Potter contended that a new discipline is needed to address the basic problems of humanity. This discipline should combine two types of knowledge: biological knowledge, or the science of living systems (hence 'bio'), and knowledge of human value systems (hence 'ethics'). Integrated knowledge is the only way to tackle the major problems of overpopulation, war and violence, pollution, poverty, shortsighted politics, and even the notion of progress itself. These issues should have priority because they jeopardize the survival of humankind. [1]

The problems identified are not typical medical challenges. Rather, they refer to the social and environmental context in which they arise as global concerns. Of course, there are connections with health and disease. Pollution and poverty, for example, are linked with deterioration of health. Reduction of disease and mortality will lead to an increase in population. War has severe medical consequences. But, though medically relevant, these major problems cannot be properly addressed by medicine or healthcare alone but rather demand a broad ethical approach. Potter did not systematically elaborate such an approach. He argued that a new discipline was necessary and suggested a name for it without presenting a comprehensive theoretical and practical program. Nonetheless, he articulated three components that are crucial for elaborating a comprehensive ethical system, all of which assume the central role of the environment in ethical considerations.

INTERCONNECTEDNESS

The first component is interconnectedness and interdependency. Potter recognized that human beings cannot be separated from the surrounding world. They are embedded not only in society and culture but also in the natural world of animals and plants. This means that the starting point for ethics is different. It is not, as in biomedical ethics, the individual patient or professional who has to decide how to cope with the increasing range of scientific and technological possibilities offered by contemporary medicine. While respect for individual autonomy is an important moral principle, it should be recognized that persons are always embedded in a broader context. Individuals flourish when conditions exist that promote and sustain autonomous life. What is true for individual persons is equally true for humankind as a whole. Human beings as a species can survive only when the natural world in which they live is preserved. Health for the human species is "part of health for the biosphere." [2] Healthcare and earth care cannot be separated. Finally, interconnectedness is manifested in global threats such as nuclear warfare and ecological disasters. Survival within a country is impossible without addressing the global context. In other words, bioethics should go beyond the level of individual ethics.

THE FUTURE

The second component is concern for the future. Ethics should assume a long-term perspective, since what is ultimately at stake is not individual survival but the survival of humanity. This survival cannot be guaranteed. The population explosion, ecological catastrophes, widespread pollution, weapons of mass destruction, and diminishing natural resources endanger the ecosystems upon which humanity is dependent. Potter argued that instead of searching for short-term solutions, we must explore various scenarios, in particular how decisions today will impact the future world that we leave to our children and subsequent generations. This explains why he regarded politics as a priority topic. In political systems with regularly changing governments, politicians usually neglect long-range goals, since these efforts will not generate votes in the next election. In his emphasis on the future, Potter was influenced by the work of the famous anthropologist Margaret Mead. [3] Because the future will not take care of itself, how it will develop depends on our decisions and actions. That requires, as Mead reiterated, that we have sufficient knowledge about what is possible and what is probable. Only then can we make a normative assessment about what is preferable. Mead argued that the future should become a subject of scientific study; universities should establish "Chairs of the Future." Potter proposed setting up a Council on the Future, an independent and interdisciplinary institution "above politics" whose mission would be to predict the consequences of the application of new knowledge. [4] It should also be a democratic forum, initiating public debate and engaging all citizens. Potter's proposal intended to create a platform to address the interests of future generations. (Countries such as France, Finland, and Hungary would, following Potter, establish institutions to represent future generations.)

EVOLUTION

As a biological researcher, Potter was intrigued by the evolutionary perspective, particularly issues of survival and extinction. In his view, natural selection does not guarantee long-term survival. Its aim is more limited: survival through adaptation to the present environment. It does not take into account that the environment itself is continuously changing so that adaptation now does not rule out vulnerability later. Natural selection is "incapable of seeing into the future." [5] This flaw in the process of evolution is not often noted, since it is generally assumed that evolution means progress. Potter found the concept of progress to be problematic. In nature, progress is not intrinsic and inevitable. Natural selection can lead to extinction of species. Human beings cannot be sure that they will continue to exist. But as a species they are unique; because they are conscious of the process of evolution, they can take steps to influence it. They do not merely adapt to their environment as do other species but anticipate that the environment itself is continuously changing, and they can influence and create environments that future generations will face. The unique feature of human beings is adaptability rather than adaptation. From Potter's perspective, adaptability requires scientific know-how of evolutionary processes, but more important, a framework of values to determine in what direction these processes need to go, what future goals they must accomplish. On the basis of knowledge of nature and evolution, human beings should attempt to influence and modify society and culture.

This refers to the third component of bioethics, that is, the analogy between natural and cultural evolution. As the basic units of cultural evolution, ideas have the same role as genes in biological evolution; their diversity generates variety and a range of future possibilities. Human beings can and should use biological science to accomplish cultural progress. Knowledge of the life sciences should guide the development and application of ideas aimed at the survival of humanity. In this respect, Potter was influenced by the French geologist and philosopher Pierre Teilhard de Chardin. They both shared the concern that human beings are the crucial actors in evolutionary processes. Because of their reflective capacity, humans can give direction to their own development—becoming, so to speak, "collaborators" in the process of evolution. The perspective of evolution therefore is not deterministic but implies human freedom and responsibility; it is focused on future possibilities.

Environment

The three building blocks of Potter's new bioethics (the interconnectedness of humans and the surrounding world, the concern for the future, and the analogy of natural and cultural evolution) can help to broaden the usual ethical approach to science, technology, and healthcare. The common thread of all these components is a focus on the environment.

During the 1960s, when Potter wrote the articles that were later compiled in his book *Bioethics*, public and political concerns about the environment increased rapidly. Rachel Carson published *Silent Spring* in 1962, illustrating the harmful impact of pesticides. The wreck of the oil tanker *Torrey Canyon* in 1967 caused an environmental disaster, polluting the southwest coast of the United Kingdom. Paul Ehrlich's best seller *The Population Bomb* (1968) warned that, owing to over-population, millions of people in the 1970s would starve to death. In 1969, the Cuyahoga River in Ohio, then one of the most polluted rivers in the United States, caught fire. Air pollution created severe smog in cities like Los Angeles and London. The first report of the Club of Rome in 1972 showed to the world the rapid depletion of natural resources. These events and publications made the general public and policymakers aware of the significance of the environment. Legislative action was taken with the first modernized environmental laws; for example, the United States instituted the Clean Air Act (1963), the Water Quality Act (1965), and the Endangered Species Preservation Act (1966). There was also an upsurge of public campaigns. Environmental activists have organized a nation-wide Earth Day since 1970. In the same year, the US government established the Environmental Protection Agency. One year later, the United Nations proclaimed the annual celebration of an international Earth Day. Greenpeace started its protest campaigns, also in 1971.

In this context of environmental concerns, Potter's writings give the environment a central place in bioethics. His journal articles discuss the limits of the natural environment, the connectedness of humans and world, ecology, and the question of what is an optimum environment. The publication of the book *Bioethics*, however, provided an opportunity to convey his central message more clearly: the prefix 'bio' in *bioethics* refers not simply to human life but to all life. This message is communicated in two distinct ways.

First, the book cover displays the famous photo taken from the *Apollo 11* spacecraft in July 1969. The near full earth is visible, showing Africa, the Middle East, Europe, and western Asia. [6] Using this photograph to illustrate a book on bioethics affirms a different vision from the one that underlies the usual books on medical and healthcare ethics. Historically, the photos of planet Earth taken from outer space played an important role in promoting environmental activism. They produced a global consciousness. Showing the globe in the middle of the vast darkness of space demonstrates the vulnerability of the earth. The home of humanity ("spaceship Earth") in fact is rather small, isolated, and finite—and thus fragile—within an infinite universe. This sense of vulnerability was further strengthened by an even more spectacular series of pictures taken by *Apollo 17*, the so-called blue marbles. They became the iconic symbols for environmental concern. [7]

Second, Potter locates his bioethical discourse within the environmental movement and the new discipline of ecology. Both emerged as a new political consciousness and normative orientation in the second part of the nineteenth century. [8] Potter associates the notion of bioethics with the "Wisconsin tradition" of environmentalists such as John Muir and Aldo Leopold. [9] He dedicates *Bioethics* to the memory of Leopold, who showed that human ethics cannot be separated from ecology. Nevertheless, this memorial seems to be a last-minute addition. The book itself makes no references to Leopold, and in his later monograph *Global Bioethics*, Potter admits that he discovered Leopold at a late stage. [10] Between 1940 and 1948, when they both worked at the University of Wisconsin, they never met; Potter was not even aware of Leopold's existence. [11]

Furthermore, in *Bioethics*, Potter draws on the principles of ecology, arguing that "we should look upon earth, man, plants and animals, seas, and atmosphere as a balanced ecological system." [12] Ecology is the scientific study of the interrelations between plants, animals, human beings, and their surrounding environment. The term was first used by German biologist Ernst Haeckel in 1866. It is a value-laden concept, as it expresses the vision of a balance between humankind and the natural world. Potter often refers to this idea of balance. [13] He argues that we should be careful not to disturb the "fragile web of nonhuman life that sustains human society." [14] The main risk, in his opinion, is that owing to a growing human population, the balance of the ecosystem will be destroyed. If the ultimate goal of bioethics is the survival of humanity, fundamental ethical considerations should address our obligation to future generations. But his focus on survival was not only the result of increasing interest in ecology. It was also promoted by a fusion of biological and economic perspectives in the 1970s. New awareness of scarcity and the need to conserve nonrenewable resources advanced an economic discourse, especially in regard to energy, that shared similar interests and values with the ecological discourse. Additionally, it assumed that nature is characterized by stability and balance. Scarcity of resources could endanger the future of humankind. From this point of view, it is argued that long-term effects rather than short-term profit motives should drive decision making. This fusion of ecology and economy explains the emphasis on conservation. The stability of the ecological system requires sustainable management of scarce resources.

Global Bioethics

In the 1980s, especially after his retirement in 1982, Potter adopted a different approach. He had to acknowledge that the development of bioethics was exclusively concentrated on medical and individual concerns, narrowing his broader conception to the perspective of biomedical ethics. In the rapidly growing bioethics literature, Leopold was not referenced, [15] nor were any links to environmental issues established. Furthermore, environmental ethics, as it developed later in the 1970s, had taken its own, separate path without connections to bioethics.

In a landmark publication in 1987, Potter distinguishes more clearly than before two kinds of bioethics: medical bioethics and ecological bioethics. [16] The first is focused on the moral problems facing individual physicians and patients; it concentrates on personal autonomy and individual rights; and it takes a short-term view related to treatment options. The second is concerned with the survival of humanity; it concentrates on responsibilities and future generations; and it takes a long-term view associated with the preservation of the ecosystem. Both are major areas of contemporary bioethics. The problem, however, was that the term *bioethics* had become restricted to the first area only. To Potter's embarrassment, no attention was paid to ecological issues. In order to save his initial conception of bioethics as a broad approach, he developed two strategies. The first was to elaborate the background credentials of the new term. This strategy is fully deployed in Potter's second monograph, *Global Bioethics*, published in 1988. Emphasizing in the subtitle of the book that bioethics builds on Leopold's heritage, Potter devotes several chapters to Leopold's work. He argues that it has been his interest in ecology that has made him coin the term *bioethics*. As a colleague at the University of Wisconsin, Potter saw himself continuing the line of Leopold's thought, extending the scope of ethics beyond the ethics of individuals. Leopold should therefore be called, according to Potter, the "first bioethicist," since he proposed a new and broad ethical approach, contrary to what is promoted as 'bioethics' in contemporary discourse, which is nothing more than resuscitated medical ethics focused on individual interactions. [17]

In this second book, Potter not only criticizes the limited perspectives of medical bioethics but first of all elaborates the environmental dimensions of bioethics. His basic concern is the risk of extinction—the possibility that humankind will not survive in the future. [18] Human beings can only escape the fate of many other species if they mobilize the ethical resources in their traditions and societies, and combine them with scientific knowledge. Preservation of the biosphere or a healthy ecosystem is the guarantee for human survival. Future disaster can hardly be avoided unless we change our ethical perspective. Leopold's famous statement about right and wrong ("A thing is right when it tends to preserve the integrity, stability, and beauty of the biotic community. It is wrong when it tends otherwise.") expresses the same ethical imperative. [19] These concerns are the basis for Potter's second strategy. Throughout, *Global Bioethics* presumes that medical and environmental bioethics are complementary (especially chapters 3–7). On the one hand, ecological dilemmas concerning deficient water supplies, toxic waste, pollution, and acid rain seriously affect human health. On the other hand, medical ethics dilemmas such as teenage pregnancy, abortion, and contraception cannot be examined in isolation from social factors like poverty, lack of education, and environmental conditions related to population growth, fertility control, and carrying capacity of the planet. The conclusion is that ethical problems require the coupling of medical and ecological bioethics.

If the need for environmental bioethics and its complementary relationship with medical bioethics is better understood, it is evident that both areas of bioethics ultimately share the same goal of human survival. They should therefore not be separated but rather harmonized in one comprehensive approach. This is what Potter in his 1988 work calls *global bioethics*. The broad notion of bioethics encompassing medical, social, and environmental concerns, as initially envisioned, can thus be supported and revitalized under a new name. Bioethics is global in two senses: it is unified, bringing together the various concerns mentioned, and it is worldwide, taking into account not just local and regional but also planetary problems and solutions. [20]

Having launched the new concept of global bioethics, Potter's activities in the next decade concentrated on its promotion and elaboration. He repeated the message that the current crisis is not simply the problem of extinction of species but the possibility of *human* extinction. Survival of humankind will require fundamental cultural change. This demands a new approach to ethics. The goal of global bioethics is "developing a morality that places long-range goals of human survival ahead of short-term economic gains." [21] He argues that scientists have a special responsibility in this endeavor. [22] Public health professionals also can play a role in bridging the gap between medical and environmental ethics, because they are not merely concerned with patient autonomy and individual ethics. [23] In his quest to identify how and where the idea of global bioethics might be disseminated, Potter engaged with religious discourse, especially the efforts of Catholic theologian Hans Küng to develop a universal world ethic. [24] It inspired him to argue that global bioethics provides "an ecumenical common ground" in healthcare and earth care. [25] In his very last publication, appearing a few months after his death, Potter summarizes the mission of this expanded concept of global bioethics: "Global survival in the long term will not be possible if the world population is not brought under control and possibly reduced—will not be possible if the environment is not protected—will not be possible if human health is not improved—will not be possible if biodiversity is not protected—will not be possible without a transformation of society—will not be possible without a sense of the meaning of community." [26]

Bioethics as Biomedical Ethics

At the end of his life, Potter acknowledged that his books were unknown to most bioethicists. [27] Reflecting on the development of the new discipline, bioethicists confessed that they had never heard of Potter or his work. [28] Not only were his publications not used or known, but his contribution was not even recognized. An international congress in 2010 on the founders of bioethics included scholars such as Daniel Callahan, Edmund Pellegrino, and Robert Veatch, but none of them mentioned Potter. [29] In well-known bioethics textbooks, his name is absent. [30] Surprisingly, even histories of bioethics contain scarce or no references to Potter. [31] Albert Jonsen in his overview of the development of bioethics has a few lines about Potter but only within the context of Warren Reich's thesis of bi-located origins. [32] This thesis contests Potter's claim to priority, suggesting that the term *bioethics* was in use among people who established the Kennedy Institute as the first academic institution in the field, before Potter's publications. However, there is no written evidence, so these claims, based on hearsay, seem to justify in only retrospect the use of *bioethics* for a more narrow medical approach. Potter's book was published in January 1971, while it was not until June 1971 that the term *bioethics* was included in the institute's original name: the Joseph and Rose Kennedy Institute for the Study of Human Reproduction and Bioethics. In the meantime, *Time* magazine had used the word in the spring of 1971 in reference to Potter's book, so the new term might have become known to the general public. Strangely enough, and uncritically, the bi-location thesis has been accepted as official history. [33]

This particular construction of history does more than simply neglect Potter and his work or weaken his priority claim of inventing the term. In fact, it legitimizes the common use of the term for a narrower approach than envisioned by its originator. In a few years' time, the neologism became widely used but without any links to its conceptual inspiration and motivation. Therefore, the historical issue is not so much recognition or marginalization of Potter but justification for a specific ethical approach under the name of 'bioethics.' [34] This is obvious in two rhetorical strategies. The first is the argument that bioethics is "the same old ethics" but applied to new dilemmas. Under the heading "Bioethics" in the first edition of the *Encyclopedia of Bioethics*, Danner Clouser argues that bioethics is not a new set of principles; as applied ethics it is just the application of general moral rules to new issues in a specialized area of concern that he labels the "bio-realm." In other words, bioethics is not a new ethics. It does not have a specific "essence" that demarcates it as specific or innovative; it is just a "de facto list of issues" without a common thread. [35] So there is no need for a new approach as advocated by Potter. The new challenges do not require a discontinuity with traditional ethical expertise.

The second strategy argues that the new issues that elicited the development of bioethics are primarily biomedical challenges. They present themselves immediately in the daily lives of individuals. Patients and health professionals have to make decisions concerning technological interventions such as resuscitation or transplantation, and they expect moral guidance from bioethics. Social worries concerning civil rights, overpopulation, or environmental degradation are in Clouser's opinion merely a "contributing factor" to the growth of bioethics. [36] This argument is supported by Reich: the focus on concrete medical problems is understandable, because it attracts greater interest among physicians, the public, and policymakers. [37]

The chief aim of both strategies was apparently reassurance. Confronted with challenges such as lifesaving interventions, genetic engineering, or in vitro fertilization, the general public, the medical profession, and policymakers could rest assured that they were facing nothing new. There was no need for a new discipline, a new ethics, or different and unusual policies. But this reassurance obscured the fact these issues were indeed something new. First of all, bioethics is not only characterized by new issues; it also proceeds on the basis of a new methodology, particularly the set of principles developed at Georgetown University and articulated in the *Belmont Report* (written by the US National Commission for the Protection of Human Subjects of Biomedical and Behavioral Research), suggesting that there is consensus on at least some ethical starting points for addressing dilemmas.

Second, principlism emphasizes that bioethics is the professional domain of philosophers. While traditional medical ethics was the business of healthcare professionals, a new category of experts are now the significant drivers of the bioethical debate, namely, experts in general normative theory. A classic article by philosopher Stephen Toulmin explains why: medicine saved the life of ethics. The new issues in bioethics provided an opportunity for moral philosophy to overcome its analytic abstractness and meta-ethical preoccupations that had made it irrelevant until the 1960s, at least in the Anglo-American world. [38] (This was exactly Potter's complaint: bioethics had been hijacked by "professors of philosophy" or "firmly established ethicists in departments of philosophy" who were using the new term to re-brand their traditional ethics.) [39] As a consequence, it is not necessary to reexamine moral traditions themselves. Also, the need for inter-disciplinary collaboration is limited; while it is necessary for philosophers to co-operate with scientists, a clear demarcation between science and ethics should be maintained.

Third, the focus on medical priorities reflects a specific orientation. While Potter considered bioethics to be the search for wisdom, the new bioethicists aim at solving problems and providing solutions. Such practical orientation does not consider the broader context (of human survival or global injustice) but concentrates on individual rights and duties to facilitate treatment decisions. Furthermore, it ex-presses a specific conception of ethics, not as a way of living and thinking, not as a form of life that respects life in general, but as a pragmatic exercise focused on analysis, argumentation, and justification. Potter readily admitted that he was not a professional ethicist. In a video message to the World Congress on Bioethics in Gi-jón, Spain (2000), he called himself "an amateur in biological philosophy." [40] Early on, his work was swiftly dismissed by Clouser, who claimed that what Potter advocated is applied science and that it should not be called ethics. [41]

These different orientations explain why the initially broad notion of bioethics came to designate a narrower medical-ethical approach. Besides practical reasons (as a retired professor, Potter did not have the resources and networks available to the Kennedy Institute), the limited impact of his expansive viewpoint might also be due to Potter's eclectic approach. His 1971 book is a compilation of articles that does not develop a coherent theory of the new discipline. The ethical component is underdeveloped. He does not present a deep analysis of problems from a norma-tive point of view or elaborate ethical principles such as future-related responsibil-ity. Ethics in his view is intuitive, suggesting that argumentation and justification do not have priority. [42] Potter also seems to argue that the naturalistic fallacy in fact should be the naturalistic principle. We first should determine what is before we can determine what we ought to do. [43] In regarding bioethics as a kind of hyperscience (the science of survival), Potter builds the new discipline on the science of biology (rather than philosophy) and argues that it requires experience and testing instead of introspection and speculation. Doing so, the specific con-tribution of ethics as normative enterprise is lost. When science is the source of value judgments, bioethics is "biologized"; the risk is that it will only reproduce values already implicit in biological science itself. This is exactly the point made by those who want to demarcate ethics from science. Our biological nature is relevant for the choices we can make, but it does not determine what choices we ought to make. For example, ecological and evolutionary sciences provide data about the risk of extinction of the human species. But whether extinction is desirable is another matter; why should we prevent extinction? Why is human survival a good? [44] These are not scientific, but ethical, questions.

Environmental Ethics

The reduction of Potter's broad notion of bioethics to biomedical ethics is only one side of the story. The other side is that environmental ethics did not connect with the broad notion of bioethics either. It developed as a new discipline in the 1970s, although somewhat later than biomedical ethics—with its first scholarly conference in 1971, its first systematic book in 1974, and its first specialized journal in 1979. [45] Generally, major textbooks, introductions, and overviews in this new area do not include any references to Potter or bioethics. [46] Obviously, the idea that the ethical study of environmental problems should be part of the broader discipline of bioethics has not found advocates among the promoters of environmental ethics. This is curious, since the orientation of this new field is close to Potter's notion, advocating an ethics of life in all its manifestations; it is focused on human survival rather than individual well-being, on long-term perspectives, and on the connectedness of human beings with the natural world. At the same time, the orientation of environmental ethics is theoretical. Reflecting on the moral dimensions of the relationships between human beings and their surrounding environment has generated two different ethical approaches: a human-centered one and life-centered one. [47] The first (anthropocentric) approach proceeds from the fundamental principle of respect for persons; it identifies duties to human beings, employs the notion of future generations, and emphasizes the usefulness of the biosphere for human beings. The second (nonanthropocentric) approach proceeds from the fundamental principle of respect for nature; it delineates the duties owed to living things and emphasizes that they are entities that have intrinsic value or a good of their own. This second approach has two variants: biocentrism and ecocentrism. The first assumes that all life has inherent worth. This is, for example, the view of Albert Schweitzer, who espoused an ethics of reverence for life. [48] Ecocentrism is the normative theory that species and ecosystems (such as forests, lakes, deserts, and wetlands) have moral standing independent of their component individuals. Leopold's "land ethics" was such a holistic view, claiming that preserving the integrity of the biotic community was the most important ethical obligation.

Environmental ethics is defined as that area of ethics focused on the moral relations between human beings and their natural environment. [49] This definition is compatible with Potter's notion of bioethics. However, the theories developed in environmental ethics often lead to a radical rethinking of the relationship between humans and nature, expanding the boundaries of ethical discourse beyond anthropocentric concerns. Potter's position is not clear. In his writings, he refers to Leopold's extension of ethics, emphasizing the importance of the ecosystem and suggesting that bioethics is the realization of this third stage of ethics. Yet in a 1996 article, Potter does not accept the radical new approaches of biocentric or ecocentric ethics. He argues that environmental issues should be included in ethical considerations and that the focus should be on long-term perspectives and safeguarding survival. But for Potter, the interests of human beings remain paramount. Bioethics combines biological knowledge and *human* values; it is *human* survival and improvement in the quality of *human* life that requires new wisdom. Environmental ethics, in Potter's view, suffers from the same distortion as biomedical ethics. It has been appropriated by philosophers. They engage in scholastic disputes, making false distinctions between anthropocentrism, biocentrism, and ecocentrism. In what he calls "real bioethics" there is always a mix of approaches, an enlightened anthropocentrism that acknowledges the central role of the biosphere for human survival. [50] Potter does not see an opposition between theories; the ecosystem is valuable precisely because it has intrinsic as well as instrumental value. [51] In this analysis, Potter's view is supported by Roderick Nash, who characterizes most of the activity in environmental ethics as "academic and esoteric." [52]

Its theoretical orientation makes environmental ethics different from bioethics in the narrow conception of biomedical ethics. In the context of healthcare, emphasizing values such as saving life or the sanctity of life presupposes an anthropocentric point of view; life is interpreted as human life. Ethics is also interpreted differently, not as philosophical discourse but as practical, applied ethics; it is an activity that takes place in the interaction between human beings, focusing on individual decision making regarding health and disease. The nonhuman world, such as the environment and other forms of life, can be taken into account but only in relation to human needs and interests. Environmental ethics therefore is not only a more philosophical discipline but also has less practical impact. One reason is the diversity of theoretical approaches. Each has different policy implications. Anthropocentrism aims to preserve species that have the highest value for human beings, while ecocentrism preserves endangered species that contribute to the survival of ecosystems. Anthropocentrism argues that the human population should be stabilized, while biocentrism and ecocentrism focus on reduction. Another reason is that there is a gap between ethics and policymaking. In biomedical ethics, ethical discourse and policy were connected from the beginning. The *Belmont Report* (1979) formulated the principles of respect for persons, beneficence, and justice as the ethical framework for medical research. Policy documents in environmental ethics, such as *Our Common Future* (1987) or the Rio Declaration on Environment and Development (1992), use moral concepts but do not discuss and elaborate ethical principles; they do not translate the moral concepts into practical consequences and policies. [53]

Two other differences between environmental ethics and biomedical ethics are worth mentioning. One is that the sense of crisis, even imminent apocalypse, is often more urgent and deeper in environmental ethics. Overpopulation, pollution, depletion of resources, deforestation, desertification, and climate change make the planet uninhabitable. Human beings are destroying ecosystems in a suicide course for the entire species. The second difference is that the source of ethical problems is perceived differently. In biomedical ethics, it is the power of science and technology that infringes on individual rights and values. Environmental ethics more often criticizes the economy and the culture, in particular the human lifestyle. Technical interventions will not be sufficient to save the planet; only a change of values will do. Curiously, these two differences highlight arguments that Potter used to introduce the broader notion of global bioethics: a focus on priority problems and a criticism of economic systems. For Potter, global bioethics should act as a counterweight to the sanctity of the dollar. [54] Today's economies, as he points out, have "failed to serve the social good and the goal of sustainable global survival." [55] Environmental ethics, therefore, introduces a major point of critique that is missing in biomedical ethics.

The conclusion thus far is that (a) bioethics was introduced as the name of a broad approach that combines individual, social, and environmental ethical concerns; but (b) has been narrowed and reduced to biomedical ethics, focusing primarily on individual ethical concerns; and (c) has been almost completely ignored in a new area of ethics with an exclusive focus on environmental issues. Although biomedical ethics and environmental ethics emerged around the same time that Potter coined the word *bioethics*, there are hardly any connections between these two new areas of ethics. How can it be argued that these disconnections and fractures will be healed in the emerging field of global bioethics?

Reconnecting Biomedical and Environmental Ethics

Three arguments make the point that biomedical and environmental ethics should be reconnected in an encompassing ethical approach.

EMERGENCE OF NEW PROBLEMS

A growing number of scholars have called attention to the interconnections between health and environment. For example, Jessica Pierce has pointed out that environmental degradation has negative impacts on health. [56] Environmental problems that have raised increasing concerns since the 1960s are persistent; they have not diminished but rather have grown in intensity. The chemical disaster in Bhopal, India, in 1984 and the nuclear disaster in Chernobyl in present-day Ukraine in 1986 not only killed a substantial number of people but also had serious long-term health effects. Nowadays, it is undeniable that pollution, climate change, and scarcity of water have significant consequences for health and disease. On the other hand, healthcare itself can have an impact on environmental degradation. Potter used the global argument that better medical care leads to population growth; this produces habitat destruction and pollution, which leads to environmental degradation. The argument can also be specific and local. Medical centers produce toxic waste. Current practices of healthcare, especially in more developed countries, inefficiently deplete scarce resources. The notion of sustainability should therefore be applied to healthcare. [57]

The idea that the environment is important for health is not new. It has been a crucial element in social medicine, occupational healthcare, and public health, as the next chapter shows. But its significance has changed recently, transforming the environment from one among many determinants of health into a crucial, global component in the pursuit of good health. This transformation will require the development of "green bioethics": a broader ethical perspective that takes into account that human beings and nature are connected. [58]

CRITICISM OF CURRENT REDUCTIVE APPROACHES

The second argument is that biomedical ethics and environmental ethics as currently conceived have a reductive approach that is too narrow for today's ethical problems. Jonathan Moreno, for example, argues that the destruction of New Orleans by Hurricane Katrina in 2005 shows the need to revive Potter's broader concept of global bioethics. In natural disasters, medicine and public health cannot be separated from environmental challenges. [59] Threats to health arise in a global context of social injustice and environmental damage. Biomedical solutions are therefore often inadequate as long as the social, economic, environmental, and political determinants of health are not addressed. Bioethics discourse should do more than focus on individual problems; it should develop a collective, global vision of how human existence can be improved. [60] Bioethics should move beyond the bedside and concern itself with public health, population issues, animal health, and certainly the environment. The challenge for contemporary bioethics is to promote health and at the same time justice and sustainability. This will require a global concept of bioethics that includes social and environmental concerns. [61]

Environmental ethics, on the other hand, has been increasingly criticized for being too theoretical. It should be more focused on practical problems in environmental policy, ecological research, and conservation management. Such a turn to "environmental pragmatism" could be inspired by a biomedical ethics that is interdisciplinary, problem-oriented, and focused on specific cases. [62] While its critique of anthropocentric approaches and its holistic orientation toward populations and ecosystems provides a broader ethical framework, environmental ethics can itself be an applied ethics that combines theoretical pluralism with practical decision making. For example, recent policies and practices in conservation biology now put emphasis on the needs of human beings and not merely on the intrinsic value of nature and ecosystems. [63] At least some environmental scientists have argued for a broader conception of bioethics that does not focus solely on medical issues. [64]

Need for a New Ethical Framework

Ever since Warren Reich acknowledged that Potter's legacy should be recognized, the idea of global bioethics has been gaining ground. [65] A broad conception of bioethics is now developing that not only addresses emerging global ethical problems but also articulates a framework of ethical principles that can be applied to those problems. [66] Such a broad focus will necessarily introduce environmental considerations into the bioethical debate concerning health and disease. [67] This extension or broadening of ethics is not surprising; the circle of moral concerns will further expand as ethics continues to evolve, as it has done throughout human history. [68]

Conclusion

Although Potter maintained a broad conception of his new term *bioethics*—one that embraced medical, social, and environmental concerns—biomedical ethics and environmental ethics have been disconnected for decades. The recent emergence of global bioethics in the face of global problems provides the opportunity to bring these two areas of ethics together in an encompassing framework of thought and action. The convergence of disparate areas of bioethics into an overarching notion of global bioethics will extend the agenda of bioethical debate; focus critical analysis on the larger, global context of problems; and redefine relationships of human beings with each other and also with the biotic community. This is one of the lessons of ecological thinking and environmental concerns: human beings are necessarily connected with other forms of life, and they are embedded in nature. This means that ethics, although practical, reflects a way of life, a worldview. It is not simply an instrumental approach that regards the environment as a useful tool for human beings. The connectedness of human beings furthermore implies that ethical discourse cannot merely focus on individual interests; it should take into account social concerns such as justice and solidarity. The next chapter argues that the convergence of biomedical and environmental ethics into the larger concept of global bioethics is specifically advanced by the new emphasis on biodiversity. This term was introduced in public and scientific debates in the mid-1980s. While the word *environment* directs attention to the external surroundings of human beings, and to interactions between human and natural worlds, the notion of biodiversity assumes that human beings are part of the biosphere itself.

2 | Biodiversity

B iomedical ethics and environmental ethics have been disconnected from the beginning, although both originated from the same inspirational source in Van Rensselaer Potter's concept of bioethics. How is it possible to reconnect both areas of applied ethics in the conception of global bioethics? This chapter argues that such reconnection is possible because of the growing attention to biodiversity. This relatively new term has several advantages over the emphasis on the environment that may benefit the development of a broad bioethical perspective. However, the recognition of the significance of biodiversity not only opens up possibilities; it provides strong arguments that today a broad notion of bioethics is necessary and inescapable. There is a need for convergence of biomedical and environmental ethics, and thus a more extended agenda of bioethics, because biodiversity is essential for health, human well-being, and survival. This need is furthermore underlined by the global character of contemporary bioethical problems. Loss of biodiversity is a global problem that requires global answers. This chapter examines the emergence of biodiversity as a new way of thinking about the interconnectedness of human beings and the natural world. It then analyzes the ethical implications of biodiversity, as a moral good itself and as a moral appeal to action. It concludes by explaining how the focus on biodiversity can bridge the divide between biomedical ethics and environmental ethics through promoting global bioethics as originally envisioned by Potter.

Biodiversity: A New Focus

May 22 has been proclaimed by the United Nations as the International Day for Biological Diversity. [1] The neologism *biodiversity* was introduced as shorthand for biological diversity. The occasion was the National Forum on BioDiversity, a scientific conference organized in Washington, DC, in September 1986. The proceedings were published as *Biodiversity* in 1988. The editor, Harvard professor Edward O. Wilson, is now known as the "father of biodiversity." In the literature, the new term became widely used, especially after 1992 when the World Conference on Environment and Development in Rio de Janeiro adopted the UN Convention on Biological Diversity. [2] The convention defines biological diversity as "the variability among living organisms from all sources including, inter alia, terrestrial, marine and other aquatic ecosystems and the ecological complexes of which they are part; this includes diversity within species, between species and of ecosystems." [3] While it is generally accepted that diversity is manifested at three levels—species, genes, and ecosystems—the term is regarded as vague and not easy to define. In practice, it has become synonymous with all living things, with the richness and variety of life on earth. There are also problems in measuring biodiversity, which is often limited to counting species as the basic unit of diversity and to making lists of endangered species. Biodiversity in this approach is interpreted as the richness (or loss) of species. It is still disputed whether the focus should be on the number of species or on the degree of difference between species. Regardless of the theoretical and practical controversies, it is remarkable how quickly the notion of biodiversity was disseminated in popular and scientific discourses. [4]

Environmentalism

The 1960s saw a rise of environmentalism, particularly in developed countries. Inspired by Rachel Carson's *Silent Spring* (1962), a broad social movement emerged with deep concerns about the deterioration of the natural environment caused by pollution, waste, chemicals, and pesticides. [5] These concerns were often associated with criticisms of consumption patterns, overpopulation, urbanization, industrialization, and technological progress. Grassroots activism advocated protection of nature and limits on human interventions. In the United States, it resulted in the adoption of the Endangered Species Preservation Act (1966) and the establishment of the Environmental Protection Agency (1970). In Australia and Europe, environmentalism led to the founding of "green" political parties. As a social movement, environmentalism could draw on a long tradition of concern for the environment in many countries. In the United States, the ideas of Henry David Thoreau (1817–62) and John Muir (1838–1914) had inspired a veneration of wilderness and wildlife. [6] Wilderness was no longer regarded as a dangerous space to be avoided but as a place of balance, harmony, and order that should be protected against humans. [7] It provided the opportunity to experience trans-human values: beauty, creativity, a sense of unity with nature, contemplation, health, and relaxation. In an era where everything is humanized, wilderness presents the "other" of civilization and thus the possibility of eluding the dangers of modern life. [8] It is the realm outside of human control; that which escapes human dominance. Confronted with the landscape of Yosemite Valley, Muir felt the same sense of humility, human insignificance, and vulnerability of the natural world as did later generations that gazed at the "blue marbles" images (photographs of planet Earth taken from outer space). This sense of vulnerability and the need to protect nature from humans also inspired modern environmentalism—even more so, now that science and technology, and especially the atomic bomb as their contemporary product, can destroy life on the planet.

The emphasis on wilderness preservation and protected areas is contested. Indian scholar Ramachandra Guha argues that from a global perspective the preoccupation with wilderness is devastating for people in developing countries. [9] Establishing national parks and reserves in these countries has resulted in displacement of villages, removal of poor peasants, and transfer of resources to the rich. Wilderness areas are designated for tourists, disregarding the needs of the local population. Environmental issues directly harmful to the majority of people (such as lack of safe drinking water, inadequate waste disposal, soil erosion, and air pollution) are often not considered. In this view, Western environmentalism justifies practices that are unjust. Its distinction between human beings and nature, and the ethical theories based on it (anthropocentrism, biocentrism, and ecocentrism) are not very helpful. This distinction, according to Guha, is "of very little use in understanding the dynamics of environmental degradation." [10] It diverts attention away from the basic ecological problems, namely, overconsumption and militarization (arms races and wars). The distinction frames the fundamental ethical problem in such a way that the focus of debate is not on the economic and political structures of modern societies that support a destructive lifestyle of individual consumers in some parts of the world. Therefore, protection of wilderness and endangered species is not the same as protection of the environment. The fundamental problem is not the human misuse of natural resources; it is the imbalance in use, the inequalities within human society based on structures of social injustice, exploitation, power, and domination. Thoreau and Muir may be icons for American environmentalism, but they are not relevant for other cultures. Similar criticism refers to the fact that biodiversity itself is not equally distributed. Most biodiversity "hotspots" are in developing countries. [11] Policy discourse often suggests that problems with biodiversity predominate in the global South, while conservation solutions are offered in the North. [12] The implication of this argument is that the ethical debate should be expanded: the emphasis should be on global inequality and injustice.

The focus of environmental concerns is contested not only from a global perspective but also within Western countries themselves. Particularly in Europe, environmentalists argue that the social and economic system needs to be changed and that issues of power and justice must be addressed through a transformation of values. [13] Within the United States, approaches are not consistent. In the first part of the twentieth century, concerns about wilderness were mainly driven by commercial interests. Wildlife was regarded as a commodity. Policies were based on conservation, regulated use, and scientific management, rather than preservation and protection. After 1945, the focus shifted toward preservation, and Aldo Leopold played an important role in this change. [14] He argued that an ecological point of view should combine an economic approach with an ethical one. The role of humans should change from master to member and citizen of the earth. Human beings are part of a biotic community. The emphasis on the economy was misplaced: nature is not a "stockpile of raw materials." [15] However, in the middle of the twentieth century, the pendulum swung back in the other direction. Ecology again was dominated by the bioeconomic paradigm. Economic thinking was applied to the study of nature as a "storehouse of exploitable material resources" that required rational resource management. [16] It is clear that these environmentalist approaches depend on different ethical visions of nature. If nature is regarded, in Max Oelschlager's words, as "the source of human existence, rather than a mere resource to fuel the economy," it will be respected because it has value in its own right. [17] But if it is a means to satisfy human needs, it will be used for economic purposes. These diverging ethical perspectives produce different practices: arcadianism and imperialism. [18] The first is based on harmony—peaceful coexistence with other organisms and symbiosis between human beings and nature. The second is based on exploitation and utilitarian use of resources, justified by humanity's dominion over nature. These different ethical views and the reemergence of the bioeconomic approach are the backdrop for the interest in biodiversity.

Popularity of Biodiversity

Why has the interest in biodiversity so rapidly increased since the 1980s? Three factors can be identified: globalization, a sense of crisis, and the need for policies. [19]

GLOBALIZATION

Biodiversity was in fact recognized because of globalization, at least for those outside the scientific community. People around the world became increasingly aware of the amazing diversity and beauty of the natural realm. The aerial photos and documentaries of Yann Arthus-Bertrand impressed millions, especially his film *Home*. Another example is *Ushuaïa*, a weekly television program (since 1987, and television channel since 2005) animated by journalist (and former minister of ecological transition) Nicolas Hulot in France. Spectators could feel sympathy for threatened species in Africa or concerns for the rain forest in Brazil, conscious that biodiversity is not merely of local significance. Anti-whaling actions of Greenpeace and campaigns of Friends of the Earth against the World Bank for financing projects causing deforestation in developing countries attracted a lot of supporters. The 1988 murder of Chico Mendes, a Brazilian rubber tapper and environmentalist, led to an international outcry and made him a symbol of the global environmental movement. Biodiversity thus illustrates the global nature of environmental problems as well as the need for global solutions. [20]

At the same time, biodiversity is reduced or lost as a result of the same processes of globalization. Dissemination of the use of pesticides contaminates local groundwater and thus drinking water in a much larger area. Global traffic is moving thousands of species around the world, for example, chytrid fungi that destroy amphibian populations everywhere, from Panama to Tasmania. Amphibians are now the most endangered class of animals. [21] Global movement also increases the proliferation of emerging diseases such as avian flu and the Ebola and Zika viruses. Globalization of agriculture and the search for more efficient production of food leads to reduction of the genetic variety of crops, and therefore, increased vulnerability to crop diseases. Local problems of pollution and environmental degradation are aggravated by global problems of acid rain, ozone depletion, and global warming, which all impact biodiversity. Multiple threats to biodiversity converge and reinforce each other. Global warming will lead to heat waves, natural disasters, and extreme weather events; new emerging diseases and changed distribution of epidemics; extinction of species; bleaching of coral reefs; and transformation of ecosystems. [22]

The environmental crisis therefore affects everybody, regardless of location. It is also not limited in time; loss of biodiversity has intergenerational effects. Biodiversity is the heritage of all human beings and will be essential for future generations. The notion of biodiversity furthermore illustrates that a biological perspective is insufficient. Loss of diversity has global effects beyond the realm of biology. An early example is the 1846 potato famine in Ireland. Because a single variety of potato was cultivated, the genetically uniform crop was vulnerable to blight. The disease devastated the crops and produced massive famine, killing one million people and producing mass emigration, changing the history of countries.

SENSE OF CRISIS

The 1986 conference book, edited by Wilson, clearly sounds the alarm. Wilson himself issues an urgent warning that we are destroying "the greatest wonder of this planet." [23] He is supported by other authors, all predicting the worst imaginable catastrophes: before the end of the twenty-first century, we will experience the same effects as those produced by a nuclear winter; "the greatest single setback to life's abundance and diversity" in four billion years; and a real "biocrisis" beyond repair. [24] The crisis discourse focuses on the extinction of species. Everybody knows about lost species: dinosaurs, dodos, mammoths, and passenger pigeons. But few are aware that the present rate of biodiversity loss is unprecedented and exceeding anything in the past. Each day one species was lost in the 1970s; one species per hour in the 1980s. Wilson calculated that in the 1990s, each day 74 species would be extinguished—three each hour. In 2007, the situation turned out to be even worse: in that year, 150 species were lost each day. [25] This loss is tragic for two reasons. It is irreversible; when a species is lost, we cannot bring it back. Extinction is worse than individual death; it is the "death of a form of life." [26] It is also tragic because we do not even know what we are losing. Today there are approximately 1,800,000 known species. But we do not know the actual number of species; it is estimated that in reality there are between 5 million and 30 million species. [27]

The focus on the extinction of species refers to a deeper concern: the extinction of humanity itself. Biodiversity, as Wilson argues, is "the key to the maintenance of the world as we know it." [28] Loss of biodiversity is the most fundamental global environmental problem because it endangers the survival of humanity. All

other environmental problems contribute to this danger: pollution, overexploitation of natural resources, depletion of the ozone layer, soil erosion, and climate
change. Because extinction is definitive, biodiversity loss is a disaster worse than
other disasters such as nuclear war or economic meltdown; it will take millions
of years to recover. [29] This catastrophic scenario implies a moral appeal. Extinctions nowadays are no longer natural events but occur as the result of human
activity. During the past 600 million years there have been five major mass extinctions. The most recent of these was 65 million years ago, when an asteroid impact wiped out the dinosaurs. But now, in what has been called the Anthropocene
(the geological age in which humans influence the natural world), all kinds of species are dying faster than before. We, as human beings, are destroying other
species—and ultimately ourselves. [30] Humans are undermining their own existence. The big ethical question is how such destructive and self-annihilating
behavior can be justified.

Loss of Biodiversity

The International Union for Conservation of Nature (IUCN) Red List (2014)
assessed 73,686 species; 22,103 are threatened with extinction (30%).

- One in every four mammals is threatened; one in every eight birds.
- Among threatened organisms are 41% of amphibian species, 33% of
  reef-building corals, and 30% of conifers.
- The total number of known threatened animal species has increased
  from 5,205 to 8,462 since 1996.
- A total of 8,457 threatened plants are listed (about 2% of the world's
  described plants).
- Every year, 7.3 million hectares of forest are lost.
- Half of the tropical forests have already been cleared.
- In the past 35 years, biodiversity has been reduced by more than a
  quarter. [31]

NEED FOR POLICIES

The invention of the term *biodiversity* signals a change of perspective that focuses more than before on effective policies. Instead of elaborating theories or initiating more research, it is urgent to undertake action before it is too late. Wilson
makes it clear that, faced with the biodiversity crisis, we know the causes (habitat

destruction and fragmentation, invasive species, overexploitation, diseases, and climate change), and we know what to do. [32] The new concept shifts the perspective in three ways. First, biodiversity becomes a wake-up call to save species or ecosystems from extinction; rather than theorizing about stability among species or determining the health of the ecosystem, preventing extinction is a vital task. [33]

Second, biodiversity changes the perspective from preservation to conservation. Preservation means that nature needs to be preserved for its own sake; it should be protected against present and future consumption. Environmentalism traditionally focuses on the preservation of endangered species and the creation of special reserves and national parks. Conservation means that nature should be defended for the sake of humans and should be saved for future consumption. It implies a more prudent use, not absolute exclusion from human use. [34] The emphasis on biodiversity is associated with conservation as a more feasible goal for policies. The new discipline of conservation biology points out that a holistic and dynamic approach is needed to ensure biodiversity, safeguarding first of all evolutionary change. It is impossible to prevent all extinctions. Rather than saving individual species, it is important to protect diversity of species. And instead of trying to save flagship species (such as elephants and pandas), efforts should focus on habitats and ecosystems. [35]

Third, the notion of biodiversity moves attention from protection to sustainable use. It is a conceptual and practical instrument to solve the long-standing tension between conservation and development. The notion links ecological and economic views. Considering biodiversity as a global resource that generates significant benefits, especially in developing countries, it will, if properly managed, reconcile conservation with economic development. It is from this perspective of bioeconomics that Wilson calls for a "new alliance between scientific, governmental, and commercial forces." [36]

Biodiversity and Bioeconomics

Biodiversity is an important ideological tool in the revived environmentalism of the 1980s—the so-called New Environmentalism that focused on the practical value of nature. [37] Biodiversity is viewed as an economic asset and can be quantified so that it is measurable. Like monetary entities in the economy, species can be considered as equivalent entities that are comparable and exchangeable. The new science of ecological management can then assess the impact of policies. Furthermore, biodiversity can be regarded as a "portfolio of natural assets" [38] or "biological wealth"; each species is a chemical factory that has potential benefits for medication, food, and human survival. [39] The new discipline of ecological economics demonstrates the value of ecosystem services that provide life support to humanity. [40] The bioeconomic perspective of these new approaches has quickly become influential in environmental policies. In 1987, the United Nations' World Commission on Environment and Development introduced the notion of sustainability (linking environment and development). The Convention on Biological Diversity, adopted by the Earth Summit in 1992, is based on the assumption that biodiversity is commercially valuable and that an economic approach is the most effective way to conserve biodiversity. The underlying idea is that market mechanisms can make people interested in caring for the environment.

From Environment to Biodiversity

The use of the term *biodiversity* in popular and scientific discourses provides a more favorable opportunity for the application of a broad perspective in bioethics than does a focus on the environment. Potter published his book *Global Bioethics* in the same year that Wilson's conference book appeared (1988), but in it he does not mention biological diversity or biodiversity. At the same time, his concerns are similar to the ones associated with the crisis of biodiversity: issues of extinction and human survival. Bioethics as a new discipline is necessary because it should recognize, first, that the human, plant, and animal worlds are interconnected and, second, that the natural world is necessary for human survival. Potter's orientation is not merely toward the interactions between human beings and the environment; he proceeds from a broader vision. [41] Inspired by Leopold, he assumes that humankind is completely dependent on the natural environment; healthy ecosystems are the precondition for the survival of the human species. This precisely echoes Leopold's argument for an expanded ethics that human beings belong to a biotic community of soil, water, plants, and animals. For Leopold, and also for Potter, nature is a self-regulating system. In their writings, both often draw upon notions of harmony, balance, homeostasis, and equilibrium. Leopold emphasizes the connection between structure (ecological diversity) and functioning (stability) of the environment. He is fascinated with the concept of diversity, though he does not use the term *biodiversity*. In his view, human efforts should be directed at preserving the capacity of the ecosystem for healthy functioning rather than preserving individual animals, plants, or trees. [42] Leopold's emphasis on the community ("the land") and the embeddedness of human beings in nature has two ethical consequences, shared by Potter. First, interconnectedness implies concern for all living beings. Humans are no longer conquerors of the land-community, as

Leopold famously claims, but its members and citizens. [43] Second, land, food, health, security, and community cannot be separated. Nature is a resource for agriculture, food, and medicines, as well as for beauty, recreation, and moral imagination, but it cannot be reduced to this; it also has value in itself. Leopold characterizes the economic perspective as "hopelessly lopsided." [44]

Against this theoretical backdrop, biodiversity as a concept provides a more fertile ground for global bioethics, and thus for reconnecting biomedical and environmental ethics, than does a focus on the environment. The first reason is that biodiversity implies a different type of relationship with nature. Talking about the environment refers to the surroundings, the context in which human beings are living. Usually, this is the natural world around us: the forests, lakes, and mountains. In this sense, environment is synonymous with nature, but nature is often regarded as apart from human beings. [45] However, the environment can be wider or smaller, global or local. For example, we refer to the intrauterine environment. The discourse of biodiversity has a different emphasis. It refers to a wider context in which human beings are included. Instead of interacting with the environment, human beings themselves are part of the natural world. The notion of biodiversity stresses that we are dependent on natural ecosystems; without them we cannot survive. This focus on interconnectedness and interdependency is one of the characteristics of global bioethics. [46] The second reason is that the concept of biodiversity is connected to cultural diversity. Biodiversity is not static but dynamic and changeable. Changes are often related to human interventions. Natural conditions determine how human cultures and civilizations emerge and flourish. On the other hand, cultures transform natural conditions and influence biodiversity.

Since people are part of biodiversity, natural and cultural evolution cannot be separated. Biodiversity is a better vehicle through which to articulate an evolutionary perspective than environment. The connection between natural and cultural diversity was also highlighted by Potter when he was developing his notion of bioethics. Knowledge of the living world ('bio') is indispensable for cultures that aim at survival. Potter argues that cultural diversity or plurality is essential for imagining survival strategies but that not all cultural approaches are acceptable. [47] The same perspective is later reflected in UNESCO's Universal Declaration on Bioethics and Human Rights: respect for cultural diversity is a principle of global bioethics, but not all diversity is equally acceptable.

Third, biodiversity is a value-laden concept. It is assumed a priori that it is a good that must be conserved. Like the environment, it can be a threat (harmful for health, and a source of new, emerging diseases), but in contrast to the environment, it also provides opportunities and benefits (for example, potential medicines and food). The environment has an impact on health and therefore requires attention as one of the determinants of health. Biodiversity is more fundamental, since it is a source of health; without it, health would not even be possible. This additionally implies that biodiversity has a value that needs further specification. The notion of biodiversity invites an ethical discourse that connects individual, social, and environmental perspectives in the single approach of global bioethics.

Finally, biodiversity introduces a different ethical setting. Seeing the environment as biological and physical surroundings or the natural world assumes that the individual human being comes first. He or she relates to and interacts with the environment. From an ethical point of view, the environment should be well cared for, while threats to human interests should be prevented or eliminated. This view is consistent with the emphasis on the autonomous person in biomedical ethics. The notion of biodiversity reverses this ethical thinking. What comes first is the unity and interdependence of the living world. Human beings are first of all embedded and contextual beings. This is also the basic message of global bioethics: ethics goes beyond individual perspectives; it is social ethics.

Values and Ethics

Subsequent chapters on topics such as health and disease illustrate how biodiversity is approached from different perspectives: as scientific, economic, and ethical challenges.

THE SCIENTIFIC PERSPECTIVE

The science of ecology that inspired Potter has the same etymological root as economy (οἶκος, "family house" or "household"), referring to the proper administration and management of resources. It focuses on the relations between living organisms and their environment, presupposing the ideas of balance, harmony, and stability of nature. [48] Concepts such as natural order, equilibrium, self-regulation, and homeostasis were often used, for example, by Potter and Leopold. In this perspective, interventions must be directed toward the whole system. Preserving the biotic community should be more important than economic benefits, even when biodiversity is often aligned with bioeconomics. The problem is that ecology in practical settings is connected to economy. Especially after the emergence of conservation biology as a new discipline, the role of science in biodiversity has been defined as environmental management. [49] In this

approach it is, first of all, important to identify and describe as many species, genetic varieties, and ecosystems as possible before it is too late. But the mission of
ecology is not just categorizing nature; it is protecting, saving, and rescuing biodiversity as much as possible. Ultimately, the scientific challenge is value-laden.
It assumes that biodiversity is good (for various ethical reasons and because it
has intrinsic or instrumental value) and that action is morally required to protect
it. This morally driven mission can lead to interventions that conflict with the respect for nature that traditionally motivated environmentalism. Restoration ecology, for example, reintroduces animals in areas where they have died out, hoping
to reverse biodiversity loss. [50] To save species from extinction, they are conserved
ex situ (in zoos, aquariums, botanical gardens, and seed banks). Breeding programs and hybridization with other species are used to increase the threatened
population, and then animals are released from captivity into the wild. Scientific
management is also applied in other areas of conservation. Forest trees can be
improved through gene transfer and clonal propagation. New scientific and technological tools are available to reduce environmental degradation and loss of biodiversity. It is argued that they should be used, now that we live in the Anthropocene, the
epoch where fundamental environmental changes are caused by humans. Simply
respecting nature is not sufficient to reverse biodiversity loss. The scientific perspective offers a biocentric version of the trans-humanism discussed in biomedical ethics. Enhancement techniques have long been applied to produce transgenic animals
and genetically modified crops. Now the promise is that genetic engineering can
lead to de-extinction: the resurrection of extinguished species such as the passenger
pigeon or the Tasmanian tiger. [51]

   This kind of Promethean environmentalism engenders two discussions that are
essentially ethical. [52] One concerns the idea of progress. The assumption is that
there always will be a solution for problems; new tools and technologies will be
invented, such as geo-engineering to reduce global warming or genetic engineering to re-create species. Potter had already criticized this assumption. The basic
worry is that it takes the root cause of problems as their solution. The idea underlying this approach is that the world can be managed; it is the ethical perspective
of imperialism. However, it is exactly this idea of dominion and planetary management that has produced the biodiversity crisis. [53]

   The second discussion concerns the idea of nature. Promethean environmentalism assumes that the idea of wilderness as a world beyond and untouched by
humans is gone. The idea of nature has become meaningless, because everything
is of our own making. There are no places unaltered by human beings, so the dis-

tinction between nature and culture, between original and wild versus artificial and constructed, no longer exists. Human beings have "ended nature as an independent force." [54] The implication is that in situ conservation (national parks and wildlife refuges) cannot be distinguished from ex situ conservation; the entire world has become a zoo. [55] The perspective of the end of nature raises the question of whether there is any difference between naturally occurring biodiversity and human-made biodiversity. Does only the former have independent moral value, or is value dependent on the human perspective? It further articulates the importance of values; if nature is constructed and thus requires perpetual management, what will be the values that guide this management? Human beings have always influenced nature, whether or not it was regarded as an independent force. But the overall record of these interventions is mixed, to say the least. Agriculture has facilitated the growth of the human population but at the same time has destroyed forest habitat and biodiversity. The emphasis on productivity has created monocultures, making crops vulnerable to pests and diseases; increasing use of pesticides and fertilizers has contaminated aquifers. Given this history, the promise of Promethean environmentalism cannot be trusted. [56] We already have control over nature; how do we know that we will have better control? The conclusion is that the scientific perspective on biodiversity in its traditional or Promethean form entails presupposed values; it is founded on an ethical perspective that should be clarified and critically analyzed.

The Economic Perspective

Ecology and economics were directly connected in the 1987 United Nations' report on environment and development. Prosperity, justice, and security were linked in a new approach "in which all nations aim at a type of development that integrates production with resource conservation." [57] Along with ecology, economics shares the discourse of scarcity. Often-used metaphors are the lifeboat and spaceship. [58] Biodiversity loss means depletion of resources that are finite and not renewable—hence the need to set limits. Also shared is the focus on utility. Things must be categorized (and this is the task of ecology), before they can be utilized. Edible plants must be differentiated from inedible and toxic ones. Ecological knowledge therefore is targeted at use or exploitation. Subsequently, economic expertise determines what use is wise. This determination rests on two theses: nature is a resource, and nature has value. Biodiversity is the "environmental wealth" of humanity. [59] It has always been used to produce food, to make clothes, to reduce diseases, and to create shelter and a safe environment. It will continue to be used as a resource to satisfy legitimate human needs. In this perspective, the use value of biodiversity can be economically calculated. If the total market and nonmarket value of all ecosystem goods and services is estimated, the global value of biodiversity is approximately $33 trillion (while the total GNP of all countries is $18 trillion) according to a study in 1997. [60] Ecosystems provide services that are indispensable for global life support, such as climate control, nutrient cycling, and atmospheric quality. On the basis of these theses, Wilson in 1988 could proclaim that our goal in the biodiversity crisis is clear: "making biodiversity a source of economic wealth." [61] This is not a scientific conclusion but an ideological program founded on specific assumptions about the nature of human beings and society. One is that the focus on market-based mechanisms for conservation will make more people interested in the environment because it appeals to self-interest. Potential benefits will be a strong incentive to conserve biodiversity. The emphasis on the utility and benefits obtained from ecosystems will be a more powerful motivator than appeals to the intrinsic value of nature. [62] This ideology of "green neoliberalism" can be translated into practical policies. Ownership and markets are fundamental notions for environmental management. Biodiversity should no longer be regarded as the common heritage of humankind; it is the property of states. They should facilitate access to biodiversity by creating a market for biological and genetic resources. Nature can thus be exploited without being destroyed because it "can pay for its own survival." [63] These policy ideas were soon implemented in the Convention on Biological Diversity. The conclusion is that the economic perspective on biodiversity also proceeds from specific moral values. A property- and market-based approach prioritizes private interests and individual autonomy. It neglects to recognize that the basic problem with biodiversity is not lack of ownership or free access but the imbalance in utilization. Even if there are "owners," mostly in developing countries, they are not benefiting because they have no access to the technologies to develop products from "their" biodiversity; they also cannot profit because the international property rights regime primarily benefits developed countries. This imbalance raises ethical issues of global justice and inequality. Furthermore, the neoliberal ideology overlooks the fact that in a global context biodiversity is a common good that is essential for humankind as a whole. A healthy ecological system is more fundamental than an economic system; soil, air, water, forests, and oceans are irreplaceable and indispensable for human survival. [64] Biodiversity is a common good that should not be divided into private transferable commodities. It requires a global public health approach.

THE ETHICAL PERSPECTIVE

Because the scientific and economic perspectives are both value-laden, biodiversity ultimately presents an ethical challenge. All scientific studies and economic calculations notwithstanding, the final decision about how to address biodiversity, as Wilson has emphasized, is one of ethics. [65] The ecological crisis facing humanity is an ethical crisis. [66] Science and economics describe, examine, and analyze biological diversity, but how to relate to biodiversity is an issue of values: should it be protected, conserved, preserved, managed, or commercialized? An ethical perspective will be required to determine what to do because biodiversity is a normative concept. [67] It is often assumed a priori that conserving it is good and losing it is bad. However, the explanations for the normative force of the concept differ. Two basic views can be distinguished: biodiversity presents a moral good, and it implies a moral appeal for action.

A powerful argument, particularly from the economic perspective, is that biodiversity is a good because it has instrumental value. Ecosystem services provide essential components of human well-being. Biodiversity harbors genetic and biological resources for medicines. It is a resource for food, drinking water, and industrial materials such as oil, rubber, and fibers. Furthermore, it is important for recreation and tourism. [68] Beyond immediate economic advantages, biodiversity also promises as yet unidentified benefits, since most species are not even known. There is enormous potential for future use. For example, most of the food supply is based on 12 plant species, whereas there are thousands of edible plant species. Biodiversity therefore is a kind of insurance; it is valuable because it includes a range of options and opportunities that we may need in the future.

This type of argument associates the value of biodiversity with its actual and possible uses. It is regarded as a resource that is beneficial for human beings. A different argument presents biodiversity as a moral good in itself that needs to be respected whether or not it is useful. Biological diversity is an ecological legacy and heritage that human beings have received, not so much from previous generations but from the natural evolution of life itself. [69] The least they can do is to make sure that this legacy is transferred to future generations. This is crucial, not because biodiversity is merely beneficial but more essentially because it is the precondition of human life; it represents "the world in which the human spirit was born." [70] This is the type of argument elaborated by Paul Wood. Biodiversity is not a resource itself but the source of resources. It is a necessary precondition that maintains the biological resources crucial for global health and survival. We should therefore distinguish between the value of biological resources and the value of biodiversity itself. Biodiversity is an "essential environmental condition." [71]

Biodiversity furthermore has intrinsic value because diversity in general is valuable. Diversity means difference, variety, variability, alternatives, and multiple perspectives. It allows us to escape from domination, homogenization, and monoculture. Many scholars have argued that biological diversity is important for the stability and productivity of ecosystems; it guarantees adaptation and survival. [72] In this positive sense, biodiversity is associated with cultural diversity, which includes a range of value systems that allow for various paths of development. [73] Since globalization often leads to homogenization—for example, the loss of languages—respect for diversity is more often articulated as an ethical principle in the critical discourse of globalization.

The second explanation of the normative force of biodiversity is that the concept is not only a moral good, but it also inspires moral action. We must do something to conserve it if we can do so. Biodiversity cannot simply be described as a natural phenomenon; it calls for a moral responsibility to protect and conserve. [74] The reason why the concept motivates and inspires action is that the contemporary biodiversity crisis is the result of human behavior, of cultural evolution. We ourselves are responsible for the current extinction of species and loss of biodiversity, contrary to earlier extinctions. This presents a moral dilemma. We are the only species that can save other species. The moral urgency is that now humans themselves are becoming an endangered species. This explanation requires rethinking the relationship between humans and nature. If loss of biodiversity is the result of human activity, we have to change our relationship with the natural world; we have to change our ways of thinking and acting in a finite world. [75] The future is determined by culture, not nature. This is exactly the point that Potter reiterated to underline the need for another approach in ethics. But even if we do not agree that we live in an era in which nature has ended as an independent force, in a post-natural world, nature can no longer be separated from humans; it necessarily includes human beings. [76] That means that how we relate to nature will not be different from how we relate to humans and other beings. If there is no longer a strong division between nature and culture, the distinction between anthropocentric and nonanthropocentric perspectives loses much of its relevancy. The focus can be on human concerns without denying that other forms of life are valuable on their own. The basic question is, as Holmes Rolston puts it, "how we humans belong in this world, not how much of it belongs to us. The question is not of property, but of community." [77] From a global perspective, we are citizens of the world, belonging to the earth. In our connection to biodiversity we continuously make trade-offs: between benefits for humans versus other forms of life, between current benefits and future costs (current versus future generations), and between benefits to a few and costs to many (individuals versus society). [78] If this is a correct interpretation of the fundamental ethical challenge, the basic ethical problem will be that of global justice and responsibility.

Broader Bioethics: The Need for Different Ethical Discourse

The previous chapter has argued that Potter introduced *global bioethics* as an umbrella term to cover medical, social, and environmental concerns. *Global* means not only that it is worldwide but also that its theoretical and practical approaches are widely inclusive. Global bioethics, as it is currently developing, proceeds from perspectives that differ from the ones that dominate mainstream bioethics. [79] First, it takes a broader view of the human person—not as a free-standing and autonomous decision maker but as an embedded and socially connected entity. In global bioethics, personal autonomy is just as important as in mainstream bioethics, but it recognizes that the context of individual existence should necessarily be considered. Because of this connectedness to other human beings as well as surrounding nature, human beings are vulnerable. Second, global bioethics has a positive notion of society. Human beings can only flourish in relationship to others. Bioethics should focus on the social, economic, environmental, cultural, and political conditions that make flourishing in health possible. The role of government is to provide for those conditions that determine the health of the population; it is not the role of the market. It is necessary to direct ethical debate toward the notions of cooperation (rather than competition), social responsibility (rather than individual responsibility), solidarity (rather than private interests), as well as global justice (rather than increasing inequality). Third, the focus of global bioethics is on the common good. Living together means that human beings share heritage and common goods. They interact with each other in the public sphere. Common interest is not simply the sum of private interests. Global bioethics rejects the neoliberal assumption that the human person is primarily driven by self-interest. It does not regard individuals as consumers who will act on the basis of benefit and profit but as citizens who will be concerned with the common good. People might care about biodiversity even if they themselves will never benefit from its protection, because they want to protect future generations. Fourth, global bioethics emphasizes collective action. Individual action is important but cannot bring about social transformation. Engaged individuals planting trees and eating organic food will not change the world. The logic of neoliberalism emphasizes individual responsibility and consumer choice. Nongovernmental organizations, civil society, and media promote the idea that personal choices will make a difference. The problem with this view is that it will not change the distribution of power. Global environmental problems, for example, are not the consequence of destructive individual choices that should be altered, although these choices are not irrelevant. Forms of collective engagement will be necessary to influence social and economic conditions that produce global problems. But this is precisely what the logic of neoliberalism tries to prevent. It argues that while the scale of problems seems overwhelming, there is no reason for fatalism or radical change, since at least individuals can do something. Making this point obscures the fact that governments, corporations, and policies are equally, if not more, responsible for environmental degradation.

These global bioethical perspectives produce a moral discourse that directs attention to the root causes and underlying value framework of ethical problems. The formulation of a new bioethical principle to protect the environment, the biosphere, and biodiversity in the Universal Declaration on Bioethics and Human Rights demonstrates that the separation between environmental and biomedical ethics can no longer be maintained within bioethics. The usual arguments against such separation are utilitarian: ecology has consequences for human health. Countering environmental degradation such as toxic waste and pollution is therefore a public health concern. However, the significance of biodiversity provides at least four additional reasons to include environmental concerns in bioethics.

Global Character of Problems

Environmentalism is not a phenomenon exclusive to rich and more developed countries. A global perspective reveals different varieties of environmentalism, many not associated with affluence. Guha discusses movements against deforestation in Thailand, the protests of Ogoni People against oil pollution in Nigeria, the Green Belt Movement in Kenya, and the Chipko movement against deforestation in India. These are examples of what has been called "environmentalism of the poor." [80] This global view illustrates that all human beings are intertwined in the same processes of biodiversity loss and climate change. Environmental problems are trans-boundary problems that require international cooperation. These global challenges have also instigated various responses at local levels, depending on the cultural and religious context. Responses differ because the challenges are not scientific but ethical. Contrary to what Wilson argues (that the causes of biodiversity loss, such as overpopulation and food insecurity, are biological problems), conservation is not a biological or technical issue. [81] Loss of biodiversity affects not just biologists but all people. It entails choices among values. Since it is impossible to protect all biodiversity, how do we determine what needs protection? [82] Furthermore, the value of biodiversity is not dependent on individuals. [83] Its value is located at the supra-individual level of communities, ecosystems, and populations, and therefore requires a global evaluative perspective. [84]

Source of Problems

Wilson's interpretation of the causes of biodiversity loss as biological problems is not uncommon. It presents a specific definition of the problem, assuming that overpopulation is a challenge (since it is interpreted as a biological phenomenon) but that overconsumption and inequality are not (since they are social and political issues). This definition does not recognize that there is a fundamental inequality at the root of biodiversity loss. In the real world, responsibility for such loss is not equally divided. Developed countries are responsible for environmental degradation that primarily damages developing countries. Also, blame is not distributed equally. Developing countries that harbor the most biodiversity, and especially species richness, are regarded as the source of problems (because of corruption, poverty, and inefficient governance), while solutions are offered by developed countries. [85] These interpretations disregard the fact that destruction of biodiversity is rooted in processes that are mainly driven by developed countries. From a global perspective, some members of the world community are engaged in social practices that harm other members of that community. [86] For example, Larry Lohmann points to Thailand, where forests have rapidly disappeared because of the expansion of agriculture. The driving forces were business interests, state bureaucracy, and the military, supported by international agencies such as the World Bank, and with a focus on export for non-local needs. Ordinary farmers and rural village groups defended biodiversity, but they were often overruled by commercial interests. [87] The logic of economic value is frequently prioritized over the logic of conservation. The source of biodiversity problems is therefore not different from the cause of global bioethical problems. One-sided, ideologically driven processes of neoliberal globalization produce many of today's bioethical problems (e.g., brain drain, organ trade, commercial motherhood, pandemics). The same processes underlie biodiversity loss.

GLOBAL ETHICAL FRAMEWORK

Global bioethics works within an ethical framework that combines individual, social, and environmental perspectives. This is also the goal of environmental ethics: it argues that the circle of moral concern should be extended to include animals, plants, species, and ecosystems. This is the moral extensionism advocated by Leopold and seconded by Potter; it is a new ecological sensibility that not only includes nonhuman animals and other living beings, but also present and future generations. [88] For Potter, such extension is a crucial step from natural to cultural evolution. This is not a radical move; only the scope of moral considerations is "extended." Though critical of the notion of progress in science, Potter (and others) make the assumption that ethics undergoes development. Moral consciousness has expanded over time and will continue to expand.

Ethical concerns regarding biodiversity are furthermore articulated with the same ethical principles that are prominent in global bioethics. The first one is the principle of justice. The most deprived and vulnerable regions of the world are the most polluted and degraded ones. Also, people in these regions consume many fewer natural resources than those in developed countries. At the global level, human beings in 2009 consumed the equivalent of 1.2 "Earths" (indicating how much land, sea, and other natural resources are needed to produce what is consumed, measured as "global hectares"). Current global consumption therefore creates a significant ecological deficit, but there are major differences between countries. In the United States, every person consumes 6.0 "Earths"; in Cambodia, 0.4; and in Costa Rica, 1.1. One "Earth" per person would be a just share. [89]

Earth Overshoot Day

By August 2, 2017, human beings had used more from nature than the earth could renew in that entire year. By 2030, we will need two planets. The "overshoot" day has moved backward rapidly. In 1985, it was November 5. Australia and the United States consume, respectively, 5.2 and 6.0 planets per year. [90]

Another principle is respect for vulnerability. This principle applies not only to individuals (e.g., children and the elderly) but also to species and ecosystems, to populations and countries, and even the planet itself. [91] Because living beings are interdependent, human well-being cannot be separated from biodiversity; damages in some components will make the entire ecosystem vulnerable. Global responsibility is another prominent principle. If we are in the Anthropocene and in the process of the sixth extinction, perpetual management of nature is required. This does not imply that nature is human-made. But it implies that human responsibility has expanded; nature is in need of extended care, just as many human beings require continuous care. Finally, concern for future generations is a shared principle, as is the principle of benefit sharing.

The principles that are shared in the ethical discourse of biodiversity and global bioethics do not primarily refer to individual interests and goods that dominate the concerns of mainstream bioethical discourse. On the contrary, they refer to commonalities, global commons, and common goods. Biodiversity is regarded as the common heritage of humankind. It is the legacy that the present generation has received from the earth that its members have in common. The principles also refer to the idea of a global moral community. This community has expanded in space by including not only human beings but all forms of life, and in time by including various generations. Being part or member of a community implies a different positioning of human beings in regard to nature. As citizens of the earth, humankind is no longer master. The pervasive dualism in Western culture between human beings and nature and the view of nature as a mechanism or machine that can be used for human benefit is problematic from this community perspective. The basic problem is how human beings relate to other members of this global moral community: other human and nonhuman beings and other living things. [92] The idea of the community of life presents the challenge of "conviviality": how to take care of the environment of which we are a part. [93] Relevant ethical questions therefore belong first of all to the domain of social ethics. They arise from a culture of domination where anthropocentric interests always have priority and are contrasted to nonanthropocentrism. Power, justice, and equality are then the matrix for ethical reflection. By the same logic of domination, the unjust use of power produces the social problems of poverty, inequity, and environmental degradation. [94]

GLOBAL PRACTICES

The fourth argument against separating biomedical and environmental ethics is that healthcare implies earth care. [95] The emphasis on biodiversity demonstrates that concern for the environment is part of healthcare itself. Traditionally, indigenous people have taken care of biodiversity, without making distinctions between healthcare and earth care. [96] Nowadays, global citizenship means care and responsibility across humanity and nature. [97] Being global citizens and not merely consumers demands the cultivation of human sensibilities for the vulnerability of all forms of life and for environmental injustice. This citizenship will direct attention toward the structural setting in which problems emerge and to the need for collective action to address these problems. As in global bioethics, a common denominator in this culture of care for the world is the critical analysis of the dominant ideology of neoliberalism. Ethics should not be dictated by economics. Current efforts to address environmental problems therefore combine ecological protection, development, and justice. The notion of sustainability reconciles economic, environmental, and social dimensions. But it is true that in practice, the economic perspective often dominates. Development is interpreted as economic growth. The economic system is taken for granted, so environmental concerns have limited impact. In fact, sustainability introduces a rhetoric of reassurance: there is no need for drastic changes and fundamental choices, since economic growth, environmental conservation, and social justice can all be realized. [98] Green neoliberalism is also founded on the dualism between nature and society. Nature is a resource to be used and exploited; but the use should be wise so that this resource will not be depleted for generations to come. The Convention on Biological Diversity illustrates the problem. On the one hand, it articulates that biodiversity has intrinsic value. On the other hand, justice implies sustainable use and fair benefit sharing. Despite the rhetoric, in practice this leads to a neoliberal approach. Biodiversity provides resources that have an economic value. Property rights are the best way to conserve biodiversity. Rather than regarding biodiversity as global commons or common heritage, the convention holds that it belongs to states. Countries in the global South can commercially exploit their biodiversity, while countries in the North can patent products derived from it. The idea is that biodiversity can be conserved and used at the same time. The results are disappointing. Biodiversity continues to decline. Use of biodiversity does not deliver much for developing countries. [99] Critical ethical analysis would reveal that the neoliberal processes of globalization that are responsible for environmental degradation and global injustice cannot at the same time be employed and promoted as the solution of these problems.

Conclusion

Potter introduced the concept of bioethics as an appropriate name for the extension of biomedical ethics from individual to social and environmental issues.

He was influenced by Leopold's idea of the community of life, which emphasizes that humans and nonhumans are interconnected. Their well-being, flourishing, and survival cannot be separated. This chapter has argued that the concept of biodiversity provides an ethical discourse that connects human beings and nature, as well as natural and cultural diversity. Such encompassing discourse is especially relevant for health and healthcare. Health nowadays is interconnected; human health, animal health, and healthy ecosystems are linked. Subsequent chapters examine how biodiversity puts essential notions of bioethics within a broader perspective: health, disease, medicines, food, and water. This examination shows how concerns for the environment have significantly shifted. While in Potter's time the focus was on pollution—emphasizing risks and agents that directly harm individuals—it is now broader. [100] Climate change, for example, produces global systemic change and a permanent crisis of biodiversity, thus worsening the living conditions of all beings. While environmental pollution and toxic substances have traditionally been included in the domain of public or occupational health, biodiversity should clearly be the professional concern of healthcare in general.

As this chapter has argued, there is another reason why biomedical and environmental issues can no longer be separated: globalization. Biodiversity loss is a global phenomenon. But health is now also a global endeavor. The idea of the community of life therefore extends the scope of health challenges; it demands a global bioethics. [101]

The emerging area of global bioethics takes into account the global character of contemporary ethical problems. It furthermore implies that different practices are needed. Trans-boundary problems require international cooperation, solidarity, and collective responses—and thus an ethical discourse beyond the usual emphasis on individual autonomy and personal responsibility. Global bioethics also critically addresses the sources of these problems and provides an ethical framework to scrutinize them. This critical approach is important because this chapter has indicated (and subsequent chapters elaborate) that biodiversity talk often derives from an economic point of view. Bioethical analyses usually do not pay much attention to this economic framework in which problems emerge and multiply. Bioethicists sometimes argue that ethical concerns should be separated from social, economic, and political issues; these issues are either not within the domain of ethics or are too complicated to address. [102] Global bioethics, on the other hand, focuses on the social, economic, cultural, and political context of ethical problems. It attempts to overcome a mere economic point of view. It argues that economic approaches pretend to be neutral and objective but in fact depend on how the human and natural world is valued. Neoliberal policies of globalization often conceal ethical and political judgments. Global bioethics not only criticizes the underlying ethical assumptions of neoliberalism but also gives priority to the basic challenge in the current era of the Anthropocene, which is identified by Jedediah Purdy as "what kind of world to make together." [103]

Finally, global bioethics promotes a global consciousness focused on shared identity, a sense of global solidarity, and concern with what all beings have in common. Highlighting commonalities is important, since processes of globalization are increasingly associated with inequalities. Biodiversity is not equally distributed across the world. The same is the case for responsibility and blame for loss of biodiversity. Access to natural resources, like access to healthcare and medication, is not the same for everybody. Controversies abound over the use of resources. From a commercial point of view, searching for resources in biodiversity is promising bioprospecting with many potential bioeconomic benefits. From the point of view of indigenous and local populations who have accumulated traditional knowledge about using such resources for centuries, it is biopiracy. Such inequalities continuously raise issues of global justice. Environmental degradation leads to separation between those who suffer its consequences in disease and poverty and those who have the means to escape from these threats. [104] The focus on justice and fairness therefore provides an overarching ethical framework to address health as well as biodiversity. A recent call for action in the *Lancet* advocates a new social movement for planetary health. [105] Global bioethics should consider the health of ecosystems; conserving biodiversity is necessary, because human health depends on it. Collective action is needed to address the unjust global economic system that has been created. Bioethics as global bioethics should concentrate on people and equity. It should focus on the needs of populations across the world, especially the most vulnerable. It can only do so if it unites individual, social, and environmental concerns in a comprehensive response to ethical challenges.

A World of Dignity for All?

Data from the United Nations *Millennium Development Goals Report 2015*:

- More than 800 million people live in extreme poverty and suffer from hunger.
- Among children under age five, 160 million have inadequate height for their age owing to insufficient food.

- Half of global workers are working in vulnerable conditions.
- Each day, 16,000 children die before their fifth birthday, mostly from preventable causes.
- One in three people (2.4 billion) uses unimproved sanitation facilities.
- Over 880 million people live in slumlike conditions.
- Only 36% of the 31.5 million people with HIV in developing countries receive antiretroviral treatment. [106]

3 | Health

This book argues that contemporary healthcare is in need of a broad conception of bioethics that combines medical, social, and environmental concerns. Confronting new and emerging global problems that cannot be adequately addressed by mainstream bioethics with its primary focus on individual decision making will require a global bioethics as advocated by Van Rensselaer Potter. The previous chapter has proposed that the new notion of biodiversity provides a major impetus for developing and applying such a comprehensive model of global bioethics. Subsequent chapters further elaborate why this is the case. This chapter focuses on health. The central purpose of healthcare is the maintenance and improvement of health. This is impossible without considering biodiversity, which provides significant determinants of health such as nutrition and water. It is also a source for new medication. Loss of biodiversity is responsible for diseases and disasters. But biodiversity is more than a positive resource or threat to health; health itself is unthinkable without the existence of biological diversity. The implication is that healthcare can no longer ignore environmental issues if it wants to accomplish its goals.

Health concerns associated with the environment are not new. Present-day healthcare has lost its connection with the environment because it is primarily oriented toward technological interventions and medical treatments of individual patients. A similar orientation is noticeable in mainstream bioethics. It is driven by the question of how individuals should make decisions regarding the use of available treatments and interventions. However, such specific orientation is rather new; during most of its history, Western medicine was largely concerned with the natural and social environment in which human beings were living.

Health and Environment

Health is determined by environmental influences. That is the thesis of one of the oldest Hippocratic writings, *Airs Waters Places*. The author argues that anybody who wants to pursue properly the science of medicine must proceed as follows: first, consider the effects of each season of the year; then, the winds and properties of the waters, the geographical conditions and places where people live, the soil, the mode of life; and finally, the weather and climate. [1] Careful attention to environmental factors has for a very long time provided the medical framework for health and disease. These basic medical concepts have been related to the context of human existence rather than to individual persons.

The idea that health and disease are associated with atmospheric and ecological circumstances inspired ancient Greek and Roman medicine. What has been called "meteorological medicine" assumed that physicians can anticipate and predict diseases of populations if they know the climate, soil, water, air, and geographical conditions. [2] This assumption is based on a particular worldview, inspired by Ionian natural philosophers such as Empedocles. The world is regarded as a cosmos, a regular, balanced, and beautiful order constituted of four basic elements: water, earth, fire, and air. The constitution of the human body corresponds to this order; it is made up of four humors. Health is the perfect and harmonious mixture of these constituents. Diseases emerge when the balance between bodily humors is disturbed under the influence of changes in the environment. Within this cosmology, the human being is not only dependent on the macrocosm but part of it, as a microcosm. [3]

This cosmology has determined medical thinking and practice for centuries. Living in harmony with nature is regarded as the pinnacle of health, as it has been in most traditional worldviews. Nowadays one would argue that human beings are part of the biosphere; they do not simply interact with it and are dependent on it, but belong to it. This highlights the idea of interconnectedness of humans and the surrounding world that is one of the building blocks of Potter's idea of bioethics.

The central role of the environment in Western medicine has been particularly powerful in explanations of disease. Beginning in antiquity, these explanations were characterized by two "thought-styles," that is, theories about the nature of medical problems and possible ways to resolve them. In the past, there was no way to know whether these ideas were true or false, but they were useful in gathering scientific evidence. [4] The first, contagionism, attributed diseases to minute living organisms (*contagia*, or germs) transferred through contact between individual persons. This conception was important in medieval times (establishing quarantine and pest houses) and into the fourteenth and fifteenth centuries (during the bubonic plague). But until the nineteenth century, it was impossible to confirm the existence of *contagia* and their role in transmitting diseases. For most of the time, the dominant theory was miasmatism. Disease was attributed to exhalations, impurities, and odors (*miasmas*). These toxic substances emanated from insalubrious conditions in the environment such as bad drainage, lack of ventilation, polluted water, crowded housing, dirt, and waste. Under specific conditions, it was believed, the poisonous miasmas are propagated through the air (often noticeable by smell) and affect populations rather than individuals. Disease is the product of the environment and not the result of an individual agent or microorganism. [5] Miasmatism as first articulated in Hippocratic meteorological medicine could explain the emergence of epidemics as well as common diseases such as malaria ("bad air" attributable to marsh miasmas). However, it could not prove the association between disease and environmental determinants, especially the quality of surrounding air. For a long time, there was no practical implication; place and climate, soil and air seem unalterable.

Miasmatism revived in the eighteenth century and was reformulated as an environmentalist disease theory. This new interest was driven by the conviction that environmental conditions could be modified. Physicians undertook scientific observations and collected statistical data on the habitat of human beings. They also initiated measures to avoid diseases, such as drainage of swamps and standing waters, cleansing of streets and public areas, and ventilation. [6] A new discipline of medical geography studied the impact of place and natural situation on health so that specific environmental conditions could be related to miasma and disease. The arrival of cholera in Europe in the 1830s provided another boost to miasmatic theory. [7] Traditional measures such as quarantine, closing borders, military blockades, and isolation of patients (based on contagionism) did not stop the epidemic. Most physicians related the disease to an unhealthy environment. They argued that it was important to modify the environmental conditions that facilitated the spread of cholera rather than to have precise knowledge about the causes of the disease. [8] Practical approaches were urgent because of the deterioration of the environment. The Industrial Revolution had significantly changed the natural and social environment. Deforestation and enclosure of common rural areas had increased urbanization in many European countries. Population growth and poverty were associated with overcrowding, miserable living and labor conditions, and general lack of hygiene and sanitation, especially in cities. [9] Against this environmental and social background, the new challenge of cholera motivated the emergence of the "sanitarian movement." Uniting physicians, lawyers, philosophers, teachers, and politicians, the movement advocated for improved hygienic conditions through sewer systems, waste removal, potable waterworks, and better nutrition. [10] The basic idea was that human beings are determined by environmental influences, not merely natural ones but increasingly also by social circumstances: dirt, polluted water, high humidity, and bad air as a result of the urban habitat. It was recognized that diseases such as typhus, tuberculosis, and cholera are first of all diseases of poverty. [11] Living conditions are unhealthy because human activity has deteriorated or even destroyed the previously healthy natural environment. This transformation of the environment is jeopardizing human health.

The sanitary point of view implies a specific perspective on disease and the role of medicine. Disease is not considered primarily an individual problem but the result of human interventions in the environment; it is a social problem. The same view applies to health; it is a public challenge originating from the conditions in which people live. That means that coordinated efforts are necessary to sustain a healthy environment. Although health is a value and blessing for individuals, in this view it should basically be effectuated at a population level, requiring collective action, preferably by strong governmental intervention. "Sanitizing" the environment, and thus safeguarding the health of citizens, is the duty of society and state authorities. Ultimately, influencing the intimate connections between health and environment is a political challenge. This is epitomized by the views of Rudolf Virchow, the German founder of scientific pathology. When poverty, hunger, social misery, disease, and environment are closely linked, then medicine, he argued, is a social science; politics is medicine on a grand scale. [12]

Meteorological medicine, miasmatism, medical geography, hygiene, the sanitary movement, and social medicine show that during most of the history of Western medicine, health has been considered within an ecological context. Medical theories as well as healthcare practices have been based on inescapable linkages between health and environment.

Health and Biodiversity

Nowadays there are numerous examples of the interconnection between health and biodiversity. [13] As in the past, emphasis can be placed on threats and diseases, as well as on the positive contributions of biological diversity. Loss of biodiversity can lead to emerging infectious diseases. It can change the distribution of existing disease vectors so that, for example, malaria can resurge in areas where it has been eradicated in the past. It can also give rise to new diseases owing to dissemination of viruses such as Ebola and Zika. Biodiversity decline is related to the introduction and spread of invasive species that further affect biodiversity. Reduced access to clean water and food shortages can be the result of disruption of ecosystems. It is well documented that environmental degradation will have serious consequences for human health. These threats testify to the interactions between human beings and their environment. An example is the depletion of ozone levels over the Antarctic and Arctic, reported in the 1980s and 1990s, resulting from chemicals (chlorofluorocarbons) used in aerosol cans. The ozone layer of the atmosphere shields the earth against ultraviolet radiation, and its depletion reduces the protection it provides. DNA and protein of living beings is damaged, increasing the incidence of melanomas, cataracts, and skin cancers in humans. In this case, often presented as a successful policy intervention, the damage has been diminished through global action. [14] However, in many other cases, conflicts about how to protect health as well as the environment are not easy to solve. Chemicals may have a range of toxic effects, but harms are often difficult to prove.

Agriculture is one of the major destroyers of biodiversity (because of deforestation and destruction of habitat), but it is necessary for the production of food. Malaria is a major global cause of mortality and morbidity; it can be prevented with pesticides that eliminate mosquitoes but that at the same time produce environmental harm. [15]

Declining Sperm Counts

Compared with 1971, men in Western countries today are producing 60% less sperm. Pesticides, hormone-disrupting chemicals, diet, stress, smoking, and obesity are blamed for this decline. The negative impact of the modern environment on male health is reflected in the sperm count. [16]

Despite these negative interactions, more and more attention, especially since the adoption of the Convention on Biological Diversity (CBD), has been directed toward positive relations between biodiversity and health. [17] Biodiversity is considered a major source of useful medication. Common drugs like aspirin, digitalis, and taxol are extracted from trees and plants. The medicinal use of these natural substances has been traditional knowledge among indigenous peoples for centuries. Certain species of frogs in the rain forest, for example, produce deadly toxins used by native Indians to poison arrows. Scientific study of these toxins has led to better understanding of electrical transmission in nerves and muscles. Many species can provide models for studying and understanding human physiology and pathology. These opportunities for science and medicine are lost because of the current extinction rates of many species. Biodiversity is furthermore an important source of food. It is estimated that there are 75,000 edible plants; humans use only a few of them. Genetic diversity protects crops against diseases and therefore contributes to food security. Here again, biodiversity is invaluable for science. Genetic diversity particularly is regarded as a library for the life sciences. Biodiversity also provides mechanisms to control pests and pathogens, and to protect against disturbances. For instance, insects are biological indicators for accumulation of chemicals in the environment. Finally, biodiversity is important for mental health; as a source of recreation, creativity, and spirituality, it enhances the quality of life. [18]

In 1995, the American Medical Association urged healthcare professionals to become more aware of the important relationship between human health and biological diversity. [19] Since then, awareness about the crucial role of biodiversity for health has greatly increased. Ten years later, a World Health Organization (WHO) report warned that approximately 23% of the global disease burden (the number of years of healthy life lost to disease) and 24% of all deaths can be attributed to environmental factors. [20] These factors, such as unsafe drinking water, air pollution, occupational hazards, destructive land use, and ecosystem changes can all be modified so that improvement of health is feasible. A recent report underlines the essential role of biodiversity: it is "a key environmental determinant of human health." [21] Nowadays, a wide range of research activities, professional publications, and policy initiatives demonstrate that human health and biodiversity are undeniably interconnected and that environmental issues have become a growing concern for contemporary medicine and healthcare.

What Is New?

Clearly, health concerns related to the environment are not new, but today's global focus is. In earlier times, physicians could take into account local conditions and direct their efforts at "sanitation" of the immediate environment; nowadays polluting activities in one part of the globe will affect countries in other parts. Another novel factor is awareness of the relational context. Human beings are not merely facing their environment and possible threats coming out of their surroundings; they themselves are part of this environment that constitutes their lifeworld. In practice, humans and nature, nature and culture, humans and society can no longer be strictly separated. The third new dimension refers to governance: how should we deal with biodiversity loss when it is a global phenomenon, connected with social and economic impacts? The three new dimensions are elaborated below, as they illustrate how environmental concerns are incorporated in global bioethics; they also indicate which ethical principles are at stake.

Global Health

That medicine and healthcare today are global activities that cannot ignore environmental changes throughout the world is demonstrated by the enormous interest in global health. [22] Academic institutions, mainly in developed countries, have set up new programs in global health; growing numbers of students spend some time working in healthcare in other countries. In the United States, the Obama administration launched its Global Health Initiative in 2010. [23] Financial resources for global health have substantially increased. New philanthropic organizations have been created and new global partnerships launched that focus on eliminating malaria and tuberculosis, developing vaccines, and promoting childhood immunization.

Previously, *international health* was the term used to refer to cooperation between countries in the area of healthcare. [24] European countries, for example, worked together to control the spread of infectious diseases, such as cholera epidemics. In the nineteenth century, they convened international sanitary conferences. A number of European powers had extensive colonial empires and were concerned about "tropical diseases" that could be imported. In the twentieth century, several institutions were established to collect data, study epidemiology, and implement measures, culminating in the creation of the World Health Organization in 1946. International health was often focused on infectious diseases and involved only a few disciplines; it included cooperation between a limited number of countries; and it often implied unilateral relations of providing expertise and aid to developing countries. The perspective has become much broader over the past two decades. The term *world health* already indicated that health is an issue for all people everywhere, and not simply in the relations between separate countries. Health transcends national boundaries, but more important, its determinants, such as social and environmental conditions, stretch beyond borders. Addressing health concerns therefore requires global cooperation and the involvement of multiple disciplines (medicine, nursing, epidemiology, statistics, social sciences, health policy, health management, economics, and political science). Also, examining health at a global scale exposes enormous differences in health between and within countries and leads to concerns about the explanation and justification of such health inequalities, particularly when effective interventions and treatments are available to reduce or eliminate them. This broad perspective is now expressed by the term *global health*. [25]

Global Health

The concept is defined by all of the following:
- Focus on the health of people everywhere (health for all)
- Recognition that health transcends national boundaries
- Understanding that the determinants of health reach beyond borders
- Identification of common needs and vulnerabilities
- Need for global cooperation
- Involvement of multiple disciplines
- Concern about health inequalities

Global health is an appropriate notion now that globalization has significantly changed the context of healthcare and bioethics, as described in the preceding two chapters. Global health refers to common problems faced by global citizens. Our own health cannot be separated from the health of everybody else. When a new epidemic disease caused by Zika or Ebola viruses breaks out, it is a reason for concern independent of how far away we are located from the source. It is also a reason to provide support and share expertise—the sooner, the better. Global health focuses attention on the commonalities we share, our mutual vulnerability, but also our shared potential of technology and knowledge that is at the disposal of humanity and that can be applied if we cooperate and coordinate actions. This presents an entirely new context for the interactions between health and environment. Rather than being sensitive to the impact of environmental factors on their individual patients, health professionals are now confronted not with local conditions but with global challenges such as global warming and ozone depletion that are increasing health risks generally. Previously, air pollution was localized (as smog in certain cities); toxic waste was dumped in specific sites and produced illnesses in people living near them; water sources were contaminated by chemicals from specific industries. These localized risks and hazards still exist, but the environmental problems are globalized: air quality is deteriorating everywhere; safe waste disposal is not available for millions of people; and water is becoming scarce in many areas of the world. That means that a sanitary movement within a town, region, or country will not be sufficient to address problems. Action is required at multiple levels by a range of actors, and not just states. The challenge goes beyond cleaning up the environment, although that will be required continuously. More is needed than remediating and mitigating localized health threats and sources of disease; threats have now become systemic because the environmental context that sustains human life and health is breaking down. [26]

Relational Context

The last argument refers to the awareness that human beings are not just coping with the environment but are part of it. More than ever in the history of healthcare, interconnectedness of human beings and the surrounding world has become a crucial consideration. It had been emphasized by Potter (chapter 1), studied by the new science of ecology, and expressed in the notion of biodiversity (chapter 2). Now it determines the relationships between humans and their world, thereby incorporating not only the social but also environmental context into the new framework of global bioethics. The growing awareness of interconnectedness can be illustrated by many examples (see box below). It is also shown in several recent contributions to the debate and practice of global health, especially in regard to determinants of health, the right to health, and planetary health.

Vultures, Rabies, and Burials

Asian vultures, once abundant in India, Pakistan, and Nepal are now critically endangered. The cause is kidney failure induced by diclofenac, a painkiller used in veterinary practice and accumulated in animal carcasses. Vultures used to scavenge on carcasses of dead animals as well as human corpses left exposed to the elements in so-called celestial burials. They can eat diseased animals without getting sick, but they also ingested the drug by eating animals that had been treated. The decline in vulture populations caused a population explosion of feral dogs and rats that normally compete for the carcasses. These animals, unlike the vultures, are vectors of diseases such as rabies. India now has a huge human rabies problem. Burial ceremonies in which the human body is left to be consumed by vultures needed to be changed, because this traditional practice has become more dangerous, attracting rabid rats instead. [27]

Determinants of Health

Human health is determined by the conditions of daily life more than by medical treatment and healthcare services. This is the basic message of WHO in the report of its Commission on Social Determinants of Health, published in 2008. [28] The determinants are very broad, referring to "the circumstances in which people are born, live, work, and age," including such negative factors as unemployment, unsafe workplaces, urban slums, risky built environments, lack of education, gender discrimination, and food insecurity, but also "the quality of the natural environment in which people reside." [29] The view that health is a comprehensive and social rather than a narrow biomedical concept is clearly connected to environmental concerns. The report links healthy places and healthy lives in a separate chapter. It discusses the impact of climate change and environmental degradation, analyzing examples of air quality and sustainable agriculture. [30]

This broad view of health is not new. It has been expressed in the Declaration of Alma-Ata, adopted by WHO in 1978, whose aim was "Health for All" by the year 2000. [31] Following the lead of the Lalonde report, one of the first policy documents relating environmental issues to health, the declaration recognized that health should entail more than providing clinical services and building medical infrastructure. Marc Lalonde, minister of national health and welfare in Canada, argued that the level of health cannot be equated with the quality of medicine; health is a field with four principal elements: human biology, environment, lifestyle, and healthcare organization. [32] In the 1970s, the broad view of health was furthermore supported by demographic studies. Thomas McKeown's thesis—that since the end of the seventeenth century health has improved without significant contributions of medicine—became widely known. Gains in life expectancy in Western countries were not the result of medical interventions but rather of environmental influences, particularly better nutrition, hygiene, and standards of living. [33]

Despite these initiatives and notwithstanding the broad definition of health in WHO's constitution (health as "more than the absence of disease"), the focus of global health activities for most of the time has been on diseases, biomedical approaches, and technical solutions, rather than on the social context of health. From its beginning, WHO has been marked by a tension between vertical (disease-oriented) and horizontal (health system–oriented) approaches. [34] However, the overwhelming majority of global initiatives have applied biomedical interventions (vaccines and drugs) and campaigns to eliminate specific diseases. [35] Recently, experiences with Ebola (discussed in the next chapter) and the work of Michael Marmot are reemphasizing the need to strengthen health infrastructure and address underlying determinants of health. Marmot argues that a focus on social determinants is now more relevant than ever. Inequalities in health are growing, not diminishing. Life expectancy is increasing, but the gap between the longest- and shortest-living populations remains the same. [36] Commitments to address inequalities are fragile; only a few countries take action. Furthermore, there is political resistance; paying serious attention to social determinants of health places the neoliberal policy outlook in question. [37] In fact, austerity politics further undermines the determinants of health. In most countries, policies of social protection are reduced so that vulnerability and insecurity are growing because the social and economic conditions that negatively impact health are no longer mitigated. [38]

Rights to Health

The idea that health is primarily determined by the conditions in which people are living is often associated with another powerful idea: health is a fundamental human right. The WHO constitution, adopted in 1946, states that "the enjoyment of the highest attainable standard of health is one of the fundamental rights of every human being." [39] The International Covenant on Economic, Social and Cultural Rights, entered into force in 1976, affirms this right (article 12). States have the responsibility to respect, protect, and fulfill the right to health. The covenant implies the improvement of "all aspects of environmental and industrial hygiene." [40]

Although the right to health is contested, and its concept and applicability criticized, its international acceptance has increased. [41] Many countries have included the right to health in their constitutions. [42] The UN Committee on Economic, Social and Cultural Rights monitors the implementation of the covenant. It issues comments on the normative implications of the right, specifying the obligations of states. Special "rapporteurs" examine country situations, publish annual reports, and investigate individual complaints. The result is a growing body of legal interpretations and evidence. Citizens and nongovernmental organizations have appealed to this right to pressure governments into providing better access to treatment and care. [43]

Connecting the discussion of social determinants of health with the right to health introduces a new political and ethical context. [44] The emphasis on health as a basic human right is a counterforce to the neoliberal ideology that regards health primarily as a commodity and driver of economic growth. Health as a right articulates a discourse in which health is not chiefly a resource for economic purposes but a common good that requires respect and protection by states, which in turn have the responsibility for providing basic services that meet human needs. [45] Health is also not a local concern but a universal right that should apply globally, even if the circumstances are different. This discourse has normative implications.

Reports on social determinants convincingly show that major differences in health exist within and among countries. [46] It is evident that in theory everyone has the right to health; in practice, this does not result in equal health outcomes. Implementing the right is the responsibility of states, but they clearly do not all provide sufficient or appropriate measures to do so. They often use apolitical language that does not refer to the neoliberal ideology, thus masking the root causes of health inequalities. The emphasis on the right to health supports the position that health disparities should be remedied and that it is the social responsibility of governments to address their root causes. [47]

PLANETARY HEALTH

A third development that articulates the interconnectedness of social and environmental issues in global health is the concept of "planetary health." It is a rallying cry for new approaches in science, policy, and governance. The focus on the planet intends to redirect attention from the virtual global sphere down to earth, to the reality of the Anthropocene, where human beings are rapidly degrading the conditions under which they live and thus endangering their health and civilization. [48] The new term expresses precisely the interconnectedness of human beings, societies, and biodiversity. Human health is inseparable from health of the earth. Earlier, the notion of health was applied metaphorically to land and ecosystems. Aldo Leopold talked about "land health," regarding health as "capacity for internal self-renewal"; health refers to integrity, stability, balance, and harmony. [49] Another application was "ecosystem health." Ecosystems are like organisms; they are healthy when they are stable and sustainable. [50] The concept of planetary health goes beyond this metaphorical use. It is based on undeniable evidence that human beings cannot be healthy if the planet is not healthy. While the notion of global health conveys the idea that the health of all human beings is interconnected and that human health is linked to healthy social conditions, the concept of planetary health ties individual health and social determinants more explicitly to environmental concerns.

Focusing on the health of the planet is policy-oriented. The basic question is: How can planetary health be safeguarded when biodiversity is rapidly lost? For example, how can a growing world population have sufficient food when agriculture already is a major cause of biodiversity loss? [51] This emphasis on policy reiterates two issues that Potter emphasized in his conception of bioethics. First, it is concerned with the future of humanity. Humans need a healthy environment to flourish and survive. Second, it expresses the need for collective action. Bioethical discourse should present a new vision that will inspire a social movement to sustain health for all.

The Sustainable Development Goals (SDGs) adopted by the United Nations in September 2015 combine the global problems of health, society, and environment into a practical framework of international cooperation and action. Among the "17 goals to transform our world" are "Ensure healthy lives and promote well-being for all at all ages" (goal 3); "End poverty in all its forms everywhere" (goal 1); "Ensure access to water and sanitation for all" (goal 6); and "Take urgent action to combat climate change and its impacts" (goal 13). [52] For the next fifteen years these goals will guide the international community in addressing global health, biodiversity loss, and social inequalities.

Governance

The connection between biodiversity and health has furthermore stimulated concerns about governance. This has become a major challenge in global bioethics. [53] The issue is not merely that addressing global health problems requires inclusion of social as well as environmental dimensions, as illustrated in the SDGs. It also demands a horizontal rather than vertical focus, impacting the healthcare system rather than specific diseases. This implies intersectoral action to overcome departmentalism and governmental silos. It furthermore requires transdisciplinary cooperation. It is estimated that approximately 75% of new emerging diseases (e.g., Ebola) in humans are transmitted from animals. Cases of rabies are increasing in Asia and Africa. These challenges require human, animal, and environmental expertise so that integrative research and practices can be developed. The concept of "One Health" is promoted to effectively cover subject areas such as zoonoses, agriculture, and food and water safety. [54] For example, the global problem of antimicrobial resistance demonstrates that human, animal, and environmental health are linked. Animal agriculture in the United States uses almost 80% of all antibiotics sold in the country. Antibiotics are used as growth promoters, especially in the pork and poultry industry (see chapter 6). The animal waste pollutes the environment and infects humans, who develop resistance against the antibiotics used. At least 23,000 Americans die each year from antibiotic-resistant infections. This problem can only efficiently be addressed when medical, veterinarian, and environmental experts cooperate. [55]

Most important, this broader scope of governance will have implications for the ethical debate on health policies. If environmental concerns are relevant and inescapable for bioethics, a broadening of the ethical framework is required. For example, draining swamps or using pesticides can extirpate mosquitoes as vectors of malaria or Zika virus and therefore promote health, but this may have negative consequences for the environment and will reduce biodiversity, affecting health in the longer run. What will be the right approach? A characteristic of global governance is its articulation of a range of responsibilities—not merely individual responsibility, as is often the case in mainstream bioethics, but also that of governments, businesses, and organizations. It furthermore aims to influence the root factors underlying ethical problems—for example, the political and economic order as well as power structures—as upstream determinants of health inequalities, poverty, and biodiversity decline. [56] The emphasis on planetary health demands transformation of the global economy. [57] Finally, in terms of global governance, scientific evidence and normative viewpoints often cannot be separated. Evidence is either limited or not decisive. For this reason, the power of policymakers is frequently based on intellectual and moral appeals. What is needed, then, is moral reasoning alongside evidence. [58] So if medical, social, and environmental dimensions of global problems are interconnected, moral reasoning should be expanded into global bioethics to provide the ethical principles and viewpoints that are appropriate for global governance. [59] The domain of bioethics therefore needs extension and redefinition. [60]

Ethical Implications

Growing awareness of the essential connections between health and biodiversity, as well as the new attention paid to global health, social determinants of health, the right to health, planetary health, and finally the challenges of global governance all point to the need for a different and more encompassing ethical perspective—one that (in contrast to mainstream bioethics, which separates itself from environmental ethics) features a global scale, a wider set of issues, and a conceptual scope beyond individual autonomy. In global bioethics, where health is connected to the natural and social world, different ethical principles become relevant. Globalization of healthcare over the past few decades shows why. Global health has enormously improved. Overall life expectancy throughout the world during the final two decades of the twentieth century increased by eight years. Child mortality has substantially decreased. [61]

But the progress in global health is associated with two paradoxes. The first is that while human health has never been better in the history of humankind, there is increasing disparity. Hundreds of millions have not benefited from new therapies, diagnostic technologies, and better healthcare. Regrettably, many poor people are still dying from diseases that have existed for centuries. Second, although global health has improved, the natural world has deteriorated. The growth of agriculture, necessary to feed a growing human population, is a major cause of biodiversity loss. Expanded and more productive food systems have led to better human health but at the same time have worsened the global ecosystem. These paradoxes in the connections between health, environment, and society demonstrate two things: the basic ethical problem is inequality, and a broad range of ethical principles is needed to address ethical challenges.

INEQUALITY

One of today's major social, political, and ethical challenges is inequality—manifested in income and wealth, gender, race and ethnicity, class, power, and social status. Inequality is not a new problem. Globalization, however, has significantly increased inequality within and between countries. [62] The 85 wealthiest individuals in the world are as rich as 3.5 billion mostly poor people. In the United States, 95% of the economic growth since the financial crisis of 2009 has been captured by the wealthiest 1% of the population, while the bottom 90% have become poorer. [63] This economic inequality is not only rapidly growing but is also associated with other types of inequality. In the United States, 46 million people are currently using food stamps. One out of six Americans is in poverty. [64] Incomes have been stable for three decades for average citizens, while annual compensation for CEOs was 243 times higher than that of average workers. Minorities have been particularly hurt, as Joseph Stiglitz shows: "Between 2005 and 2009, the typical African American household has lost 53 percent of its wealth . . . , the average Hispanic household has lost 66 percent of its wealth." [65] As illustrated earlier in this chapter, studies of social determinants of health have demonstrated the connection between wealth and health. People living in the poorest neighborhoods in England, for example, die on average seven years earlier than those living in the richest areas; during this shorter life, they spend 17 years longer with disability. [66] Economic inequality and health inequality are furthermore associated with inequality in environmental degradation. An example is toxic waste: it is often dumped near poor and minority neighborhoods. Traditional societies are most vulnerable to biodiversity loss, since they often rely directly on ecosystem services. For example, they use herbal drugs; with progressing deforestation, the sources for their traditional medicine diminish while the medication derived from the herbs is often patented by foreign companies and unaffordable for this population. Widening inequality, finally, leads to migration; poor people move to richer countries, sometimes as health professionals (causing brain drain or care drain in their home countries), or other times as illegal immigrants. Similar determinants underlie health inequalities, environmental change (especially climate change), and social conditions (especially poverty). [67]

The present concern about inequality is related to a negative view of globalization. Few deny that globalization has major advantages and completely reject it. Criticism, however, is directed at the specific politics of globalization driven by neoliberalism. This ideology assumes that human beings are first of all self-interested individuals, rational actors maximizing their own benefits. Society, in this view, is the aggregate of individual interests. Governments should only take care for the proper functioning of the market. Social safety nets should be removed or reduced. Taxes should be lowered for companies and environmental regulations minimized in order to facilitate the free trade of consumables, including healthcare and biomedical technology. This ideology promises growth and wealth. Inequality is deemed acceptable, since in the end it will lead to wealth for all. Benefits for the most wealthy and entrepreneurial individuals "trickle down" to the rest of the people. Privatization and commodification of healthcare will lead to improved health of the general population. The same approach goes for biodiversity: therefore, policies should first focus on economic growth, then alleviation of poverty, and then environmental protection. [68]

Growing inequality demonstrates that the promises of neoliberal policies are false. They are based on economic myths such as the trickle-down effect and the beliefs that wealth is the result of hard labor, that the market is efficient, and that market forces are autonomous ("the invisible hand"). In fact, most wealth is inherited; the rest of the population does not benefit from increasing wealth at the top; and markets are shaped by deliberate government policies and are also skewed through monopolies, subsidies, and limited competition. [69]

Neoliberal globalization is now one of the major sources of global bioethical problems. Globalization potentially can benefit everyone. In practice, however, it is associated with rising inequality, vulnerability, marginalization, and exclusion. This has created a different context for bioethical problems. Elsewhere it has been argued that ethical problems in healthcare during the second half of the twentieth century were generated by scientific and technological advancement as well as professional power and paternalism. [70] In response, mainstream bioethics emerged in the 1970s. It has developed an individualistic paradigm focused on respect for autonomy, principles of beneficence and non-maleficence, and justice. The moral horizon of global bioethics, however, differs. It focuses on the situated and embedded individual rather than the autonomous person disconnected from other persons, the community, society, and the environment. It addresses as sources of ethical problems the social, economic, and political powers at work in globalization rather than advances in science and technology. Global bioethics critically analyzes neoliberal policies and practices, arguing that market-driven approaches do not enhance the quality of healthcare and do not increase access to it. On the contrary, decreasing public expenditures for health and reducing the social welfare infrastructure makes healthcare less accessible. [71] Acknowledging that present-day ethical problems often emerge within a different context and from different sources than in the past leads to the search for another, broader ethical framework than mainstream bioethics currently provides.

This is also the lesson of the new approaches to the concept of health, discussed earlier. The enormous differences in mortality and morbidity within and among countries are the result of differences in social, economic, environmental, and political contexts. WHO mentions three fundamental drivers or underlying structures: constellations of power, money, and resources. [72] Addressing these root causes of inequality in health requires that the focus of ethical analysis should go beyond the emphasis on personal autonomy and individual responsibility.

RANGE OF ETHICAL PRINCIPLES

One conclusion to be drawn from the global nature of contemporary bioethics problems, as well as their origin in neoliberal ideology, is that this forces the ethical discourse to go beyond an individual focus. Neoliberal policies affect individuals all over the globe, but individuals cannot settle global problems. Global ethics needs a wider ethical framework, as advocated by Potter. It needs to bring individuals together in common activities and actions. The final reference is humanity, not the autonomous individual. That requires an ethical point of view that is shared by as many others as possible and at least aspires to apply universally to all the world's people, assuming of course that in practice full agreement will never exist. Global problems ultimately require cooperation and solidarity based on shared perspectives. This broader view also recognizes that the inequality associated with globalization makes justice one of the core issues of global bioethics. [73] The significance of biodiversity for health and healthcare is furthermore addressed by new ethical principles related to vulnerability, future generations, and sustainability. Although global challenges such as Ebola and Zika pandemics require agreement on basic ethical principles, it is also important to respect the diversity of moral views. Another conclusion is that the social-ethical nature of global bioethics initiates a continuous search for common perspectives and values that are shared among many cultures and religions. One implication is that global bioethics is necessarily a work in progress. It will never be a finished package of ethical principles or practice guidelines (such as those articulated by "principlism" in mainstream bioethics) that can be applied all over the globe. The challenge for global bioethics is to take differing moral views into account while at the same time searching for convergence, as Robert Veatch has argued, toward commonly shared values. [74] For this to happen, a moral vocabulary of interaction, dialogue, participation, and cooperation is essential. Common ground needs to be cultivated through ongoing interaction and communication.

A Different Normative Grammar

In her famous best seller, *Behind the Beautiful Forevers*, journalist Katherine Boo describes the slum dwellers in Mumbai. Their misery is not just lack of money. In a marshy area near the international airport and a sewage lake where garbage is dumped, they are trapped in a system that works against them. Every day they are confronted with structural violence, social injustice, and pervasive corruption. In Mumbai, 42% of the 27.4 million inhabitants live in slum conditions. Rapid population growth begets increasing pollution, inadequate sanitation, and contaminated water supplies. Boo's stories show how poverty, disease, hunger, violence, and environmental degradation are interconnected. [75]

The inescapable relation between individual, social, and environmental dimensions that is demonstrated in the concept of health requires ethical principles that extend beyond those used in mainstream bioethics and that can envision all relevant dimensions. Reflecting on the importance of biodiversity and its fundamental associations with health and healthcare therefore leads not only to the conclusion that environmental issues can no longer be separated from healthcare, and thus bioethics. Introducing biodiversity concerns into bioethics also gives the debate a wider scope, reinforcing the development of bioethics into an ethical discourse in times of globalization. Including environmental concerns in bioethics is not just an extension of bioethics but a real transformation. Since these principles go beyond the individual point of view that so often characterizes bioethical debate, they provide not only a broader theoretical perspective but also more practical opportunities to address global bioethical problems. The following chapters discuss these principles in relation to basic health issues connected with biodiversity, in particular principles of justice, vulnerability, diversity, future generations, and solidarity.

JUSTICE

The evidence for disparities in health between and within countries is often associated with a normative distinction between inequalities (differences in health) and inequities (inequalities that are unfair). Findings that poor people die earlier than rich people because they lack access to healthcare or that some people die from starvation while there is sufficient food in the world are regarded as unacceptable. These outcomes are avoidable through lowering the price of healthcare and medication, and through better distribution of food. Because they can be avoided, the inequalities are unfair. [76] As Marmot points out, "health inequalities that could be avoided by reasonable means are unfair. Putting them right is a matter of social justice." [77] Similar arguments are used to criticize environmental inequalities. Poor and marginalized populations are disproportionately exposed to environmental harm. It is cheaper to ship waste to poorer countries, disregarding the harmful consequences, than to dispose of it in richer countries. [78] These populations are also discriminated against because neoliberal policies prioritize economic growth over environmental protection. This priority destroys biodiversity as the basis of life. The poor are told to delay caring for the environment, while at the same time the environment is destroyed by their own efforts to survive. [79]

Against this backdrop of inequity, the principle of justice provides a powerful lens through which to analyze and criticize bioethical problems. Since the global context is marked by economic, political, and social power differences, it is essential to foreground the perspective of justice in any attempt to understand, scrutinize, and rebalance the individual and social dimensions of global ethical problems. It is also indispensable for addressing the environmental dimension. [80] Emphasizing the principle of justice will focus bioethical analyses and debates on underlying structures and power constellations that determine the context in which ethical problems arise.

VULNERABILITY

Vulnerability is a relatively recent notion. It has been used since 2000 in a variety of disciplines such as nursing science, social sciences, disaster studies, environmental sciences, and public health. The use of the term in bioethics is even more recent. This growing interest and use are related specifically to processes of globalization. Since 2005, when UNESCO member states included the principle of respect for human vulnerability in the Universal Declaration on Bioethics and Human Rights, it has been regarded as one of the fundamental ethical principles of global bioethics. [81] However, vulnerability is a broad notion; it can be applied not only to humans but also to biodiversity, species, ecosystems, and countries. In mainstream bioethics, vulnerability is primarily related to the principle of respect for autonomy. Vulnerability refers to lack of self-management and self-control. [82] Being vulnerable means not being able to exercise individual autonomy, either because of impaired autonomy or not being autonomous (anymore or not yet). From this perspective, vulnerability is a weakness; the proper response is protection. In global bioethics another perspective has emerged. The discourse of vulnerability has particularly expanded in the context of global phenomena such as natural disasters, pandemics, and environmental degradation. Vulnerability results from the damaging impact of neoliberal logic. Economic growth is associated with biodiversity loss. But the impact of biodiversity loss is unequal; poor and traditional societies are most vulnerable, since they rely directly on ecosystem services. [83] For example, when medicinal natural products are lost, indigenous populations cannot afford to purchase drugs developed from their traditional knowledge. Loss of biodiversity therefore exponentially exposes vulnerable groups and further underscores inequality. Vulnerability is, therefore, not an individual concern but is socially produced, since society itself is affected. From this perspective, ethical analysis should focus on the broader context that produces vulnerability—on what Powers and Faden have called "systematic patterns of disadvantage." [84] This is only possible in an ethical framework that goes beyond the individual perspective and that includes justice, solidarity, care, and social responsibility. The principle of respect for human vulnerability is closely related to the principle of justice. Global health is frequently regarded as an ethical commitment to the most vulnerable people and populations. [85] Everybody should have an equal opportunity to have equal health status. That implies that efforts should be specifically directed to the most disadvantaged persons and groups. Those vulnerable groups with the poorest health should receive priority. Meeting the basic sanitary and nutritional needs of the world's poorest people is not an impossible task: it costs slightly more than the annual consumption of perfume in Europe and the United States. [86]

Diversity

Chapter 1 described how processes of globalization produced global consciousness, an awareness of the world as an interconnected network. This consciousness has a double effect. On the one hand, it interprets globalization as a means to provide equal opportunities that can benefit everyone over time and that can promote the convergence of cultures. But the risk is that such equalizing processes may lead to homogenization and uniformity, especially in the cultural domain. On the other hand, global consciousness also articulates the diversity of the world; its shows the differences and varieties. It is argued that this diversity needs to be respected because it has a special value. Biological diversity, for example, needs to be conserved. [87] The first normative postulate of conservation science is that diversity of organisms is good; extinction of species is bad. It is essential for the stability and productivity of ecosystems. It is fundamental for the capacity to adapt and survive. [88] Biodiversity is therefore regarded as an ethical concept. [89] Diversity generally means variety; it offers humankind a range of possibilities and opportunities to face future challenges and has been developed and exploited by human beings since their origin. Biodiversity has particular value because it is fundamental to human health, well-being, and the survival of the species. It is the precondition for the flourishing of human beings—the life-support system for hu-

manity. But it also has value because it provides many resources to human beings. The CBD mentions various reasons to protect biodiversity, based on a wide
spectrum of values: ecological, genetic, social, economic, scientific, educational,
cultural, recreational, and aesthetic. [90] Biodiversity thus provides services. The
risk of this view is that the value of biodiversity is reduced to economic utility.

Similar debates about the value of diversity occur in regard to other forms of
diversity, especially cultural diversity. While in the past, diversity has been regarded as a source of difficulties, insecurities, conflicts, and discord, it is now
considered a positive—contributing richness, authenticity, uniqueness, and toler-
ance for difference. [91] Given these favorable connotations, many now argue that
biological and cultural diversity cannot be disconnected. This leads to the contention, discussed below, that conserving biodiversity is not possible without respecting and protecting cultural diversity. The emphasis on cultural diversity had
been fostered within the discipline of global bioethics to alleviate the fear of global
homogenization; at the same time, it risks creating polarization when it primarily articulates differences. Globalization also implies hybridization and cross-
fertilization of cultures. The recognition that other people are different leads at
the same time to an awareness that everyone shares the same humanity. That the
idea of diversity combines relatedness as well as difference is evident in biological
diversity. All species are related to other species (through evolution), yet the range
of different species (as studied in ecology) is vast. The image of the "tree of life"
immediately demonstrates connectedness plus difference. It is less obvious that
cultural diversity links differences and commonalities. As a response to the equalizing and homogenizing tendencies of globalization, it stresses that cultures are
not equal. As argued above, globalization in fact increases inequality, and inequality in many cases induces a discourse of inequity, injustice, and vulnerability. Emphasizing cultural diversity means that biological, social, economic, and
political differences are considered to be cultural attributes. A vision of difference
is provided without inequity. Differences should be respected instead of eliminated. While diversity discourse has now become pervasive and highlights concerns about globalization, it is nevertheless a rather weak counterdiscourse to
neoliberalism, arguing, for example, that the poor deserve respect but that redis-
tributing wealth is going too far. As Walter Benn Michaels puts it, "celebrating
diversity . . . is now our way of accepting inequality." [92] The Universal Declaration on Bioethics and Human Rights (UDBHR) incorporates the principle of respect for cultural diversity as one of the principles of global bioethics. [93] But

given its tendency to focus only on differences, it is a weak principle that is restricted even in the text of the declaration: it cannot infringe on other principles such as equality, justice, and equity.

FUTURE GENERATIONS

One of the principles of global bioethics particularly relevant for environmental issues is the principle of protecting future generations. It expresses an attentiveness to the future that was one of Potter's deep concerns. The responsibility of the present generation for posterity was endorsed in the United Nations Conference on the Human Environment in 1972. [94] It was emphasized in the Rio Earth Summit in 1992 and incorporated in the texts of the CBD and the Rio Declaration. [95] An important further articulation was the adoption of the Declaration on the Responsibilities of the Present Generations towards Future Generations in 1997, followed by its inclusion as a separate principle of bioethics in the UDBHR. [96] The discussion of the principle is influenced by the notion of sustainability, since the World Commission on Environment and Development defined the latter in terms of the former. [97]

Faced with erosion of biodiversity, references to future generations are often made in connection with the paradoxes of progress mentioned earlier. Our children and grandchildren will do much better in terms of health, education, and life expectancy. But their future quality of life will be worse in terms of pollution, climate, ravaged environment, and depleted resources. In other words, they will live longer but in dire circumstances. This discrepancy is regarded as an issue of intergenerational injustice. What are the rights of current generations in the consumption of current resources? Biodiversity is the common heritage of humankind; it belongs to humanity as a whole, bequeathed from our ancestors. The argument then follows that the present generations have moral responsibilities to posterity, at least not to leave a world that is irreversibly damaged so that our descendants cannot survive. However, the argument is contested. Can we have duties to people who do not exist? Do we know the needs and interests of future people?

Notwithstanding the theoretical difficulties, it is undeniable that future generations have become a moral concern. [98] These people yet to come exemplify the extension of ethics, advocated by Leopold and Potter. One reason for growing concern about the future is the power of technology, which reaches far ahead. Another is the global consciousness discussed previously. The awareness that we are interconnected is not only spatial but also temporal. Progress in genetics has made us attentive to what we have inherited and what we transfer to our offspring. Sensibility to cultural differences includes historical development but also the conservation and transmittal of culture to distant generations. The emergence of a global community implies concerns not only for present generations but also past and future ones. These concerns were reinforced by the sense of fragility and vulnerability produced by the photos of planet Earth taken from outer space. [99]

Solidarity

The focus on future generations is often connected with the notion of intergenerational solidarity, as is the principle of justice. Interconnectedness of people sharing similar goals and a common identity can produce and manifest solidarity at local, national, and global levels. [100] Appeals to solidarity are often a response to injustices, discrimination, and inequalities suffered by vulnerable groups and individuals. Solidarity is characterized by collective action. Disquiet and indignation bring people together, and they unite in social action. Sometimes this results in structural arrangements such as political movements, social services, and state retirement funds. Since solidarity demonstrates that one belongs to a community, it is committed to the common good. It differs from charity and care because it is based on mutuality. Everybody is supposed to contribute according to his or her capacities, and all may benefit according to their needs. There is no distinction between giving without receiving and taking without giving. Compassion and reciprocity go together. [101]

Its emphasis on common good and collective agency makes solidarity an important conceptual tool for criticizing and resisting neoliberal policies and arrangements. It introduces a moral vocabulary of interdependence, community, and commons that transcends the neoliberal discourse of self-interest, rational actors, and maximizing gains. Significantly, the African Charter on Human and Peoples' Rights (1981) was the first international human rights document to refer to solidarity. [102] Furthermore, the notion of solidarity expresses commitments, because when human beings recognize their shared humanity and vulnerability, it engages them in moral action. With the emergence of global bioethics, the challenge is whether forms of solidarity can be constructed and cultivated at the global level. Traditionally, solidarity has had a limited perimeter. It was restricted to specific vulnerable groups based on gender, ethnicity, or religion, thus including some and excluding others. With growing global consciousness, solidarity is expanding. Concerns for future generations have produced a sense of solidarity across generations, while global justice concerns have expanded solidarity spatially. The question is whether this global perimeter can cover not only solidarity with fellow human beings but also other forms of life. There is no reason to answer negatively. As argued in this chapter, the loss of biodiversity is impacting human health. The root causes of environmental degradation are structural; they are associated with economic, social, cultural, and political inequalities; they are therefore difficult to change via individual agency. Care for biodiversity and protecting the environment requires engagement and collective action, and thus global solidarity. This means not only caring for fellow human beings but also for other beings who are in disadvantaged, weak, and vulnerable positions in the same global community.

The Lesson from Biodiversity and Health

The current state of debate in environmental ethics makes it easier to introduce and apply the above principles into a broader discourse of bioethics. [103] For a long time, the theoretical orientation of environmental ethics has been dominated by a divergence between anthropocentric and nonanthropocentric approaches (see chapter 1). The emergence of the notion of biodiversity introduced another ethical perspective. In contrast to the term *environment*, it articulates that human beings are situated and contextual—not just surrounded by an environment but included within a diverse and living world (chapter 2). The notion of health, as discussed in this chapter, points in the same direction. It is a transversal notion, bridging the distinction between individual and context. The health of human beings cannot be disconnected from healthy biodiversity. [104] If the planet is not healthy, humans cannot be healthy. Biodiversity and health therefore stress practical interconnectedness and relationality rather than theoretical dichotomy and dualism. Two such dualities are especially criticized.

Nature versus Culture

Maintenance, resilience, and restoration of biodiversity are associated with cultural diversity. Human beings have always used and modified nature, while cultures have adapted to the natural environment. Traditionally, cultures have also been dependent on natural resources; they have built knowledge and practices in relation to biodiversity; they have developed value systems, as well as languages to identify and classify other species and substances. [105] The connections between biological and cultural diversity are expressed in the concept of "biocultural diversity." [106] The concept is useful as an attempt to overcome the dualism between nature and culture. It first shows how the same mechanisms are impacting both types of diversity. Studies in this new research area show how traditional knowledge regarding nature is disappearing under the influence of globalization. For example, people have often no local knowledge of the native flora. Schoolbooks for children refer to flora such as roses and palms rather than the plants that grow in their region. [107] Since 2007, more than 50% of the world's population lives in cities, often with limited exposure to their natural habitat. Indigenous populations, as "guardians of the land," have been driven from their traditional habitats owing to deforestation and mining. [108] Erosion of cultural diversity is furthermore reflected in the disappearance of languages. More than half of the world's populations speak one of seven dominant languages (Mandarin, English, Hindi, Spanish, Russian, Arabic, and Bengali). This is only 0.1% of the approximately 7,000 languages spoken. [109] The majority of these languages will not survive. It means that humanity's cultural domain of symbolic expression will be seriously reduced. [110]

The new concept of biocultural diversity also demonstrates what adaptive solutions are available. Diversity provides a range of options, while monocultures increase vulnerability. [111] If culture and nature cannot be separated, as is furthermore highlighted by the idea that we now live in the Anthropocene, nature cannot be conserved by separating it from humans. [112] Policies aimed at preserving wilderness and protecting national reserves from human interventions will not work. [113] If humans are integrated in nature, they should use their cultures to enhance biodiversity. Biocultural diversity studies can show what kind of strategies humankind has developed in the course of its history to address environmental and social problems.

The ethical implication of biocultural diversity is that it presents a conceptual and practical framework for overcoming dualism. Ricardo Rozzi has highlighted another ethical implication: it connects habits and habitats. [114] Modern ethics has separated human habits from the places where human beings live. Etymologically, the word *ethics* has two sources in ancient Greek: customs or habits (ἔτος) and dwelling place or habitat (ηθος). According to Rozzi, attention to the latter etymology has faded away over time. Recognizing the link between habit and habitat, however, underlines an ethical approach that connects human beings, culture, and nature. If it is true, as Rozzi points out, that "inhabiting a particular habitat generates recurrent forms of inhabiting," then moral character is cultivated within an environmental setting and within a social and cultural context cohabited by human and nonhuman beings alike. This provides an ethical perspective on human beings embedded in a biological, social, and cultural environment. [115] Embeddedness makes the human body permeable but also vulnerable, as discussed in the next chapter.

HUMAN BEINGS VERSUS NATURE

Biodiversity loss continues, species protection delivers few results, and environmental management is often ineffective. Policies fail since they do not recognize that biodiversity cannot be disconnected from human beings and their social and cultural context. A wider focus is therefore required that combines concern for the environment with human and social concerns about justice, vulnerability, diversity, future generations, and solidarity. An integrative approach is needed that takes into account issues of justice and human rights. [116] This is the lesson of reflections on biodiversity and health. Health is a relational concept; as the result of a balance between humans and nature, it is the equilibrium between organism and environment. Between human beings and nature there is not an abstract dualism and indifferent interaction but a fundamental interrelationship, since humans themselves are part of biodiversity and can only be healthy thanks to healthy ecosystems.

The usual point of view is that only human beings have value. Nature has no value in its own right; it has value only insofar as it satisfies our preferences and contributes to human well-being. Biodiversity is valuable, and thus worth conserving, because it is useful: it is a source of food, medication, industrial materials, and recreation. This view is associated with the Western imperialistic perspective that human beings exist to control and dominate nature and to use it as a resource to benefit their interests. It implies a diminished notion of nature as a marketable commodity that exists to be produced and consumed. Biodiversity is regarded as a system of ecosystem services—providing renewable resources and functioning as regulating machinery to produce human well-being by cleaning air, purifying water, and controlling diseases. [117] However, nature also has non-use values. For example, it has option value. Most of biodiversity is presently unknown. It has potential for future use. For this reason, it should be conserved as resource for future generations. Furthermore, whether or not biodiversity has a current or potential use, the existence of certain species such as elephants and whales is valuable. A contrary point of view, occurring especially in many non-Western perspectives, but also in the CBD, argues that not only human beings but also non-human species and even ecosystems have intrinsic value. The value of biodiversity is not dependent on human value judgments or human interests. [118]

Both views are no longer satisfactory. They assume that human beings and nature are disconnected as separate entities. There is no balance between managing nature and meeting human basic needs. The only choice seems to be between protection and exploitation. In a relational approach, when dynamic interaction and interconnectedness are determinative, the distinction between anthropocentric and nonanthropocentric views is reevaluated. Philosophers Bryan Norton and Sahotra Sarkar have introduced the concept of "transformative value" as a way to go beyond the distinction between intrinsic and instrumental values. [119] They argue that the utility of biodiversity is an insufficient value to justify conservation. Some components of biodiversity have no (detected) use for human beings. Other parts (such as the Ebola virus) have negative value. In general, the use value of biodiversity (for example, food and shelter) is too important to trade in the marketplace. [120] Attributing value to biodiversity because using it satisfies individual preferences is too simple and static a view. The point is that experiences with nature influence our preferences. These experiences encourage us to evaluate and critique our preferences; they may stimulate us to form new preferences. Biodiversity can transform our values. It has transformative value. [121] This means that biodiversity has value beyond mere utility, whether or not one agrees that it has intrinsic value. The ethical framework presented by Norton and Sarkar transcends the dualism between humans and nature. Biodiversity has value and needs to be respected and protected, not merely because it is a useful resource or has intrinsic value independent of human beings but because humans and nature are included in an interactive value context. [122]

Relational and Inclusive Ethics

The comprehensiveness of biodiversity, aptly demonstrated in the notion of health, has inspired efforts to develop an ethical perspective that transcends the dichotomies between humans, nature, and culture. Dirk Lanzerath, for example, argues for an "anthroporelational" ethics. [123] The fate of nature and human beings cannot be delinked. Rather than attributing value to either humans or nature, it is the relationship that should be valued. This point of view is appropriately expressed by Marie-Hélène Parizeau: "It is not biological diversity that is a value in itself, but the relationship that we have to the environment through our culture and daily lives." [124] This statement reiterates Leopold's vision that human beings are members of a biotic community, an interconnected web of life, who are in an ethical relation to the ecosystems. Such a turn to a relational ethics goes beyond the notion of intrinsic value, whether applied to humans only or also to nonhumans. It means that relationships are the source of value. If, as argued earlier in this chapter, human beings and nature, people and places cannot be separated, the ethical notion of individual autonomy should be redefined and expanded. [125] The human self is always in relation: connected to others, to social structures, and to biodiversity. A relational concept of the self may break down the tendencies toward domination that characterize Western approaches to environmental and social problems. A relational ethics furthermore advances a different interpretation of biodiversity—not as a commodity for use and exploitation to satisfy the preferences of individual consumers but as a commons that is shared for the benefit of communities.

Conclusion

The ethical discourse that has developed from reflections and practices regarding biodiversity stresses the interdependence of human and nonhuman beings within a global moral community rather than contradictory values and interests between human beings, society, and environment. The growing recognition that, as a result of globalization, health should be regarded and approached as a relational and transversal concept illustrates the same point: human health cannot be disconnected from the health of nonhuman beings or from the health of the environment and the planet as a whole. The notion of health overrides the distinction between individual beings and their social and environmental context. The growing interest in biodiversity and global health therefore supports two significant features of the developing discipline of global bioethics. First, it underscores a similar conception of human beings: they are not primarily autonomous individuals who operate as self-interested, rational agents maximizing immediate benefits but, rather, they are communal and ecological beings. A human is a social being and not merely *homo economicus*. He or she will flourish in a shared context, respecting biodiversity as common heritage and as commons, sustained for the survival of humankind. Biodiversity also shows that human beings are situated in another sense: they are grounded or rooted in local places so that habits and habitats are connected. This is typical for processes of globalization. The ideals of cosmopolitanism, global consciousness, and sense of planet are at the same time linked with local and situated concerns, a sense of place and community. [126]

The second common feature is the broadening circle of moral concern. Leopold's presentation of environmental ethics actually inspired Potter to formulate his concept of bioethics. It was not taken up by mainstream bioethics. Current global bioethics, however, is expanding beyond individual ethics, taking into account social and environmental concerns. This is not merely a matter of adding more ethical principles to the mainstream bioethical ones. Extension implies transformation of bioethics. Taking nonhuman entities into moral consideration does not imply arguing that these entities have intrinsic value. The scope of moral concern is extended on the basis of relatedness, a broader sensibility because of relationships and interactions within biodiversity. [127] The primacy of individual autonomy in mainstream bioethics can no longer be maintained if the emphasis is on interrelatedness of living beings and the embeddedness and situatedness of individual human beings. This does not imply that human individuals are subjected to more powerful ethical principles. As argued in this chapter, dualistic distinctions between nature and culture, humans and nature, but also individuals and context are difficult to maintain if humans are part of biodiversity. But it does mean that bioethics should locate the individual within a global grammar of commonality characterized by justice, vulnerability, diversity, future generations, and solidarity.

This chapter has examined the notion of health. The ancient idea of meteorological medicine has been rediscovered. Some places are healthier than others, just as certain social conditions do not enhance well-being. [128] Contemporary health discourse is therefore deficient when it does not take into account the environment. Healthy human existence is unthinkable without healthy biodiversity. The concept of health therefore integrates individual and social as well as environmental dimensions. As a consequence, ethical reflection on health and healthcare needs to adopt the same broad outlook. Environmental degradation and loss of biodiversity affect basic issues that are of prime importance for health and healthcare, such as disease, drugs, food, and water, discussed in subsequent chapters. Environmental concerns should be included in bioethical discourse, as advocated decades ago by Potter. These concerns not only require a broad ethical approach but will reinforce the discourse of global bioethics. They emphasize the situatedness of human beings and will help to move bioethical debate beyond the perspectives of autonomous, self-interested individuals.

4 | Disease

Emerging infectious diseases are one of the best examples for showing that biomedical and environmental ethics cannot be separated. Pandemics of avian influenza or Ebola and Zika virus are the consequences of human interventions such as deforestation for the sake of economic development. These interventions are reducing biodiversity and bringing human beings into closer contact with wildlife that is a reservoir of pathogens. Humans have, of course, always intervened in the natural world to remove threats or improve their living conditions. They have moved species from one place to another for scientific or agricultural purposes, sometimes with disastrous consequences: diseases, loss of biodiversity, and extinction of native species. Bioinvasion is nowadays connected with the fear of bioterrorism and the risk of accidental or intentional dissemination of dangerous pathogens. The growing anxiety over emerging diseases occurs in a context in which many microorganisms become increasingly resistant to antibiotics. While in the past, nature has provided effective therapies, it remains to be seen if it will continue to be a source of natural medical products. But infectious diseases are not the only threats from biodiversity loss. Human activities have impacted the environment through pollution, particularly with chemicals, transforming it into a source of chronic diseases. In most countries, including the developing world, noncommunicable diseases are now a bigger burden than infections.

This chapter discusses the association between diseases and biodiversity, as well as the ethical challenges resulting from it. It demonstrates how the general crisis of biodiversity discussed in chapter 2 is highlighted in almost continuous specific crises within the domain of healthcare. Today, pandemics (whether natural or engineered) are regarded as the most serious and likely global catastrophic risk for the near future. [1] Ethical concerns about how we as human beings should deal with biodiversity are directly relevant to health and disease. If loss of biodiversity continues, we will not only be increasingly confronted with unexpected and resurging diseases but also with reduced opportunities to remedy these plagues with resources traditionally provided by biodiversity. Including environmental concerns in a broad discourse of global bioethics is necessary for three reasons, as explained in this chapter: diseases are global phenomena and no longer localized; diseases are real and natural phenomena and at the same time the effect of human interventions; diseases demand responses, but these responses risk being driven by what Donald Worster has called the ethical perspective of imperialism, inspired by human dominion over nature that has produced the crisis of biodiversity in the first place. [2] But if it is true that one-quarter of the global disease burden is the result of environmental influences that can be readily modified, there are more avenues available to reduce disease than often presented in bioethical discourse. [3] Interactions between diseases and biodiversity present multiple ethical challenges. When individuals are infected or at risk, questions of benefit versus harm and of respect for individual liberty versus collective welfare (for example, regarding quarantine) will arise. Global threats to health, however, require theoretical interpretations and analyses as well as practical approaches that go beyond the ethical framework of respect for personal autonomy. From a global perspective, other ethical considerations will be on the table: the ethical principles of justice, vulnerability, and solidarity, as well as the question of how to deal with global health risks.

Bioinvasion

Article 8 of the Convention on Biological Diversity declares that contracting parties shall prevent the introduction of, or else control or eradicate, those alien species that threaten ecosystems, habitats, or species. [4] Especially since the 1990s, bioinvasion has been regarded as a major threat to biodiversity. The idea is that non-native species that are introduced and spreading outside their natural ranges cause extinction of native species and change existing ecosystems so that ecosystem services are affected, with important consequences for food, water, and health. The effects of what are called "invasive alien species" are mostly presented as negative. They can produce health problems such as allergies and skin damage.

Many disease vectors have, mostly unintentionally, invaded new territories and disseminated diseases where they were not known before. [5] Invasive species reduce yields in agriculture, fisheries, and forestry, and decrease water availability. The economic damage is significant (in Europe, for example, 12.5 billion euros per year). [6] As a result, many countries have developed policies to protect native biodiversity against these species, aimed at prevention, early detection, and rapid eradication. Although it took a long time, the European Union, for example, has issued Regulation 1143/2014 on Invasive Alien Species, which took effect on January 1, 2015. [7]

Mosquitoes and Slavery

Many species are transferred from their native habitats to new environments, and this can bring diseases to all parts of the world. It is not a new phenomenon. The mosquito *Aedes aegypti*, a native of sub-Saharan Africa, is an important vector for yellow fever, dengue, chikungunya, and Zika. It was introduced to the New World after the fifteenth century, probably through the slave trade with West Africa. The first yellow fever outbreak in the Americas was reported in 1648. [8]

Recently, the phenomenon of bioinvasion has been more critically examined. [9] Containing biological invasions seems an impossible task. One European assessment concludes that the introduction of new invasive species will continue and likely increase. [10] An interesting question is why this phenomenon is presented as bad and threatening. Human beings have always brought species to other countries and continents. The earliest cultivation of grapes to make wine took place in what is now Georgia and Armenia, 8,000 years ago. From there, viticulture was introduced in many countries. Exotic species were often introduced for commercial purposes, or simply because people liked them. Rhododendrons and domesticated cats are non-native to Europe but highly appreciated. Bringing non-native species to new habitats apparently manifests a human tendency to enhance the lifeworld, that is, to increase knowledge, satisfy desires, and increase economic output. The most common motive is the desire for new food items. Transferring domesticated animals and crops over long distances has indeed benefited agriculture. Similarly, the establishment of botanical gardens in

European universities during the sixteenth century for the study of medicinal plants was a bioinvasive project. [11]

The terminology itself brings us immediately into a negative discourse. The ethical context is already given: invasion is wrong; it brings diseases. Thus, it should be resisted and attacked. Objectively, the phenomenon could be called "biological exchange," species "transfer," or "relocating" or "transplanting" life. It is true that in some cases species introduced as biological agents to manage pests have become pests themselves. Evidence makes clear that native species can also be damaging. [12] But the science of invasion biology shows that most invasions fail and that most invaders produce minor consequences. In fact, there is not a fundamental difference between exotics and natives, except for their origins. [13] If so, why are non-native species regarded as the enemy, as aliens, killer weeds or biological pollution? Why is the distinction between native and alien used, and has even become the basis for global and national policies?

Two answers can be provided. The first relates to the idea of human dominion. The military metaphor suggests that we have to act; we should attack the invaders. Nature needs protection since it is valuable to human beings. We cannot allow biodiversity to be diluted and homogenized by resilient species that are spreading all over the world and extinguishing native species everywhere. That process is endangering the productivity of ecosystem services, diminishing economic benefits, and undermining the conditions for our health. Perhaps the concern with bioinvasion is also driven by the fear that diseases may be imported and get out of control. In any case, the assumption is that human beings and nature are separated. Biodiversity occurs in natural and relatively undisturbed systems that need to be kept, as much as possible, in a state of "nativeness." From this perspective, everything that is foreign is stereotyped as "exotic" and "alien." [14] This represents a new sense of wilderness. The natural world in our own territory is the place of balance, harmony, and order that needs protection against invasive species, while biodiversity in others' territories harbors all sorts of dangers.

The focus on bioinvasion, therefore, cannot be explained on the basis of scientific arguments. They are disputable. The science of invasion biology does not have to use military language. What is at stake are values. How do humans relate to nature? When they themselves are part of the natural world, they have choices other than simple exploitation or protection. Why can they not accept the development of cosmopolitan ecosystems with their attendant hybridization and continuous new arrivals and migrants? The parallels with the implications of globalization are clear. When globalization is deeply characterized by mobility, transcending borders, and deterritorialization, bioinvasion is a global happening par excellence. It is emerging and increasing owing to global processes, like infectious diseases. But the governance of the phenomenon is still primarily local, and therefore not effective; the responses do not take the global context seriously. If invasive species are really dangerous for human health, a more robust global policy will be required.

Virus Smugglers

In October 2004, during a H5N1 bird flu epidemic in Southeast Asia, a Thai man was arrested at Brussels airport. He had tried to smuggle two rare eagles in plastic tubes in his hand luggage. The birds were killed and tested a few days later. They were positive for the H5N1 virus. The consulting veterinarian developed flu symptoms. Everybody who had been in contact with the birds was given antiviral medication. Experts warned that an avian flu pandemic in humans had just been avoided. [15]

Annually, more than three million live birds are legally traded, 10% of which are endangered wildlife (regulated by the Convention on International Trade in Endangered Species of Wild Fauna and Flora). Fearing avian flu, the European Union banned the importation of exotic birds in 2007. Approximately 25% of the trade in wildlife is illegal, estimated to be worth US$50–$150 billion per year. [16]

The second answer identifies a specific culprit: the chemical industry. When invasive plants are useless and even harmful, they must be destroyed. Kudzu, a Japanese vine introduced in the United States to reduce soil erosion, is outcompeting indigenous plants and is regarded as an invasive weed. Herbicides are recommended to control kudzu, and one of the most widely used is glyphosate, produced by Monsanto, a multinational agrochemical business.

The Forestry Commission in the United Kingdom, for example, uses glyphosate to eradicate invasive rhododendron. [17] In 2014, half of the budget that the US government spent to fight alien species was used for glyphosate. [18] It is not surprising that total use of the chemical compound has increased by almost 94% since 1995, and especially since 2005 (6.1 billion kilograms have been applied over the past ten years). [19] In this context, accusations that the chemical industry, notably Monsanto, is benefiting from the war against invasive species are no surprise. [20] The National Strategy for Invasive Plant Management, developed by the US Bureau of Land Management, with a mission to eradicate noxious weeds, is strongly endorsed by Monsanto (and DuPont). [21] The company was also involved in the establishment of the National Invasive Species Council in 1999 in the United States. Demonizing alien species thus serves commercial interests. [22] Articulating the dangers of bioinvasion will help in developing new markets for chemical products that can destroy the aliens.

Bioinvasion illustrates how diseases are connected to environmental as well as social conditions. They do not just happen, emerge, or spread as natural events but are associated with, and often the result of, human activities, intentional or accidental. [23] Also, bioinvasion is not merely a scientific or biological issue that can be examined, described, analyzed, and managed. It is always interpreted within a context of benefits and harms. Does it increase biodiversity or does it harm it? Do imported species produce new ecosystems or do they disturb the balance and order of nature? Defining "aliens" presupposes a particular view of nature: it is characterized by stability, permanence, and predictability. It is regarded as "pristine wilderness" that needs protection, rather than as a dynamic, resilient, and constantly changing system. [24] The preestablished metaphors of bioinvasion also suggest that the phenomenon is wholly bad and should be opposed. Against this background, it is difficult to imagine that nature is always a mixture of native and alien species, giving rise to the emergence of novel ecosystems ("the new wild"), and in many cases assisting in the recovery, not decline of biodiversity. [25]

Bioinvasion, furthermore, complicates the notion of globalization. Invasive species show that natural barriers have broken down, that borders are irrelevant. Differences between species are disappearing, and diversity is diminishing. This "environmental cosmopolitanism" results from the mobility of life and is analogous to cosmopolitanism in many other dimensions. [26] Bioinvasion is not an exceptional but a common feature of the processes of globalization. However, current approaches to bioinvasion are localized. The Convention on Biological Diversity regards states as guardians of biodiversity; it does not define biodiversity as the common heritage of humankind so that global governance would be required. Threats to biodiversity, such as bioinvasion, are a national concern. [27] A sense of global solidarity will not therefore be easily developed. Another difficulty for global governance is intrinsic uncertainty. Knowledge about invading species is incomplete. It is impossible to predict how new invasions will play out. [28]

Infectious Diseases

Since 1980, more than 35 new infectious diseases have emerged in humans; that is one every eight months. [29] Infectious diseases are now regarded as a growing global threat. They strongly demonstrate the impact of the environment and the interconnection between individual, environmental, and social determinants of health. According to the ancient idea of medical geography, discussed in the previous chapter, infectious diseases originate in a specific location, but owing to globalization, they rapidly spread across borders. Local and national responses will be insufficient to diminish the global risks. Emerging infectious diseases therefore raise ethical issues similar to those of bioinvasion.

DISEASE AS A HUMAN PRODUCT

First, emerging infections can be due either to new pathogens or to known infectious agents. They are "emerging" because their incidence or geographic range is increasing. Known diseases such as cholera, malaria, and tuberculosis are resurging compared with the more recent past, when they were less frequent and more controlled. Often, concern is focused on diseases caused by a new pathogen (such as HIV or Ebola virus). Although a range of explanations have been supplied, it is clear that in both cases human activities and behaviors have been crucial to the emergence and pandemic spread of infectious agents.

"Aleppo Evil"

Leishmaniasis is a disease that produces disfiguring skin lesions, especially ulcers. It is caused by a parasite that is transmitted when infected sand flies bite humans. The disease is endemic in the Middle East and has been known for centuries as the "Aleppo evil." The lesions lead to facial scars that are socially stigmatizing. The disease is now increasing and spreading to Europe, owing to the refugee crisis created by the civil war in Syria.

Leishmaniasis is classified among the neglected tropical diseases—communicable diseases occurring primarily in tropical and subtropical conditions. They affect more than one billion people, living in poor, unhygienic, and marginalized circumstances. Every year there are an estimated 900,000 to 1.3 million new cases of leishmaniasis, as well as between 20,000 and 30,000 deaths. Since 2008, the incidence has been rapidly rising because of lack of hygiene, food, clean water, and adequate shelter. The disease is treatable and preventable. [30] This category of diseases was generally considered to be restricted to developing countries. But "tropical" diseases now are moving into other areas because of climate change and migration.

War and violence, poverty, and unsanitary conditions have always been important promoters of infectious diseases. Another one is environmental degradation, nowadays more damaging as a result of biodiversity loss and climate change.

Lyme Disease

Fever, headache, fatigue, and skin rash are symptoms of Lyme disease, the most frequently reported vector-borne disease in the United States and Europe. The bacterium *Borrelia burgdorferi* is the cause; it is transmitted through the bites of infected ticks. Each year 300,000 people are diagnosed in the United States, particularly in the Northeast and upper Midwest. Lyme disease occurs mainly in countries in the Northern Hemisphere.

Lyme disease is an example of a zoonosis, that is, a disease transmissible from living animals to humans. Ticks act as vectors, transmitting the bacterium to human beings. The ticks are infected through feeding on hosts, such as deer and mice that are natural reservoirs for the bacteria. Although its causative agent was identified in 1982, Lyme disease is not a new disease. [31] What has changed is the environment. The emergence of Lyme disease is the result of deforestation and fragmentation of forests, typical for suburban residential areas. This has disrupted ecosystems and reduced species diversity. Lack of predators and competitors has benefited the white-footed mouse as the host of *Borrelia*, and this has increased the risk of exposure to Lyme disease for humans. Ironically, the human desire to live close to nature has provoked the (re)emergence of disease. [32]

New viral diseases occur as a result of ecological degradation, leading to increased contact between wildlife and humans. It is estimated that 60% of emerging infectious diseases are caused by zoonotic pathogens. Of these zoonoses, almost 72% are caused by pathogens of wildlife origin. [33]

SARS

A global outbreak of severe acute respiratory syndrome (SARS) started in Asia and caused more than 700 deaths and 8,000 cases in 37 countries during 2002 and 2003. In March 2003, the World Health Organization issued a first global alert. The causative agent, identified in April 2003, is a coronavirus. It was found in civets sold in the food markets in Guangdong in southern China, where the epidemic started. Later, the virus was also found in Chinese bats. The hypothesis is that the virus "jumped" from animals to humans.

Infectious diseases can also emerge in unexpected ways through the effects of global warming. Climate change is one of the major causes of biodiversity loss. It also has an impact on infectious diseases, since microorganisms move to warmer areas. Diseases that were initially "tropical" have become global and now occur in regions where they used to be uncommon. A rare example of the effect of climate change is the recurrence of anthrax.

Emerging from the Ice

In August 2016, a 12-year-old boy died in Siberia from anthrax, and more than 2,300 reindeer had already died from this cause. The last human case of anthrax had been recorded there in 1941. The government is now vaccinating more than 200,000 reindeer. For one month, Siberia had been experiencing a heat wave, melting the permafrost. A 75-year-old reindeer carcass emerged from the thawing ice and exposed other animals to its dormant anthrax spores. The boy who died had eaten venison from an infected animal. [34]

Anthrax has been an ancient and common disease among animals and human beings. Louis Pasteur developed the first vaccine in 1881. Human vaccines became available in the second part of the twentieth century, as well as treatment with antibiotics. Anthrax is a serious disease that is disseminated through spores of bacteria that infect human beings. Nowadays, anthrax is rare, with only a few thousand cases worldwide. Unexpectedly, in 2016, the prediction of Russian scientists became reality. They had warned that global warming would melt the Arctic areas. The permafrost contains presumably 1.5 million anthrax-infected reindeer carcasses. Northern Russia had seen widespread and severe anthrax outbreaks in the early twentieth century. Permafrost degradation could be a hidden source of zoonoses.

The above examples emphasize what the new global health threats have in common. They are associated with significant loss of biodiversity, and directly and indirectly with global warming. They do not emerge spontaneously as natural events but are ultimately the products of human behavior. In the words of Peter Daszak and his colleagues, the same background process is at work: "human encroachment on shrinking wildlife habitat." [35] The dramatic rise of zoonotic diseases demonstrates the clear link between biodiversity and health and disease. The less diversity, the more risk for diseases. Preventing and reducing diseases requires environmental policies that maintain or enhance biological diversity. [36] The consequence of this link for global bioethics discourse is that the individual point of view must be transcended. Of course, when a patient has Lyme disease or SARS, he or she needs treatment and care. But the focus of mainstream bioethics, with respect for personal autonomy as its primary concern, will be too limited to understand the phenomenon of infection and its implications. A broader, contextual analysis needs to take into account ethical considerations of vulnerability, solidarity, and precaution.

The Discourse of Contagion

A second aspect relevant for ethical analysis relates to the normative presuppositions that, as in the case of bioinvasion, guide the interpretations, debates, and policies in a particular direction. Wildlife is viewed as a source of danger, a clear threat to human well-being, since unknown viruses are "jumping" among animal species and then to humans. Because biodiversity is diminishing, reservoirs of pathogens harbored by nature can no longer be controlled by nature itself but are transformed, creating a continuous danger to human health. New diseases therefore reinforce the old antagonism between humans and nature. The adjective *emerging* suggests not only that unknown germs arise out of deteriorating biodiversity but also that developing countries are places where the dangers come from (since they often are the "hot spots" of biodiversity). Microbial movement usually is in one direction. Africa is considered the birthplace of exotic viruses such as Ebola and Zika, while China is the home of influenza. Associating epidemics with foreigners and strangers is an old practice. Epidemic diseases produce fear and anxieties about death and collapse of society. They are regarded as consequences of imprudent or unhygienic behavior. [37] Contagion is unavoidable; it refers to communication, interaction, connection, and community. It demonstrates that human contacts can be perilous. When contacts increase, more diseases are communicable. But individuals can manage their health and avoid germs as much as possible. Contagion means that the source of disease is outside of the body. [38] The emphasis on "emerging" furthermore reinforces the notion of unpredictability. It suggests that there is a discontinuity with the past; risks and threats are never a repetition of the past. Viruses that continually change, rapidly spread, and jump the species barrier cannot be controlled. Threats are always potential, even if they actually do not materialize. [39]

These connotations of the terminology set the stage for what should be done. If these diseases emerge from biodiversity, it is first of all important to know more about areas and regions where progressive loss of biodiversity is happening. Paying attention to the underlying mechanism of biodiversity loss demands careful monitoring and surveillance of these areas and regions. Furthermore, if animals are hosts and reservoirs, animal and human health cannot be separated (reviving the idea of One Health discussed in the previous chapter). If we are in a perpetual war with nature, eradication of viruses is the definitive solution. Our best weapons are technical devices such as vaccines and drugs. If outsiders are to blame, policies should focus on protection of borders as well as travel restrictions and quarantine, keeping the germs out; vulnerability is located in the potential, not actual victims. If these diseases are continually emerging and reemerging, planning and preparedness are essential.

GLOBAL AND LOCAL

The third ethical issue relates to globalization. The general feeling in the 1950s and 1960s was that the time of deadly infectious diseases was over. In the early 1980s, the new disease AIDS disturbed this complacency. The expanding context of globalization alerted the public health community about global risks. Outbreaks had happened earlier (flu epidemics in 1957, 1968, and 1976, while tuberculosis resurged beginning in 1985). The horror scenario of the Spanish flu of 1918–19, which killed between 20 million and 40 million people in 18 months was taken as an notorious example of what could happen. [40] Experiences showed that health systems were not able to cope with such massive breakouts and that international cooperation was lacking. The new pandemics demonstrated that epidemics are related to place (of origin), but that their impact is no longer restricted to particular places. Cholera is an example. In the nineteenth century, the disease moved slowly from one place to another, so that precautions could be taken years in advance. Now it can suddenly break out in any place. Global dissemination is unavoidable. Travel, especially via aviation and sea vessels, has accelerated. [41] With social media, outbreaks also are immediately visible, creating fear and panic. Nowadays, there is a remarkable analogy between globalization and pandemics; it is not possible to separate "global" and "local." As there is no globalization without localization, viruses need local hosts to produce global disease. They also are mutating, self-replicating, and transforming—just as global processes are characterized by mobility, dynamism, and hybridization. [42]

Emerging Disease Outbreaks

"An emerging disease is one that has appeared in a population for the first time, or that may have existed previously but is rapidly increasing in incidence or geographic range." [43]

1993: Hantavirus pulmonary syndrome; southwestern United States (1950–53, South Korea)

1997: Bird flu (Influenza A/H5N1 virus); first cases in Hong Kong

1998: Nipah virus outbreak; Malaysia and Singapore

1999: West Nile disease; New York (1937, Uganda; 1994, Algeria)

2003: SARS; first cases in China

2009: Swine flu (Influenza A/H1N1 virus); first cases in Mexico

2012: MERS (Middle East respiratory syndrome); first cases in Saudi Arabia

2013: Ebola virus disease; first cases in Guinea (1976, Zaire and Sudan)

2015: Zika virus disease; Brazil (1947, Uganda; 2007, Micronesia; 2013, French Polynesia)

Processes of globalization have furthermore led to a heightened sense of vulnerability. Global interconnectedness and interdependency, as well as pandemics, disasters, environmental degradation, climate change, and bioterrorism contributed to the idea that vulnerability is universal and mutual. [44] Nowadays we live in an at-risk society. [45] The contemporary world is filled with risks; there is always imminent danger. Globalization not only has multiplied threats such as bioinvasion and infection. It also necessitates international cooperation to face these threats. If many of the perils are global, it is in fact humanity that is vulnerable, rather than individual persons. The implication is that ethical debate should not merely focus on how to protect individual victims of infections. When human beings are situated within social and environmental contexts, ethical analysis and discussion should concentrate on the structural conditions that have increased vulnerability for emerging diseases.

UNCERTAINTY

The fourth ethical concern is uncertainty and incomplete knowledge. Often, little is known about the viruses causing epidemics, especially about their behavior in human beings. The influenza virus in particular exhibits "predictable unpredictability." [46] Viruses undergo small genetic changes; genetic segments are exchanged with other viruses of wild and domestic birds, pigs, and humans. The host reservoir for the virus is wild birds. They can migrate anywhere. Because of growing international collaboration, scientists can rapidly produce expert knowledge and share information; the virus usually is quickly sequenced, and vaccines are produced in relatively short time. However, scientists cannot claim to foresee the future. One of the lessons of the 2009 swine flu pandemic is that "no one can predict with confidence which influenza virus will become dangerous to human health and to what degree." [47] One reason is that responses are based on previous experiences. Preparedness plans in 2009 were based on the assumption that China is the place where influenza originates, but this time the bird flu came from Mexico. The main reason for lack of prediction, however, is basic uncertainty.

THE MYOPIA OF BIOETHICS

Until recently, infectious disease has not been a major topic of concern in mainstream bioethics. [48] Ethical debate, when it existed, focused on balancing private and public interests. Infected patients are victims as well as vectors of disease. The challenge then is how to weigh respect for individual autonomy against the common good, for example, in regard to vaccination and quarantine. With the emergence of new diseases and resurgence of ancient ones, concerns have appeared on the ethical agenda that require bioethics to move beyond the perspective of individual interests.

While in the past, infectious diseases have had enormous impact on societies and cultures and have been connected to poverty, injustice, and environmental degradation, present ethical debates continue to ignore the social and environmental contexts of emerging diseases. They focus on the individual person and what

he or she can do to prevent infection or to be treated. The emphasis is on technical solutions: the development of medication and the production and testing of
vaccines. [49] This is surprising, since it is now generally acknowledged that diseases emerge as a result of human interventions that cause loss of biodiversity. [50]
Mainstream bioethics not only does not pay attention to the significance of the
contextual setting, it also does not critically examine the underlying mechanisms
that often generate bioethical problems and that are driven by neoliberal ideology.
Economic development reduces biodiversity. Deforestation and disintegration of
ecosystems bring human beings and wildlife pathogens into closer contact so that
viruses can "spill over" from one species to another. Since fish is becoming scarce
in West Africa because of expanded activities of big, foreign fishing companies, the
population is consuming more bushmeat, increasing their risks for infection. [51]
Another example of the negative effect of trade is palm oil. One in ten products
in the supermarket contains palm oil. Production, mainly for export, is associated
with massive deforestation in developing countries. It also requires extensive
use of chemicals (herbicides and fertilizers). Palm oil plantations are enormous
sources of greenhouse gases. [52] While palm oil cultivation provides employment, food, and biodiesel, its production has significant long-term effects on
the environment.

It is generally acknowledged that most emerging infections are zoonoses. Birds,
bats, and pigs are natural reservoirs of viruses that can lead to animal diseases and
be transmitted to humans, which in some cases acquire the capacity for person-to-
person transmission. But why do animal diseases develop? The "wet markets" in
southern China are often blamed for avian flu. Growing prosperity in this region
has led to the proliferation of these live animal markets, where the close proximity
between humans and animals increases the risk of viruses "jumping" to humans.
China also is the world's largest producer of pork. [53] At the global level, 21 billion food animals were produced in 2008 to feed the world's population. [54] The
role of the poultry industry, with 6.5 billion birds, is of special importance to
avian influenza.

The rapidly increasing consumption of meat and milk in developed countries
has resulted in a "livestock revolution." [55] The global demand for meat has encouraged the growth of industrialized factory farming. Meat production is predicted to double by 2050. This demand has primarily benefited large agribusinesses. In 1990, factory farming accounted for 30% of world meat production; now
almost all chickens, turkeys, and pigs are "produced" by factory farming. [56] Animal farming has significantly changed, with large numbers of animals now

grown in an industrial manner, mostly indoors, and in high density. Antibiotics, vitamins, hormones, and vaccines are used to improve production. It is clear that this method of livestock farming makes it easier for viruses to replicate, reassort, and spread. [57] Most avian flu outbreaks occur in industrialized farms. They provide the best environment for the evolution of more virulent virus strains that can become influenza pandemics. [58] Although animal health and human health are connected, mainstream bioethics rarely criticizes the industrialization of animal farming (leaving it to animal ethics or environmental ethics, as other disciplines of applied ethics). In the perspective of global bioethics, the connection between animal and human health can no longer be ignored. When industrialized farming is one of the sources of emerging infections, bioethics should focus on the role of agribusinesses. [59]

Commerce versus Health

The poultry producer Charoen Pokphand is the largest multinational corporation in Thailand, supplying 90% of the country's chicken exports. When in November 2003 virus H5N1 was found in dead chickens at the company's plants, no action was taken. Instead, workers were ordered to rapidly process the diseased chickens and send them to market. For more than two months, they killed and processed without special protection. The government, especially the prime minister, who was a former businessman, covered up the epidemic, encouraging the population to eat Thai chicken. In January 2004, the government had to admit that something was wrong. At that time, two people in Thailand and seven in Vietnam had been reported dead as a result of transmission of the virus from animals. Hundreds of millions of birds were culled in eight countries. After the Thai government imposed stringent health and safety regulations on all farmers, many small producers had to close their facilities—and Charoen Pokphand was able to expand its business. It is clear that priority was given to commerce. [60]

GOVERNANCE

Another concern that requires a broad ethical approach is governance. As in the case of bioinvasion, infectious diseases emerge, reemerge, and resurge in the context of globalization, but responses are often not directed and adapted to this context. Evaluation of the 2009 swine flu pandemic concluded that national infrastructures are insufficient; only 10% of countries have a national office to collect and disseminate data. WHO faced structural impediments such as an inadequate budget. [61] Governmental responses to this pandemic were often criticized for overreaction. The handling of the 2013–15 Ebola pandemic, however, was censured for its slow and insufficient response. [62] The World Health Organization as the "moral voice for health in the world" [63] was weak and did not operate as a global leader. The focus of debate was on potential treatment, vaccines, diagnostic tools, and protective gear—in other words, on magic bullets rather than the health system—while it was clear to many experts that Ebola revealed the fragility of the global health infrastructure. [64] Both pandemics demonstrated the public impact of global diseases. Ebola was imagined as a horrible disease; the virus was known as a possible biological weapon and depicted in several movies. Influenza was often related to the frightening scenario of the Spanish flu and warnings about the next big pandemic. [65]

Within this context of fear and uncertainty, the most important concern about governance is an ethical one: the lack of international solidarity. When a pandemic is in its early stages, rich nations show only marginal interest until they themselves are under threat and the first domestic case is identified. The first response usually focuses on keeping the disease at a distance by controlling borders, restricting travel, and imposing quarantine. These countries react as if theirs are the vulnerable populations, not the people suffering and dying in foreign and underdeveloped countries. [66] Deficient solidarity and moral responsibility for global health reveal a paradox. While it is obvious that global problems are the result of globalization, the significance of this global context is not acknowledged in governance. It is not recognized that control of communicable diseases is a global common good—in other words, that it is in the shared interest of all citizens in all countries that information be shared, that surveillance systems are interconnected and public, that medication and vaccines cannot be private property, and that interventions require cooperation. Deficient global governance is the result, on the one hand, of the emphasis on national sovereignty, where states take primary responsibility for epidemic diseases, and on the other hand, of multinational businesses practices that give priority to global trade while not sufficiently subjecting the companies to global regulation. Both governments and companies have limited interest in long-term approaches that improve healthcare systems. They have no interest in enhancing social and environmental conditions that give rise to diseases. Often the focus is on the "politics of disease": interventions target particular diseases—frequently not in proportion to the burden these diseases impose on the population but determined instead by what is feasible, based on technical, economic, and political criteria. [67] The military language of invasion and threat stimulates a search for "magic bullets" that can attack and eradicate enemies. It also underestimates the capacity of local communities to develop social resistance to disease. In Sierra Leone, for example, the rural populations learned quickly about Ebola biosafety issues, but they were not involved in Western emergency responses. [68]

The long-standing debates about eradication campaigns illustrate these issues, which are in fact ethical choices. Since WHO announced in 1980 that smallpox had been eradicated, multiple efforts have been undertaken to eradicate diseases such as malaria, polio, and Guinea worm disease. The basic idea of global health policies is that the most efficient intervention will be a short-term, disease-focused one. [69] Complete elimination of a disease (for example, by annihilating vectors such as mosquitoes) would lead to a world without infectious diseases. Eradication efforts do not require broad preventive programs and enhancement of healthcare infrastructure. The basic idea is that a technical and scientific approach will suffice (such as using chemicals) and that a similar approach can be used everywhere (because diseases do not differ in geographical settings). No need to consider infections as social and environmental problems. It is also argued that nothing is wrong with extinguishing harmful species, even when it reduces biodiversity. What is the benefit of anthrax bacteria, the mosquito *Aedes aegypti*, or the parasite *Dracunculus medinensis*, which causes Guinea worm disease? Microbes are not even included in the International Union for the Conservation of Nature's list of extinguished species. [70] It should be a moral duty to eradicate such species. [71]

The Malaria Eradication Program led by WHO (1955–69) failed. [72] In many countries, disease vectors that were initially eradicated are coming back. Outbreaks of yellow fever, for example, occurred in the 1970s and 1980s, while an effective vaccine was available that gives lifelong immunity. [73] It had, in fact, been the first disease targeted for eradication. In the first half of the twentieth century, the mosquito vector had disappeared in many Latin American countries. But the disease could not be eradicated, since pathogens are hiding in animal reservoirs. WHO estimates that in 2013 between 9,000 and 60,000 people died of yellow fever, mostly in Africa. A new outbreak was identified in 2016 in Angola, spreading to the Democratic Republic of Congo, Kenya, and China. WHO distributed 12 million doses of vaccine without halting the disease. The question is why yellow fever still is a problem. [74] Diseases such as dengue and Zika are disseminated by the same mosquitoes. Since 1970, dengue has become much more frequent, and the virus has spread through many countries. It is estimated that each year there are 50–100 million infections. [75] These more recent experiences triggered a fundamental debate about the best approach to infectious diseases: eradication, elimination (regional eradication), or control (reduction in prevalence)? [76] Nowadays, it is argued that if eradication is the goal, different approaches will be needed. Public health experiences demonstrate that a strict technical approach is of limited value unless accompanied by community participation and respect for human rights. Infections are not merely a mosquito problem, but a social and environmental one. Because knowledge is fundamentally incomplete and because humans and animals share the same habitats, there is not one method that can be applied to the world; various approaches should be applied in different places. [77] The focus of global governance should therefore shift more toward the social and environmental contexts in which infectious diseases emerge.

GLOBAL SOLIDARITY

States are learning the lesson that these global problems can only be addressed by cooperation and by overcoming short-term and national self-interest. For instance, David Fidler argues that the SARS outbreak inaugurated a new era of global governance. [78] With the establishment in 2000 of a new surveillance system (the Global Outbreak Alert and Response Network), governments lost exclusive control of information. With global alerts and travel advisories, WHO operated without authorization by member states. Countries like China initially regarded SARS as a domestic affair and state secret, withholding information from its own citizens and the outside world, and even deliberately providing misinformation. Owing to modern information technology and whistle-blowers, the Chinese government eventually had to acknowledge the outbreak and cooperate with WHO. This reversal of policy showed, according to Fidler, that governance based on national sovereignty has to make place for global health governance. Information, just like microbes, does not have borders. Fighting epidemics demands openness, transparency, and accountability. It also requires a stronger role for non-state actors. Furthermore, it demands acknowledgment that global health is a public good. That means that the national interest should be redefined: it should include the interests of other countries and non-state actors. Individual countries should act in harmony with the global community. [79]

Fidler's view is probably too optimistic. Since the SARS outbreak, weak global solidarity has been displayed in many instances (e.g., the conduct of the Thai government in 2004 and the delayed response to the Ebola virus disease in 2014).

But the need for more global governance is widely felt, now that infectious diseases more frequently emerge. More effective practices will take time to develop. Nonetheless, from the perspective of global bioethics, one dimension continues to be neglected. Critical analysis is rarely directed at the underlying social and environmental processes that generate and disseminate infectious diseases. Especially in ethical debate, the focus is on the individual person, civil liberties, and availability of medication and vaccines. This ethical focus is associated with the dominant thought-style of contagionism that considers epidemics to be the result of germs and infection, rather than social phenomena related to place and context. [80] It is also associated with the history of colonial medicine that directs the efforts of global governance toward disease-focused interventions. [81] Perhaps the major influence is the neoliberal policy orientation, emphasizing individual responsibility, efficiency, and technology. The focus of mainstream bioethics implies that the scope of ethical debate is restricted. The ideological framework of neoliberalism is never addressed or criticized. The neoliberal values driving globalization, weakening public health systems, increasing inequality and vulnerability, and privatizing health by accentuating personal instead of social and corporate responsibility are usually excluded from the range of bioethical considerations.

Nature and Humans

Finally, as in the case of bioinvasion, the bioethical debate about emerging infectious diseases is predetermined by a particular view of the relationship between nature and humans. Instead of being embedded in nature, humans see themselves as confronted by nature. Similar military language sets the stage for what needs to be done. Nature hides biohazards. Humankind is in a perpetual war with microbes. Disease is an ongoing conflict between microbes and humans. Disease detectives will identify and survey these enemies. [82] Without a continuous search for better drugs and vaccines, we are at risk of losing this arms race. We must be on constant alert and have updated biopreparedness programs. The Spanish flu is the horror scenario, and there is always fear of the next big pandemic. In this metaphorical frame, nature is the source of disease; it is not the product of human intervention. It also implies that nature can be blamed. For example, we can always refer to migratory birds or genetic mutations of viruses as sources of influenza, although neoliberal agricultural practices and domestic birds in farm factories are a much more common factor. [83]

Such view of the relation between nature and humans reflects a traditional, dualistic approach that separates the human body from the environment and subsequently focuses on the disease attacking the body and the eradication of its causes. It does not take into account the need for a broad ethical discourse (discussed in chapter 2) as a consequence of neoliberal globalization and loss of biodiversity. The result is that we are locked in a continuous struggle with nature. We have to succeed in dominating, otherwise we will not survive. But that victory will most probably further diminish biodiversity. The basic lesson from globalization is that human beings are situated and embedded in social and environmental contexts. This lesson opens up other perspectives. [84] One is that health is not merely an individual or human affair. Global interconnectedness means that human, animal, and environmental health cannot be separate concerns for ethical deliberation and practice. [85] Another perspective is that microbes are not necessarily enemies but can also be regarded as inhabitants of the same world, benefiting each other. [86] Pathogens are rare. Instead of military vocabulary, the language of cohabitation and understanding could be used. The relation between human beings and microbes in most cases is symbiotic. There are bad and good germs. Microorganisms have always been there, regardless of the current emphasis on biological threats. Most germs keep us healthy.

Biosecurity

Other perspectives will not easily open up because bioinvasion and emerging diseases are framed as security issues. [87] The term *biosecurity* was initially applied to protection of agriculture, livestock, and the environment against invasive species and diseases. Human health was included later, when infectious diseases came to be regarded as security threats. [88] The term has been widely used since the War on Terror, focusing on concerns of biological terrorism. [89]

Anthrax Letters

In late September 2001, soon after the terrorist attacks in the United States, five letters with powder containing weaponized anthrax spores were mailed. They made 22 people ill. Five of them died. The FBI investigation took a long time, working under the assumption that foreign terrorists or governments were responsible. In 2008, the FBI concluded that a biodefense expert working in the US Army Research Institute of Infectious Disease was the perpetrator, but this decision is still controversial. [90]

Biosecurity has evolved into a broad concept: the management of biological risks to the environment, agriculture, animals, food, and humans. It focuses on pathogens that can affect populations, whether by natural occurrence as in the case of emerging infections, by accidental release, or by deliberate use. The emphasis over the past ten years has been on human actors (states, terrorists, criminals, scientists) as sources of biohazards. Particularly in the United States, a biodefense infrastructure has been assembled, comprising projects like BioShield (providing countermeasures such as stockpiling vaccines and drugs against the most dangerous biological agents, increase of research to develop new drugs, and fast-track approval by the FDA); BioWatch (early warning and detection); BioSense (collecting real-time health data for the whole country); and the establishment in 2005 of the National Science Advisory Board for Biosecurity (for review of dual-use research, guidelines for publications, codes of conduct, and training programs for scientists). [91] Especially after defense against bioterrorism was incorporated into the framework of biosecurity, billions of dollars have been allocated for research. The growth of funding opportunities was initially welcomed, since it could have a significant spin-off for public health. WHO advocated an encompassing approach to biothreats because the same responses to public health emergencies will be needed, regardless of the various causes of threats. [92]

Nevertheless, there is no widely accepted definition of biosecurity. It covers heterogeneous practices reflecting different concerns. In the United States, the emphasis is on bioterrorism and laboratory safety. In Australia and New Zealand, the primary concern is invasive species, while in Europe, agriculture and food safety are priorities. [93] Practices focused on pandemics, foodborne illnesses, and acts of terrorism are not the same, even when they use similar technologies. Another difficulty is that the expected common benefits have not materialized. While the funding for BioShield multiplied, the number of emergency departments in US healthcare facilities simultaneously diminished. Grants for non-biodefense microbial research decreased. The number of epidemiologists working in environmental health shrank. Many have concluded that the emphasis on and investment in bioterrorism are counterproductive and have negatively impacted public health efforts. [94] More important, the growth of the security apparatus has in fact multiplied biothreats. The rapidly increasing number of biosafety laboratories (now 1,300 with 14,000 laboratory workers) has also raised the probability of accidental release or escape of pathogens. Between 2003 and 2009, nearly 400 accidents were reported in US labs with dangerous pathogens and toxins. [95] Biosecurity measures in other areas, such as the poultry industry, focus on protection against outside infection but do not take into account that threats may emerge within the securitized system itself, where conditions for viral emergence, infection, and transmission are produced by new methods of livestock farming. [96] Finally, many experts have become skeptical about bioterrorism. [97] Threats are rare and often exaggerated. Except for a few exceptional and small-scale cases, no biological attacks have happened. Peter Washer thus concludes that bioterrorism is a myth. [98] This may be true from the perspective of science, but the fear of bioterrorism has created, perhaps more in the United States than in other countries, a specific infrastructure and ideology with important ethical implications.

Whether or not biosecurity is regarded as an ambiguous concept that refers to a tendentious reality, bioterrorism has changed the bioethical debate on health and disease—first, by securitizing health and, second, by militarizing security. Regarding disease as a security threat is, in a way, a continuation and intensification of the military vocabulary that has always been used to characterize infections and alien species. Connecting disease and terror is also not new, since the fear and dread of past pandemics has led to social disruption, economic collapse, and cultural change. Present-day biosecurity, however, is no longer metaphorical. It has created structures, systems, and methods that survey, monitor, contain, and respond to potential threats. It has reinforced the role of states to secure their borders. It has also developed a risk discourse to deal with uncertainties, suggesting that prediction is a matter of scientific data. Talking about "risk" assumes that possible harm may result from decisions we have made. In contrast to "dangers," which arise from sources beyond our control, risks refer to threats that are the consequence of human activity. [99] Security should therefore focus on humans.

Framing health as an issue of security introduces a specific perspective. By emphasizing that infectious diseases not only have an individual impact but also endanger societies and social institutions, economic exchange, and political stability, it engages states in addressing infectious disease. But it also bypasses other frames, such as regarding disease as a medical or scientific challenge, a rights issue, or a humanitarian crisis.

Presenting disease as terror has further consequences. It leads to confusion, mistrust, and conflicting interests between developed and developing countries. For example, increased disease surveillance is useful for countries if they have a functioning public health infrastructure so that they have the capacity to respond; otherwise, they may conclude that surveillance is not in their national interest. [100]

The current framing of health and disease as a security issue moves the bioethical debate in a specific direction. The frame has certain normative assumptions that cannot be the subject of ethical examination and critique. The result is that an ethical analysis of the ideas of biodefense, bioterrorism, and biosecurity is irrelevant. [101] Another consequence is that it is doubtful whether the security framework is the best way to address current global health problems. From the perspective of global bioethics, it has at least four implications that are morally questionable.

(1) The focus on security assumes opposing sides. Threats (invasions) are foreign, coming from outside and "others." They should be contained and enclosed so that "safe" can be separated from "unsafe." Instead of enhancing relationships and interconnections with other humans, other species, and the nonhuman world, the emphasis becomes identifying, monitoring, restricting, and controlling them. At the same time, this often implies that a distinction is made between more or less valuable lives, for example, protecting human life through massive culling of animal life. Biosecurity does not mean that all lives will be secure; some forms of life are saved, but others must be sacrificed. This will become all the more necessary because life is dynamic and proliferating. While biodiversity articulates change and diversity, biosecurity is concerned with control, constraint, and containment. [102] One of the consequences of the duality of biosecurity is that it promotes antagonism rather than relationships. Security discourse may therefore hinder international cooperation and downgrade the moral vocabulary of global solidarity.

(2) A second consequence of securitization is distortion of priorities. This has already been mentioned as an effect of project BioShield in the United States. In fact, invasive species are much more common than terroristic acts. [103] The distortion is also observable at the global level. It is not simply that securitizing efforts are directed against specific pathogens (those that can be used by terrorists) or particular places from where invasions originate (usually Africa and Asia). The focus of biosecurity is on keeping the dangers out, not on addressing the conditions in which they are produced. Some dangers and threats, however, cannot be militarized. This explains why biodiversity loss and damaging neoliberal market practices are not considered security threats. [104] It is obvious that there is a conflict between the management of biological risk and the neoliberal ideology of free trade. Yet this conflict is rarely regarded as an ethical problem. [105] Efforts to reduce threats to human, animal, and plant health are regarded as attempts to create barriers to trade. The neoliberal value system is taken for granted and not

regarded as the source of contemporary global problems. [106] The ethical debate is skewed. As Peter Washer has pointed out, concerns about injustice and inequality cannot be discussed within this framework of disease. [107] As explained in the previous chapter, there is massive evidence that diseases strike more in poor and marginalized populations, especially in developing but also in developed countries. Poverty and structural violence are more important to the spread of disease than is the nature of pathogens. "Potential risks" does not mean that there is equal risk for everybody. [108] Nonetheless, inequality is not a concern for biosecurity.

(3) Another implication for the bioethical debate is that the security frame implies a specific approach to biological threats. In its perpetual war scenario, it advocates continuous surveillance and vigilance. The best response to threats is preparedness and preemption (eradication) rather than prevention. [109] In 2005, the World Health Organization launched its Global Influenza Preparedness Plan, urging countries to make national biopreparedness plans. The rationale is that the future is filled with imminent catastrophes. The assumption is that these threats cannot be prevented and that their emergence is unpredictable. Furthermore, future pandemics are inevitable. Since the risks are always potential and devastating, it is better to be prepared and take preemptive action before the pathogens are an actual threat. For public health, it means stockpiling vaccines and drugs as well as early mass vaccination.

Recent experiences demonstrate the adverse health consequences and problems connected with such an approach. After the anthrax letters, the US government spent more than one billion dollars to develop an experimental vaccine, while there was no evidence of real threats. Some time later, the government launched mass vaccinations against smallpox, even though the disease had been eradicated two decades earlier. [110] During the 2009 swine flu outbreak, Western governments spent hundreds of millions to stockpile and distribute Tamiflu, a drug that allegedly could ease flu symptoms and reduce the spread of influenza but later proved to be ineffective. [111] To prepare for a pandemic, WHO's advisory committee had recommended stockpiling and vaccination. Millions were indeed vaccinated. The pandemic was much milder than predicted. The biopreparedness recommendations not only led to waste but also to allegations that WHO experts had a conflict of interest, since many of them were associated with pharmaceutical and biotechnology companies involved in vaccine production. [112] Biopreparedness has raised ethical questions such as prioritization of vaccines and antiviral therapy. Since in many countries these will be scarce, who will then qualify to receive them? There are also questions concerning the obligations of healthcare workers. Should they have mandatory vaccinations? What risks may they take in treating and caring for patients? There are also broader ethical questions that are less often addressed. Rather than focus on a duty to care for individuals, society should take responsibility to create structures of care that support individual professionals, as Jaro Kotalik has argued. [113] The emphasis on bioterrorism and preparedness has produced fewer rather than more resources for fighting diseases. The scope of public health has been reduced to infectious diseases. While funding could have been used for providing clean drinking water, food, shelter or basic healthcare, it has been wasted to prepare for speculative and hypothetical threats.

(4) The security frame finally has social and cultural implications. It works as a mechanism of depoliticization. Its emphasis on risk assessment as an objective and scientific tool is foreclosing public debate. Evaluating risks is regarded as an expert matter. However, the aim of biosecurity measures is not zero risks (there is incomplete knowledge and thus fundamental uncertainty and unpredictability) but risks that are acceptable. The fundamental question is ethical: What is acceptable and to whom? Risk assessment therefore is not a technical issue but an ethical and political one that is related to the common good. By presenting it as technical issue, the biosecurity frame usually imposes a top-down approach of regulation and communication, not involving citizens, engaging communities, or encouraging public debate. It does not make space for dialogue, solidarity, trust, and convergence around common interests. This is affecting society and culture because of the peculiar nature of the threat. Biological threats are always possible but rarely actualized. Their probability is low, but the consequences are very serious. They are potential but permanently emerging. Biological agents are invisible and difficult to distinguish from natural sources. This context of uncertainty, urgency, and threat produce a perpetual state of emergency. Everything and everyone can be a threat. Against this background, countries have introduced new legislation and practices restricting and violating human rights, reducing and surveying the public sphere, and practically eliminating privacy. Accountability, transparency, trust, participation, and engagement as core principles of democratic societies are no longer regarded as primarily relevant in the fight against biothreats. Security demands compliance and docility, not critical citizens. [114]

The climate of mistrust and secrecy promoted by militarizing biosecurity is now focused on science. Advances in scientific research are increasingly accessible to everyone. Science, especially synthetic biology, is able to make genetically modified biological agents and thus to engineer and design pathogens. [115] This is dangerous from the security perspective because of so-called dual-use research of concern (DURC). [116] A scientific product that is beneficial may at the same time be used to harm. Though this is not a new dilemma, the biosecurity frame has reactivated it, especially since the anthrax letters in 2001. The secret bioweapons program in the former Soviet Union for making stealth viruses and superbugs resistant to treatment sounded a warning. [117] Science itself also raised concerns. Rapid advances in the life sciences instigated studies on how to make polio virus or reconstruct the 1918 Spanish flu virus, as well as how to create synthetic viruses. These cases led to debates on biosecurity and special review of research proposals and manuscripts. [118] But until 2011, no manuscript had been rejected because of security risks. That changed with the experiments to enhance the virulence and transmissibility of influenza virus.

Dangerous Knowledge

Two virology laboratories did experiments to make a mutated H5N1 virus that is transmissible by air in an animal model. Normally, humans can be infected only by direct contact with infected birds. Since 1997, 600 people have been infected with H5N1. Studying transmissibility by air, according to the researchers, can help to make faster and better vaccines against H5 viruses. In 2011, the researchers submitted reports to leading journals. At the request of the National Institutes of Health, the sponsor of the research, the National Science Advisory Board for Biosecurity (NSABB) reviewed the papers. The board recommended publishing only redacted versions, that is, without key details so that information could not be misused. It considered that the risks for public health were more serious than the benefits, not only because this pathogen could be used as a bioweapon but also because of accidental release from the laboratory. The recommendation led to strong criticism from scientists. A few months later, NSABB reversed its decision and recommended full publication. [119]

Restricting access to scientific information is a serious measure because it disregards basic values of science. [120] Open access to knowledge and freedom of inquiry are fundamental values in the ethics of science and are necessary for progress in the discipline. Knowledge may help to address problems and can reduce or prevent harm. On the other hand, knowledge may be misused in ways that seriously harm individuals and societies. This dilemma is magnified in the biosecurity context, where an objective assessment of benefits and harms is impossible. Not only is knowledge incomplete and the behavior of viruses unpredictable, but in the security framework itself, crucial information is either not available or is deliberately withheld. Bioterrorist events are extremely rare; publications that have previously raised security issues have not encouraged bioterrorism, but we do not know what efforts may have been prevented. Also, the benefits are not obvious. Scientists disagree on the value of this type of research; should it be done at all? [121] It is argued that scientists have their own interests (reputation, career) for urging publication that may conflict with the public interest. They alone cannot determine which biothreats are acceptable. [122] Assessment of risks should take place in a culture of responsibility that engages the scientific community, policymakers, and also the public. [123] Furthermore, this evaluation requires an oversight structure that prospectively screens research proposals and manuscripts for DURC. The problem is that such a structure does not currently exist and that there is no international forum to examine these issues. [124]

Many are afraid, however, that biosecurity concerns may negatively affect the spirit and practice of science. Next to distortion of research priorities, they might promote self-censorship by scientists. They may also decrease the general trust in science. It might be that some knowledge is dangerous. Openness and sharing of information may create threats at any time. Some scientists might be killers and biocriminals. The assumptions of the security frame imply that we can never trust scientific research and its practitioners. Although the security framework has mobilized financial resources and promoted political support for public health, pragmatic appreciation of these results ignores the fact that securitization misinterprets and misconstrues public health and promotes values other than health, well-being, and life. Therefore, an ethical analysis will be necessary.

The security aspect of health and disease has important consequences for the discourse and practice of global health. Health threats cross borders and require international cooperation, while biosecurity is often focused on national security, prioritizing the interests of states and the needs of their citizens (e.g., stockpiling vaccines). Healthcare, as a humanitarian endeavor, aims to provide assistance to vulnerable people and populations according to basic care needs. To accomplish this goal at the global level, global solidarity is required. However, biosecurity is not a humanitarian effort to help people with diseases but a mechanism to protect specific people and groups of people from being contaminated and threatened. It is not based on solidarity but on self-interest or group-interest. During the recent Ebola virus outbreak in West Africa, for example, some US healthcare institutions discouraged their healthcare professionals from volunteering in affected regions because they could bring back the virus. [125] One of the conclusions of a review of the global response to the epidemic is that protection is too narrowly focused. In fact, other investments should be made—namely, in primary care and public health systems that provide universal health coverage. [126] Although it has often been reiterated that everybody is vulnerable to disease, the notion of vulnerability has in fact been applied only to people in developed countries. Governance has been driven mainly by the concern to secure populations who were already protected and secured from the beginning. Potential victims, not actual sufferers have been the targets of policies. The security framework, by defining infectious disease as a major threat, has limited the scope of public health. It has concentrated on exotic threats without paying attention to the inequalities underlying those diseases and has also neglected other challenges posed by environmental conditions, such as chemicals, pollution, and chronic, noncommunicable diseases. These challenges are examined next.

Noncommunicable Diseases

Among the 10 leading causes of death in the world in 2012, noncommunicable diseases, particularly cancer, cardiovascular diseases, diabetes, and chronic lung diseases were responsible for 68% of all deaths. About three-quarters of all 38 million people who died of noncommunicable diseases lived in low- and middle-income countries. According to WHO, in low-income countries, 4 in every 10 deaths are children under 15 years old. The major causes of death there are infectious diseases: lower respiratory infections, HIV/AIDS, diarrheal diseases, malaria, and tuberculosis. [127] The association between infectious disease, environmental degradation, and biodiversity loss is therefore specifically relevant for countries that lack resources. In terms of global burden of disease, however, chronic diseases are more important in other countries. But even in poor countries noncommunicable diseases are disproportionally increasing. Some have already labeled chronic diseases as an emerging pandemic. [128]

The renewed focus on noncommunicable diseases directs attention to another type of threat related to the environment; not microbes but chemicals, pollutants, and toxic products that contribute to chronic diseases. These threats are often the result of human activities that have impacted the surrounding world, placing it in a context of harms and risks. In fact, one of the inspirations for the environmental movement was the work of Rachel Carson, who highlighted how chemical pesticides (especially DDT) destroyed the environment. [129] At that time, air pollution had already evolved into a problem, with serious smog incidents in cities such as London, Los Angeles, and New York. Legal action to reduce coal burning had already been taken in some countries (1955: Air Pollution Control Act in the United States; 1956: Clean Air Act in the United Kingdom). Carson's book also showed that pollution and toxic chemicals are no longer localized; owing to industrial expansion, they are both a national and global problem. The toxic cloud released in the Indian city of Bhopal in 1984 that immediately killed 3,800 people, initiated a global movement to better protect vulnerable populations. The nuclear catastrophe in Chernobyl in the former Soviet Union in 1986 had a similar effect. [130]

POLLUTION

Chemical pollution and its consequences for health was one of the examples that Van Rensselaer Potter used to create the new discipline of bioethics. As a cancer researcher, he was fully aware that this disease was associated with carcinogenic substances in the human lifeworld. [131] Mainstream bioethics discourse, however, rarely addresses the interrelations between chemicals, pollution, and global health. And if it does, it takes a particular substance as target without critically examining the conditions and mechanisms that produce ethical questions, leaving how to deal with it to individual risk assessment.

Air pollution illustrates that such approach is too narrow. This problem is associated with 6.5 million deaths globally each year. More than 50% of these deaths occur in China and India. [132] The main source of pollution is energy production, particularly coal power plants. It is estimated that in the European Union in 2013, about 22,900 premature deaths, 11,800 new cases of chronic bronchitis, and 21,000 hospital admissions were the result of these power plants. [133] Air pollution is a global problem. Toxic dust from countries with coal-fired power plants affects people in countries that do not burn much coal. It is also connected to strong economic interests of the energy industry. Nonetheless, it is not impossible to reduce the problem. Policies in the United States have reduced emissions of major air pollutants by almost 70% since 1970, despite continued growth of population, gross domestic product, and energy consumption. [134] The main challenge is that addressing air pollution demands global cooperation.

Economic inequalities and differences in stages of development among countries need to be taken into account, even when common interests such as health and climate are acknowledged. Proof of air pollution's negative impact on health is growing stronger, now that more data are available on disease and quality of life, as well as the economic costs of air pollution. For example, there are urgent calls to phase out and shut down the 280 coal-fired power plants in Europe. Governments (e.g., Germany and Poland) subsidize the coal industry with almost 10 billion euros annually, [135] but they are under increasing pressure to stop doing so. The United Kingdom plans to phase out coal by 2025. The Paris UN climate agreement states that coal as an energy source should be abandoned. At the local level, as in Beijing, for instance, which had endured years of heavy smog, coal power plants have become the target of citizen action and protest, putting pressure on governments to act. Seven of the ten most polluted cities in the world are in China. After decades of environmental indifference and prioritizing economic growth, the government declared a "war on pollution" and responded with a National Air Pollution Action Plan in 2013, planning to cut back coal use in metropolitan areas. [136]

CHEMICALS

It is impossible to imagine modern society and civilization without chemicals. It is undeniable that they are beneficial to humankind. The universe would not exist without chemicals. [137] Chemistry has led to many medicines, fertilizers promote agriculture, and DDT has reduced malaria. An old example is goiter. This thyroid disease is the result of deficient iodine intake. It is endemic in some areas where the water supplies and soil contain low iodine. Addition of iodine to water and salt prevents the disease. In many cases, however, chemicals can be harmful. They contribute to pollution, contamination, and deterioration of the environment and loss of biodiversity. They also can directly or indirectly damage human health and harm future generations. [138] An example is the depletion of the ozone layer of the atmosphere. Ozone filters out ultraviolet radiation. However, the ozone layer is depleted by certain chemicals, in particular chlorofluorocarbons, generated by the use of aerosols and air conditioners. Ultraviolet radiation will then lead to an increase in skin cancer (such as melanoma) and eye cataracts, and it will reduce the yield of food crops. [139]

Harm from chemicals is often difficult to detect and to prove. It frequently manifests itself over the long term. Even if chemicals seem primarily to affect the natural world, they will impact human beings. Biomonitoring shows that chemicals such as lead and cadmium show up in human tissues, or polychlorinated biphe-nyls (PCBs) in human breast milk. [140] The risks are insufficiently known for new chemical substances such as nanomaterials, increasingly used in medicines, cosmetics, and computers. Studies from Australia indicate that the general public perceives chemicals negatively, as toxic and dangerous. [141]

TOXIC WASTE

The use of chemicals has also produced the problem of waste. How do we dis-pose of products that contain dangerous chemicals? The majority of hazardous waste is produced in developed countries—the same countries that saw the rise of a broad social movement with deep concerns about the deterioration of the natural environment. In the 1980s and 1990s, environmental concerns made it increasingly difficult to dispose of toxic waste within these countries owing to reg-ulations, public debate, and higher costs. A growing international toxic waste trade developed. Africa especially was a preferred dumping site. [142] Over the years, a wide range of scandalous cases occurred. Many developing countries lacked the facilities or means to properly process or dispose of the toxic materials; they were simply dumped. Many of these exports were disguised as materials for "further use" or for "recycling." Such cases spurred an alliance between low-income countries, international organizations such as the United Nations Envi-ronment Programme, and non-state actors such as nongovernmental organizations and social movements. In the end, this alliance resulted in the acceptance of the Basel Convention, attempting to ban the waste trade with these countries. [143] The treaty in fact was weak. It regulated rather than banned the trade. Implemen-tation and enforcement measures were absent. The biggest producers and recipi-ents of waste did not even ratify the agreement. Nonetheless, the environmental alliance continues to press on. In 1995, the Basel Ban Amendment was adopted, banning toxic waste transfer to developing countries (even if it is for recycling pur-poses). This amendment has now been ratified by 89 countries but has not yet entered into force. [144] It also has no monitoring mechanism. Implementation is left to governments and companies. The European Union has taken responsibil-ity for its member states. Cases of toxic waste dumping continue within countries (e.g., in Naples, Italy, in 2016) and across countries (e.g., the toxic waste dump in Côte d'Ivoire in 2006). [145]

Waste only increases. The World Bank estimated in 2012 that the world pro-duced 1.3 billion tons of hazardous waste every year; this will almost double by 2025. [146] A growing problem is electronic waste. E-waste such as mobile telephones and televisions rapidly increases the amount of solid waste. It contains harmful chemicals such as cadmium and lead. The waste is mainly "exported" to countries like China and India for "recycling." [147]

Where Is My Old Cell Phone?

Every 22 months, on average, Americans buy a new cell phone. In 2010, they discarded 150 million "old" phones. About 80% of these phones are shipped, often illegally, to countries such as China, India, and Nigeria, where they are "recycled." The e-waste is processed to recover metals (e.g., cadmium and mercury), exposing people to serious health risks. The United States is the only developed country that has not ratified the Basel Convention. American companies can easily continue free trade in toxic waste and global dumping. [148]

Moral Concerns

For a long time, studies of pollution, chemicals, and waste have primarily examined the impact on individual human beings within a local setting, for example, the deteriorating health of people living close to a waste dump or incinerator. It is now clear that these are global phenomena that require international cooperation and global approaches. This does not mean that a focus on individuals is no longer necessary. Medicine will have to attend to the harms and diseases caused by chemicals and pollution to individual persons, and mainstream bioethics therefore will continue to raise specific ethical questions. But this focus is insufficient when it does not address the context in which such harms are produced and if it does not examine how the harm can be minimized. From the perspective of global bioethics, three ethical queries are on the agenda: assessment of risks, commercial interests, and environmental justice.

Risk assessment and management, as discussed above, is not a merely technical issue. David Resnik and Kevin Elliott have illustrated this in the case of bisphenol A (BPA), a common chemical used in plastics, especially baby bottles. [149] It is known as disruptor of hormone metabolism and is associated with several human diseases. While in Canada and Denmark baby bottles with BPA were banned, other countries and WHO concluded that there is insufficient evidence that BPA is risky for human health. Risk assessment not only involves different values (e.g., health versus commerce) but also how values of humans relate to values of the environment. [150] If serious risks to health are sufficiently evidenced, a chemical should be banned. However, in many cases the risks are unknown or not certain. When the costs to public health will be high and the expected harms are serious and irreversible, a precautionary approach is advocated. The precautionary principle justifies regulatory action even if the scientific evidence for harm to human health or the environment is insufficient. The principle is widely used, for example, in discussions about the ozone layer, climate change, food safety, infectious diseases, and bioinvasion. If scientific evidence about environmental or public health harm is not compelling, but the consequences of not acting would be devastating, especially for future generations, it is imprudent to wait to take action to avert the harm until there is complete certainty. [151] The precautionary principle is now included in many international normative instruments (e.g., the Montreal Protocol on Substances that Deplete the Ozone Layer, 1987; the Rio Declaration on Environment and Development, 1992; and the Treaty on European Union, Maastricht, 1992). It is contested, for example, by the World Trade Organization, arguing that it hinders the free flow of goods and leads to protectionism; potential hazards should only be reduced after there is strong proof that they are harmful. This controversy clarifies that the precautionary principle is basically an ethical principle. It requires a prudent and anticipatory course of action. In the case of scientific uncertainty and possible serious harm, there is no technological solution; competing interests should be carefully balanced before it is too late. [152]

As in many cases of bioinvasion and emerging infectious diseases, damage to human health is often related to commercial interests. The chemical industry is a powerful lobby in many policy debates. In 2013, European scientists warned the European Commission about the dangers of endocrine disruptors (such as BPA) and urged it to take action. They argued that there is increasing evidence that chemical exposure leads to disorders and diseases in human beings (e.g., poor semen quality, breast and prostate cancer, and developmental brain disorders) as well as loss of biodiversity (e.g., extinction of amphibians and decline of oyster and seal colonies). Scientists are afraid that commercial interests have priority over health concerns. [153] Public consultation in 2015 showed massive support for policy change. Polls show that two-thirds of European citizens favor a ban on endocrine disruptors. [154] Since then, no action has been taken except to develop a priority list for further evaluation of substances. Similar controversies surround the herbicide glyphosate. A hot debate ensued when this chemical needed to be reapproved for safety in the European Union. Seven European countries have prohibited the substance. The International Agency for Research on Cancer, a WHO working group, declared it a "probable human carcinogen" in March 2015. The European Parliament voted against the approval of glyphosate. Nonetheless, in June 2016, the EU Commissioner for Health and Food Safety decided that the harmful substance could remain on the market for another 18 months. [155] These examples illustrate that commercial interests are strong. Policymakers do not prioritize health. Chemicals and also waste are primarily regarded from a free trade perspective; they are treated as commodities that should be traded. They are externalities, considered to be by-products of production, and the negative effects on the environment and human health are disregarded. The benefits of commerce flow to free traders; the costs and harms must be absorbed by the general population. Even the precautionary principle is completely overlooked, despite pious statements of its importance in the European Union. Even from the trade perspective itself, this is not fair trade.

A third ethical concern is environmental justice. In the 1970s, environmentalism in the United States gave rise to a social movement that denounced discrimination in the location of dumps, landfills, and incinerators for toxic waste. Environmental benefits and burdens are not equally distributed. Hazardous waste is significantly more transferred and deposited in low-income and minority communities. Many scientific findings and policy reports indicated that certain vulnerable populations were disproportionally exposed to pollution. In the United States beginning in the 1980s, the environmental justice movement emerged, and many grassroots movements became active. The movement linked concerns for the environment with social justice and human rights. [156] It has led to stronger regulations and different policies—for example, the requirement for community participation in planning and decision making. But in the 1990s, waste was increasingly exported outside of the country. The same process of environmental injustice manifested itself at a global level. Waste is exported for "reuse" to countries that have no adequate recycling facilities or waste handling regulations. This is called the "circle of poison" or "toxic colonialism." Many pesticides are banned in the United States, but they are sold, exported to, and dumped in developing countries. The same double standard applies to toxic waste. While waste incinerators have become more unacceptable in developed countries, waste is exported to be burned in other countries. [157] Chemicals, pollution, and waste management reflect and contribute to global inequality. Waste is increasingly dumped in poor and developing countries. Just as local waste was originally disposed of near vulnerable communities, global trade has now shifted hazardous substances mostly produced in the developed world to vulnerable countries with less regulation. These practices are a clear example of what has been called "slow violence," subjecting mostly poor populations to "slow-motion toxicity," gradually and out of sight. [158]

Wandering Ashes from Philadelphia
The city of Philadelphia contracted in 1986 with a company to dispose of municipal incinerator ash. The landfill where it used to be placed was closed in response to political protest. The freighter *Khian Sea* was loaded with the waste and attempted to unload it in a dozen countries, including the Bahamas, Honduras, and Guinea-Bissau, but was refused entry. In December 1987, it dumped the waste, labeled as "soil fertilizer," in Haiti with the agreement of the military regime. An alliance of environmentalist nongovernmental organizations created project "Return to Sender" to bring back the waste to the United States. This finally happened in April 2000, when the ashes were deposited in a landfill outside of Philadelphia. [159]

Conclusion

This chapter has argued that diseases are now closely connected to the environment. Concerns about bioinvasion, infectious diseases, and chronic illnesses attributable to pollution, chemicals, and toxic waste are all associated with loss of biodiversity and degradation of the environment. These phenomena illustrate the interconnectedness between humans and environing world. They also demonstrate how human activities can negatively influence this world to the detriment of all living beings and future generations. The lesson is that bioethics can no longer only address the ethical interests of individual human beings or human populations. In fact, these interests are themselves closely connected and cannot be separated from planetary concerns. Disease is no longer an individual medical or biological predicament but is connected to environmental and social conditions that are the result of human activities. Disease, like health, has therefore become a global concern. It manifests itself as a symptom of the general deterioration of biodiversity. Emergence and resurgence of diseases remind us that biodiversity is a precondition of life that can easily be jeopardized. [160]

The implication is that there is an urgent need for a global ethical discourse that takes into account the interrelations between individual beings, social conditions, and environmental context. Mainstream bioethics and its primary concern with individual well-being, as well as public health ethics concerned with human populations, should be incorporated into the broader discourse of global bioethics. This chapter demonstrates that global bioethics operates at two critical levels.

First, it engages in ethical questioning of the normative presuppositions that usually frame the debates. The concern with bioinvasion divides the universe in native and non-native species. Humans are continually at war with nature, coping with alien and invasive species that travel around the world. Even more threatening are emerging viruses that change and mutate constantly and that can always lead to pandemic disasters. Both bioinvasion and emerging diseases are now encapsulated in the framework of biosecurity, including the possibility of bioterrorism. The consequence is that threats are individualized; there are dangerous species that need to be immobilized and preferably annihilated; there are frightening viruses that can suddenly create havoc and against which we need better protection (medication and vaccines). The military discourse of the security frame reiterates Western civilization's entrenched view of human dominion over nature. This reinforces the conviction that nature is to blame for diseases rather than the underlying social and political mechanisms. The frame depoliticizes the ethical debate by often implicitly assuming that human actions and decisions that lead to loss of biodiversity and that eventually produce diseases are not issues relevant for ethical analyses. Such framing of the debate, however, is highly selective.

It is undeniable that harm is produced by pollution, chemicals, and toxic waste, but this is not considered a security risk or even a biohazard. The industrialization of the poultry industry or the contamination of the living environment by agribusinesses is not subjected to analyses to explore how prevention of emerging diseases can be accomplished. The individual point of view must be transcended by an encompassing analysis. Otherwise, the separation between us and them (outsiders, aliens, countries where diseases emerge, potential terrorists) inherent in the security framework will persevere. The message that in a global era, all living beings are interconnected and living on the same planet under threat will be ignored. Ironically, it is precisely because of this global interconnectedness and human embeddedness within biodiversity that human beings are confronted with challenges such as bioinvasion, infectious diseases, and chronic illnesses. However, the usual ethical analyses and responses to these challenges continue to neglect the global processes that are at work and do not take a global bioethical perspective.

Second, global bioethics discourse directs critical analysis toward the underlying mechanisms that are at the root of bioethical problems. As argued elsewhere, the fact that many ethical problems emerge, or are aggravated and sustained, within a social, economic, and environmental context requires a different ethical discourse. [161] Globalization is ambiguous. It provides opportunities and benefits for many. However, it also produces exploitation and inequality. Particularly, it has generated more vulnerability and risks, and therefore anxiety. But, generally, the focus of ethical discourse is not on the neoliberal ideology that drives processes of globalization in the direction of inequality. Mainstream bioethics with its emphasis on individual autonomy, informed consent, rational decision making, personal responsibility, and personal ownership of the body does not take into account the value of the commons or common responsibility. Essential uncertainties about the consequences of human activities are mostly driven by risk assessments that prioritize commerce over health and the emphasis on technical, short-term solutions. Governance tends first of all to be based on the national interests of powerful states. The conceptual and practical tools that are provided by global bioethics are discussed in the last chapter. The following chapters first examine how other vital contributions to human existence and health, such as medication, food, and water, demonstrate the intrinsic connections between health, social context, and biodiversity.

5 | Drugs

The ancient city of Cyrene in North Africa was known for silphion. This medicinal plant was used and exported as a contraceptive drug. In fact, however, we do not precisely know its medical benefits. The extract was so highly valued and expensive that the plant was overharvested and became extinct. [1] The silphion story is frequently used to illustrate the beneficial relationship between biodiversity and healthcare. Nature not only is a source of diseases, as discussed in the previous chapter, but also a resource that can tremendously help to preserve health and provide treatment against a range of human ailments. Plants, animals, and microbes may be potential medication. Human beings have used them for thousands of years. At the same time, the story provides another lesson: these potential medical benefits can be lost because of overuse. Because the plant was regarded as a wonder drug and transformed into a precious commodity, it was not used in a sustainable manner and thus vanished as a species.

Previous chapters have shown how current concerns about biodiversity bring together the discourses of environmental and biomedical ethics. The ethics of human health and suffering manifests itself in individual persons—in the illnesses they have and the support and care they need—but each person lives together with other living beings in a society, culture, and environment that support his or her survival. Because human beings are embedded in biodiversity, ethical analysis should not concentrate only on individual cases. It should also focus on global threats to health. Loss of biodiversity should therefore be high on the agenda of global bioethics. First of all, because biodiversity is a precondition for healthy life; we would not survive without it. Second, loss of biodiversity will result in serious harm, for example, the emerging infectious diseases that hurt us now, but it will affect future generations even more profoundly, when rising temperatures will make human existence more precarious. This chapter discusses a third connection between healthcare and biodiversity that should inspire a broad vision of global bioethics. Perhaps one of the strongest arguments for protection of biodiversity is its medicinal benefits. Many effective drugs are products of nature or derived from them. With the current loss of species and genes, potential health remedies will disappear. They will be, like silphion, irreversibly lost. Most species of plants are not known at the moment, and even fewer have been systematically examined. On the other hand, we do not know what kind of new and emerging diseases will confront us in the future. What is certain is that with today's rates of extinction, opportunities for developing innovative new drugs are rapidly lost.

Utilizing chemicals produced by plants, animals, and microbes is not simply a matter of detecting natural products and applying chemistry. Knowledge of the effects of these natural products on human health and disease, bodily processes, and pathogens is often part of systems of traditional knowledge—for example, traditional Chinese medicine or the healing practices of indigenous people. Over time, through trial and error, along with empirical observation, people have discovered and examined the healing properties of plants; they have applied and perfected them in therapeutic practices. They have shared experiences from generation to generation. Sometimes these experiences have been codified and collected in manuals, but in other cases they are orally transmitted. In most of these systems of medical knowledge, the term *natural product* is actually misleading. A particular plant is not a medicine. Without a series of human interpretations and activities (finding out what part of the plant is interesting; what preparations should be made; what effects will occur; and how to use the "product" safely, in what form, at what dosage, etc.) plants will not transform into drugs. The "natural" product is in fact the product of human knowledge. In most cases, the result is not produced by an individual human being but through a collective effort. The medical knowledge of the drug is regarded as commonly shared, developed within a long tradition of dealing with biodiversity. It is therefore problematic to regard traditional knowledge as the property of individuals.

The modern interest in natural products, especially as a result of the bioeconomic approach to biodiversity described in chapter 2 (regarding biodiversity primarily as a commercial asset) produces ethical conflicts with the broader framework of traditional knowledge. Searching for new drugs, researchers and pharmaceutical companies undertook "bioprospecting" for biological and genetic samples, particularly in developing countries with rich biodiversity. Attention was mainly directed toward the commercial value of the natural products; the human context of traditional knowledge production was neglected. Indigenous populations, nongovernmental organizations (NGOs), and grassroots movements in the developing world labeled many of these practices as "biopiracy." Natural products from their countries were simply collected, exported, and patented by researchers and companies in Western countries, even though they had already been in use for centuries. The question of who may own products derived from nature was never asked. This question is addressed in this chapter in various ways. Often it is presented as a technical and legal issue, concerning patenting and ownership. But it has a much wider scope. First, there is the issue of ownership. Biodiversity has for a long time been regarded as the common heritage of humankind. It is a legacy that we have received from previous generations. It is a precious good that we need to preserve for the benefit of all. The chief concern is to maintain this heritage, which allows future generations to continue to survive as long as living conditions are not worsening. This also raises a second ethical concern: sustainability. Is the discourse of ownership and commercial use of natural resources helping to protect biodiversity as a future source of medication? A third broad ethical issue is global justice. The benefits of biodiversity are not equally shared. If biodiversity is a global good, and indeed can lead to innovative and effective medication, how do we assure that this good benefits humanity as a whole in a just and responsible way?

Gifts of Nature

Materials from nature have always been used for various purposes: to provide shelter, to make clothes, and to produce food and medication. [2] Human beings have detected and analyzed specific properties of plants, animals, and microbes; distinguished between therapeutic, preventive, and toxic effects; and learned how to apply them in the treatment of particular complaints and ailments. Millennia ago, herbal remedies were described in ancient Egypt, India, and China. That the leaves and bark of the willow tree relieves pain was known long before its active substance, salicylic acid, was identified in 1828, produced in a synthetic version in 1898, and then marketed as aspirin. [3] Because medication was most often derived from plants—and sometimes animals (an old example is the use of leeches)—medicine and botany have been closely connected. [4] There are many examples of the direct use of plants as remedies. More often, extracts and derivatives of medicinal plants are developed and applied. [5] Over the course of history, inventories and manuals have been published that collected the stored medical wisdom of cultures. From ancient Egypt, for example, the Ebers Papyrus lists 700 remedies (mostly plants) used in medicine. The Chinese emperor Shen Hung published a book listing 350 medicinal plants. [6] In Europe, the compendium written by Dioscorides in the first century describes 600 plants, with illustrations. It was a frequently consulted document for centuries. The establishment of universities in sixteenth-century Europe was associated with the creation of botanical gardens for educational purposes. [7] During the seventeenth century, plants were systematically collected; plant hunters gathered new species in all regions of the world. Medicinal properties were often accidentally detected in practice. [8] A famous example is the discovery of digitalis. British physician William Whitering found out that a folk healer treated one of his patients with foxglove. He initiated a long series of clinical studies, evaluating efficacy and safety. The results were published in 1785. They established the reputation of digitalis as treatment for congestive heart failure in an era when, except for opium, useful drugs were absent. [9]

More than half of the most prescribed drugs in the United States are derived from natural sources. [10] Several drugs on the best-selling list are based on natural products (e.g., simvastatin, augmentin, and cyclosporine). [11] Microbiologists David Newman and Gordon Cragg, who reviewed the utilization of natural products in the discovery and development of all new and approved drugs since 1981, show how these products play a significant role in the pharmaceutical industry. In 2014, 44 new chemical entities were approved; 25% of these drugs are unaltered natural products or derivatives of natural products. Purely synthetic drugs make up 35% of the total. [12] The reviews show that natural products are particularly important for fighting cancer and infectious diseases. Over the past 34 years, 49% of new cancer drugs are either natural products or directly derived from them. Only 17% of cancer drugs are synthetic. In the area of antibacterial drugs, 73% of the approved new agents are based on natural products. [13] It is not surprising that the authors strongly advocate expanding the search for natural products. [14] Biodiversity is not only important for drug discovery but also for medical research. It provides ideas, models, and technological methods. The discovery of a bacterium that could withstand high temperatures in a hot spring in Yellowstone National Park led to an enzyme that replicates bacterial DNA. This facilitated the creation of the polymerase chain reaction, an essential technology in contemporary molecular biology. [15] The same story applies to CRISPR (Clustered Regularly Interspaced Short Palindromic Repeats)—a revolutionary technology for changing and editing genomes (the breakthrough of the year in 2015, according to *Science*); it is not an invention but was identified in nature as a special immune system used by bacteria against viruses. It is the result of more than twenty years of studies of microbial systems that opened up completely unexpected theoretical and practical perspectives. [16]

At the moment, scientists have identified 250,000 species of plants. The majority of species, however, have not yet been discovered. Fewer than 10% of known plants have been analyzed for biological activity. [17] Of all plant species, 20% are at risk of extinction. [18] These data often provide different motivations to connect medicine and biodiversity. On the one hand, natural products that can be found in biodiversity provide a kind of genetic insurance against future unknown diseases. Since most plants have not been analyzed, who knows what promising new medication can be discovered? On the other hand, biodiversity is rapidly lost; the arsenal of potential new drugs is shrinking every day.

From the perspective of bioeconomics, this means that commercial opportunities will be lost. New valuable drugs cannot even be discovered unless more resources are invested in countering biodiversity loss. Though there are undoubtedly ethical reasons to conserve biodiversity, if nature has a commercial value, people and governments will be motivated to protect biological diversity for economic reasons too. But they should hurry.

The Lazarus Project

Australian scientists used de-extinction technology to resurrect the gastric-brooding frog. Although the frog has been extinct for decades, an embryo was created (which unfortunately survived only a few days). This species of frog was discovered in the Australian rain forest in 1973. It became a sensation because it used its stomach as uterus. Fertilized eggs were ingested by the mother and after six to seven weeks were "born" when she vomited them out. How could the progeny survive in the stomach and not be digested? Did the frogs produce certain substances or use some mechanism to inhibit the production of acids in the stomach? These questions interested medical researchers searching for better treatment of peptic ulcers. But they did not get the chance to answer them: the species was declared extinct in 2002. [19]

In the 1950s, the discovery of vinblastine and vincristine, anticancer agents derived from the Madagascar periwinkle, motivated an intense search for useful plants. [20] One major initiative was undertaken by the National Cancer Institute in the United States, which started a plant collection program in 1960 (terminated in 1982). In the 1990s, enthusiasm for studying natural bioactive compounds diminished. Many pharmaceutical companies cut projects using natural products in drug discovery and development. [21] They relied more on synthetic production methods and so-called rational drug design. [22] Sequencing of the human genome as well as new technologies (high-throughput screens; creation of large libraries of compounds) led major companies to focus on the synthesis of small, druglike molecules rather than natural products research. As a result, drug development projects based on natural products dropped 30% between 2001 and 2008. [23] Access to natural products is difficult; bioprospecting is complicated and laborious; screening and analysis of compounds takes a long time; and there are ethical and legal debates about intellectual property rights. Moreover, if a successful compound is developed from a natural product, the need for a sustainable supply becomes paramount. When a product becomes as attractive as silphion, it risks becoming extinct unless it can be made artificially. Aspirin, with 30 billion tablets consumed each year in the United States alone, would have extinguished its source, the willow, if it were not synthetically produced. [24] Cannabis, another popular medicinal plant, survives because modern technology facilitates domestic cultivation. [25]

Currently, interest in natural products is resurging. [26] One reason is that natural products have superior novelty and chemical diversity. Natural products are repositories of genetic information. They provide leads for the synthesis of new chemical entities. Medicinal plants are rarely interesting as raw materials or substances that can be applied directly as a final drug. They are intriguing because they contain promising molecules, suggest blueprints for modification, and can be the sources of novel structures. Against this backdrop, pharmacologist Alan Harvey concludes that natural products are "the most consistently successful source of drug leads." [27]

Different approaches to drug discovery and development should be assessed within the context of broader drug policy. Currently, the demand is for innovative drugs that address unmet medical needs of people across the globe. Resistance to many drugs is growing, and the fear is that soon, for several infections, no effective treatment will be available. Nonetheless, data show that many efforts and resources are directed at development of "me-too" drugs for conditions for which treatment options already exist (e.g., an eighth new type of statin in addition to the generic versions that have been on the market for years). [28] An analysis of 252 new and approved drugs between 1998 and 2007 demonstrates that biotechnology companies and universities provided 56% of all new drugs. Most of the "new" drugs (65%) produced by pharmaceutical companies are follow-on products. Biotechnology companies make most of the novel drugs that are more beneficial than existing drugs. [29] Innovative vaccines have almost all been created by public sector research institutions during the past 25 years. [30]

The second reason for new interest in natural products relates to the crisis of biodiversity. Many realize that only a tiny part of biodiversity is known and screened for potential medical benefits. Particularly the oceans, covering 79% of the surface of the planet, are relatively unexplored. Now that new technical methods have arisen to screen marine resources, they promise to deliver interesting natural chemicals. Antivirals such as azidothymidine (AZT) are derived from materials isolated from marine sponges. Potent neurotoxins are derived from ovaries and livers of puffer fish; they have promoted basic research in neurophysiology, clarifying the functioning of nerves. [31] Microbial resources promise similar benefits. Since the discovery of penicillin in 1929, microorganisms have been a source of antibiotics. But almost all antibacterial drugs from this source were identified between 1940 and 1970. The overwhelming majority of microbes, however, are undiscovered. [32]

Biological diversity is connected to cultural diversity. Cultures survive because they have intimate knowledge of the living world that surrounds them and in which they are embedded. Natural products are also parts of this world; they are not simply entities, substances, or commodities but function within a culture that learned to identify their medical effects and has integrated them as components in knowledge systems and healing practices. Degradation of biodiversity implies disappearance of these systems and practices. Because indigenous practices are under pressure, there is more impetus to study the resources of traditional systems.

Project 523

During the Vietnam War (1955–75), drug-resistant malaria was a major health problem. The Chinese government decided to help the North Vietnamese with a secret antimalaria effort (Project 523) involving hundreds

continued

of scientists. In their search for new drugs, they screened the literature of traditional Chinese medicine and examined thousands of recipes employing traditional herbs. Attention focused on extracts of the sweet wormwood plant (*Artemisia annua*), used for 2,000 years as a remedy for fevers. In 1972, the bioactive compound artemisinin was isolated, creating a new class of antimalarial drugs. Youyou Tu, the discoverer of artemisinin, received the Nobel Prize in Physiology or Medicine in 2015. In her Nobel lecture, she praised Chinese medicine and pharmacology as a treasure trove. [33]

Traditional Medicine

The experience of traditional healers has frequently been a source for new medication. Physicians and ethnobotanists discovered the therapeutic use of plants through these healers. Discoveries opened up new areas of medical research and in several cases generated enormous profits for pharmaceutical companies. At the same time, traditional medicine is an important source of care. The World Health Organization estimates that 80% of people in the world rely on traditional healthcare. That means that in developing countries this type of medicine is the "normal care" for the majority of the population, although the percentages may differ by country and continent. [34] In many African countries, there are more traditional practitioners available in primary healthcare than modern physicians. For example, in Ghana there is one practitioner for every 400 inhabitants, versus one physician for 11,500 people; in Malawi the ratio is 1:138 for traditional healers versus 1:50,000 for physicians. [35] Another phenomenon is that the use of traditional medicine in many developed countries is growing, especially herbal medicine, yoga, and acupuncture. These remedies and practices are often called "complementary and alternative medicine" because they are not part of the country's tradition. WHO estimates that in the United States, 158 million adults use complementary medicines. [36] However, despite its availability, affordability, and widespread use, traditional medicine is often not taken seriously. From the dominant perspective of medicine and healthcare, it is regarded as obscure and unscientific. Its use in the developing world can be explained because there is no alternative; nonetheless, it is argued that it provides second-rate healthcare. The common view is that, even when the majority of populations rely on it, traditional medicine cannot be trusted. Practices are variable; there is no standardization or regulation; quality control is absent; practitioners are not trained or certified; and charlatans are abundant.

Khomeini

The Institute of Elahi International Initiatives for Development and Education in Uganda launched a new drug for HIV/AIDS in 2006. The drug, called Khomeini, was advertised as an herbal drug. Hundreds of patients paid $1,659 for a 21-day cure. The government subsequently banned the drug, which was not approved by the National Drug Authority in Uganda. Testing proved that it was ineffective and simply a mixture of olive oil and honey. Nonetheless, a group of patients petitioned the president to reverse the ban. [37]

The terminology itself may have encouraged such negative views. The word *traditional* has two interpretations. One is historical: it refers to the past. It is the kind of medicine our ancestors used to practice. It presents a medical approach that is outdated; it is no longer interesting, let alone efficacious. The second interpretation is qualitative. This type of healthcare has lower quality, reliability, and safety because it is not the product of modernity. It is not the outcome of rational and professional processes. It has not been subjected to systematic scientific examination. Since traditional medicine is holistic and connected to particular worldviews and does not present objective data and evidence, it cannot even be compared with modern medicine.

In recent years, traditional knowledge and medicine have been rehabilitated. The main reason is the connection with biodiversity. [38] Changing ideas about wilderness and protected areas, especially after the adoption of the Convention on Biological Diversity, focused attention on the unjust effects of policies that protect nature from humans, rather than engaging humans with nature. Sustainable management of biodiversity is dependent on continuous stewardship of people who depend on it for their survival. Indigenous and local communities have centuries-long experience in dealing with the environment. They are dependent on biodiversity for food, water, shelter, and medicine, and they take care that ecosystem services will not be exhausted. In Ghana, for example, some animals are regarded as sacred and can therefore not be hunted; fishing is taboo on certain days according to traditional rules. [39] Traditional knowledge is preserved in order to satisfy human needs; it helps to sustain biodiversity as a source of human flourishing. Sustainability and traditional knowledge cannot be separated, especially at the local, community level. When indigenous practices disappear, more is lost than access to natural products. Sustainable management of ecosystems as practiced by local communities will also be affected. The Convention on Biological Diversity is one of the few international documents that acknowledge the significance of traditional knowledge. The text formulates three steps for proper management: (1) protect traditional knowledge ("respect, preserve and maintain knowledge, innovations and practices of indigenous and local communities embodying traditional lifestyles relevant for the conservation and sustainable use of biological diversity"); (2) promote its wider application; and (3) institute equitable benefit sharing. [40]

Another reason for the renewed interest in traditional knowledge and medicine is its importance for healthcare. In 1978, the Declaration of Alma-Ata was the first normative instrument that acknowledged the role of traditional practitioners as providers of primary care. In the next year, WHO established a program for traditional medicine, focusing initially on research and drug discovery. Later, the emphasis broadened. An important step was the Beijing Declaration, adopted by the first WHO Congress on Traditional Medicine in 2008. [41] The declaration advocates integration of traditional medicine into national health systems. It urges governments to formulate national policies so that traditional medicine can be used appropriately, safely, and effectively. Systems for quality control, standardization, and regulation should be set up. Stringent education and training should be requested for practitioners. Six months later, the declaration was endorsed by the World Health Assembly. [42]

From the perspective of global bioethics, it is remarkable that traditional knowledge has developed into an ethical concern. The Universal Declaration on Bioethics and Human Rights presents "protection of the environment, the biosphere and biodiversity" as one of the principles of global bioethics. It demands that due respect be given to traditional knowledge. Although it is not elaborated or explained, this statement could imply two different things. From a utilitarian point of view, traditional knowledge should be respected because it has practical value. It can be a source for potential medication, and it can be a basic provider for primary healthcare. In these cases, traditional knowledge needs respect and protection because it provides benefits to human beings.

From a nonutilitarian point of view, respect is required because traditional knowledge is a moral source of its own. It presents different worldviews that perhaps provide stronger ethical motivation to protect biodiversity than current views. First of all, the two interpretations of *traditional* should be critically revised. If the word refers to past experiences, every system of medical knowledge and practice has traditional medicine as a previous stage. It is wrong to assume that traditional medicine exists only in developing countries. Traditional medicine endures in developed countries as well, even if it seems to be no longer practiced. It has become history but continues to live in conceptions of health and disease. Names of diseases and therapies go back to ancient humoral theories, and the germ theory was based on popular contagionism. If, secondly, traditional medicine is interpreted in terms of quality—and disqualified as nonscientific—the distinction between past and present is not strict. It is questionable whether it has completely disappeared. Traditional local knowledge is still used in certain parts of Europe. [43] Traditional knowledge is never completed history. Several traditional remedies are scientifically validated, while many modern interventions are not evidence-based. Modern medicine has its origins in traditional medicine. Rather than being discontinuous, medical knowledge is more often than not created and transferred across generations and within a changing environment—not fixed or static but adaptive. Knowledge systems themselves are continuously evolving. Traditional knowledge is living; it cannot simply be contrasted with Western scientific medicine.

The latter interpretation of traditional medicine is particularly relevant for global bioethics. The phenomenon of globalization is frequently criticized for exporting or imposing cultural values and ethical norms, but less often for the products and commodities that it transports over the world. Modern medication, for example, is trusted because it is scientifically tested and validated. It is assumed to be effective for similar diseases in different circumstances. There is often no doubt that medical science and technology are universally applicable, even if it is called "Western" medicine. Contemporary healthcare with its focus on drugs and technology has evolved into a global enterprise with providers, patients, treatments, and facilities moving around the world. The associated values and ethical principles are controversial. The emphasis on medicinal products, diagnostic instruments, and therapeutic interventions is associated with technical and rational approaches and with the Western view that healthcare is a commodity. Traditional medicine offers alternatives, not merely in the sense that it has other treatments and drugs but another type of care—more personal, holistic, and empathic. It does not regard illness as a merely bodily event. It frequently considers healing to be a community effort. A traditional healing system explains and treats suffering, illness, disability, and dysfunction within a particular worldview.

Globalization in healthcare means that Western, scientific medicine (primarily its drugs and treatments) is almost constantly spreading around the globe. Traditional medicine, on the other hand, is determined by the local context and worldview of a specific community and culture. The same goes for global bioethical controversies such as health tourism, professional misconduct, and exploitation of vulnerable populations. Ethical issues are dispersed across the world; they do not seem to connect to a particular context. What they have in common is that they concern one type of medicine that is globalized, not so much the traditional medicine that is localized. The consequence is that for most populations in the world, experiences with healthcare are overwhelmingly different: for most patients, families, and health systems, traditional medicine is the first line of contact and has many faces. For them, this is healthcare that is nearby, accessible, affordable, and one's own. The majority of the people in Ghana, for example, use traditional healers and have very limited access to Western medical facilities. Western healthcare has globalized and is available in the country, but it does not function as an alternative to traditional medicine, at least for most people. For patients using traditional healers, traditional medicine is not simply a service system with practical value—it is more; it presents a specific worldview.

In fact, the different interpretations of traditional medicine go back to two major ecological traditions, distinguished by Donald Worster as arcadianism and imperialism. [44] The arcadian view emphasizes the interconnectedness between human beings and the natural world, the harmony of the cosmos, and the diversity of nature. The imperialistic view advocates human dominion over nature, that biodiversity can be exploited to provide for human needs. Nature is a "storehouse of exploitable material resources." [45] Nature is characterized by an integrated and rational order that can be studied, modified, and transformed in an objective, universal, and scientific way. As discussed in chapter 2, arcadianism as well as imperialism are characterized by different worldviews. While imperialism leads to utilitarian and managerial ethics, interested in the (commercial) use of ecosystem services, arcadianism implies a moral philosophy focused on relationships, interdependency, respect for diversity, common interest, and cooperation. Many of the ethical ideas of arcadianism are nowadays articulated and revived in traditional medicine as a system of worldviews.

The current WHO policy of integration and the growing interest in traditional medicine provide an opportunity to promote the ethical framework of global bioethics. From the perspective that biodiversity is primarily a resource to satisfy human needs, collecting and examining medicinal plants is ethically justified as long as it leads to safe and efficacious drugs. Mainstream bioethics has thus for a long time raised ethical questions about drug research and discovery, testing of research subjects, and safety regulations. However, as discussed in previous chapters, it neglects the wider perspective: biodiversity is not only a resource but also a precondition of everybody's life and health. In this case, sustainable use of its diverse components is everybody's responsibility. The fundamental role of biodiversity, as well as the connection between natural products, traditional medicine, and biodiversity demands this larger ethical scope. This perspective raises questions concerning access to drugs, vulnerable populations, use of traditional knowledge, and compensation. Questions about global justice have especially emerged along with concerns about global biodiversity. Biodiversity is not equally distributed throughout the world. Biological and genetic resources are harvested in economically poor countries and processed in developed countries, but the resulting products (such as drugs derived from the plants in their territory) are often unaffordable in source countries.

Comprehensive Global Medicine

Traditional knowledge and practices connected to biodiversity make two contributions to the discourse of global bioethics. First, they may lead to comprehensive global medicine. Second, they appeal to ethical principles such as justice, vulnerability, solidarity, and social responsibility, and not so much to personal autonomy and individual responsibility.

Integrating traditional medicine into national healthcare systems might lead to comprehensive medicine that is more accessible and affordable than current care. Integration, however, depends on what will be integrated. If traditional medicine is regarded as a potential source of new drugs and treatments, comprehensive medicine will mean expanding the package of drugs and treatments (adding the safe and effective traditional ones to the modern ones). Otherwise, if traditional medicine is regarded as a relational worldview with a holistic image of human beings, comprehensive medicine will refer to a compassionate healing practice on the basis of scientific approaches and evidence. In other words, integrating traditional and modern medicine can lead to distinct types of comprehensive medicine. This distinction is reflected in bioethics. Much of the literature in mainstream bioethics addresses the ethical issues of modern, scientific medicine. It does not matter much whether new medication is synthesized in the laboratory or discovered in a tropical forest. Research bioethics is concerned, for example, with the clinical testing of new medication, particularly issues of informed consent, weighing benefits and harms for the patient, and ethics committee review. How natural products are collected, how bioprospecting takes place, and whether access will be provided to the communities who used these products for centuries are not questioned. Mainstream bioethics usually regards traditional medicine as a provider of potential leads to new drugs. It is not much interested in interpretations of traditional medicine as an alternative, ethical worldview that connects individual, social, and environmental well-being, and that maintains the conditions for human flourishing.

The contribution of traditional medicine to global bioethics can best be highlighted by the term *indigenous* knowledge and medicine. [46] Indigenous knowledge is local knowledge unique to a given society and culture. It is characterized by its geographical source, that is, indigenous populations that live for a long time within a specific territorial setting and with their specific language, culture, and political and economic institutions. Simultaneously, indigenous knowledge is linked with the past generations. When the ancestors are no longer around, humans cannot survive.

There are approximately 370 million indigenous people in 70 different countries. [47] Although indigenous knowledge is very diverse, it has some general features that distinguish it from dominant and scientific worldviews. [48] Such knowledge is holistic, regarding human beings as somatic as well as spiritual entities. It is contextual, regarding human beings as embedded in their environment. It is comprehensive, considering biological diversity to be inherently linked to cultural diversity. It regards knowledge and practices based on nature as collective wisdom and a common heritage of humankind that should not be taken as exclusive individual property and used for profit.

Indigenous Knowledge Systems

The following qualities characterize the knowledge systems of indigenous peoples:

- Specific cosmology: interconnectedness of all beings
- Holism: human beings are a union of body, mind, and spirit
- Embedded in nature: human beings are part of nature
- Knowledge is developed over centuries, often transferred orally from generation to generation
- Distinct language, culture, and beliefs
- History of precolonial and pre-settled experiences
- Knowledge is collectively owned and is the legacy of ancestors
- Nondominant communities: marginalized, poor, and often exploited

These characteristics imply that in general indigenous knowledge is nonsystematic; information is not verifiable; knowledge systems and practices are variable; and its concepts of health and illness are incompatible with those of modern medicine. Features like these make the current policy of integration promoted by WHO a real challenge. How can two worlds of medicine be reconciled that have such conceptual and ethical differences? For traditional medicine, it means that drugs should be tested for safety and efficacy, that quality control mechanisms are applied, and that sustainability of medicines is guaranteed. [49] It also implies the training, certification, and professionalization of traditional healers. Efforts to make traditional/indigenous and Western/modern systems more compatible pose the ethical question of what is the best way to proceed. It is often argued that many people throughout the world have no access to treatment. But the reference here is to Western treatment. In fact, most people in the world have access to healers who are nearby and affordable, offering different forms of medicine in their own worldview. If the right to health means that everybody is entitled to healthcare and treatment, it is available for many, but it is not considered appropriate healthcare, since it is just traditional or indigenous care. There are two options. One is replacing inappropriate care with modern, Western medicine. That will be a long-term effort. The second option is upgrading and integrating traditional medicine so that most people will have at least an acceptable level of care. [50] The ethical choice is then to develop policies that replace existing care with a form of medicine that is considered appropriate (thus superior) or to intensify approaches that make existing health practices less inferior. Since traditional medicine is associated with protection of biodiversity as well as a person-centered approach, the choice should be clear, according to a recent report of UNESCO's International Bioethics Committee: from an ethical perspective, both traditional and Western medicine are essential for medical practice. Traditional medicine is no longer simply an "alternative" to modern medicine. [51] Combining two medical systems addresses the weaknesses of each one. It enhances therapeutic efficacy as well as personal care and patient-centeredness. Integration will fulfill the quest for compassionate and comprehensive care.

Global Inequality and Injustice

As demonstrated earlier, biodiversity is not equally distributed over the planet. Some areas are richer than others, and these are most often located in the developing world. Populations there are poor. The countries do not have many opportunities to technologically explore and advance their natural resources. The flow of resources is usually in the direction of developed countries. While biodiversity-rich countries provide raw materials for the production of drugs, they frequently do not benefit from these discoveries. The indigenous people who cultivated traditional medicine are often ignored, marginalized, displaced, and discriminated against. Although the sale of such material offers the producing nations commercial promises, natural product exchange practices are in fact often exploitative and unjust. [52]

Bioprospecting

Human beings have always searched for new species. Plants have been collected and introduced into other regions for the purposes of agriculture, scientific research, or setting up botanical gardens. Plant and gene hunters traveled the world. Biodiversity has long been regarded as a source of commodities, and the risks of bioinvasion and a deteriorating natural world have been downplayed. Colonial powers used bioprospecting to discover, collect, and process new products to establish commercial monopolies. Dutch seed collectors smuggled seeds of the cinchona tree out of Latin America and created quinine plantations in colonies in Asia. The rubber industry in Brazil collapsed after the British stole *Hevea* seeds and set up rubber production in their overseas territories. [53] The advancement of science and technology, the expansion of global trade, and the recognition that biodiversity is essential for human survival promoted activities of bioprospecting defined as the systematic search for biological and genetic resources in plants, animals, and microorganisms in the wild. [54] The basic idea is that this search will potentially deliver genes and chemical products that will benefit humanity—in particular, pharmaceuticals. Many of today's pharmaceutical companies have grown into multinational corporations thanks to bioprospecting and ethnobotanical research. [55] But as in colonial times, transactions have often been unequal and unfair.

Bioprospecting has two purposes: the discovery of new products (especially food and medication) and conservation of endangered ecosystems. The underlying theory is that biodiversity is commercially valuable. The collection and assessment of biological samples may lead to useful products, which thus provides an economic incentive for biodiversity protection and conservation. [56] This idea is expressed in a common definition of bioprospecting: "the exploration of biodiversity for commercially valuable genetic and biochemical resources." [57] The problem, however, is twofold. Collectors take interesting natural products from biodiverse countries and develop commercial products without compensating the source country. They recognize that biological and genetic resources have a medical (and often spiritual and cultural) value, thanks to the traditional knowledge of indigenous populations who have used these resources for decades or even centuries. The question becomes: Who owns nature and traditional knowledge? Especially in the 1990s, developing countries increasingly complained that rich countries exploited their natural resources, often without compensation.

BIOPIRACY

For a long time, bioprospecting was common and unencumbered by any concern about benefit sharing with source countries and indigenous populations. From the perspective of developing countries and environmental NGOs, it was a kind of theft, an unfair resource extraction, just as it was in earlier times. [58] The positive discourse of bioprospecting notwithstanding, the practical reality is that resources in developing countries are appropriated and monopolized by scientists and international companies in developed countries. This phenomenon of injustice is called *biopiracy*, and it is not by accident that biopiracy became a contested issue. The context of bioprospecting was significantly changed in the early 1990s by two legal documents: the Agreement on Trade-Related Aspects of Intellectual Property Rights (TRIPS) and the Convention on Biological Diversity. TRIPS is a legal agreement between member states of the World Trade Organization (WTO), in force since 1995. It introduced the notion of intellectual property rights (IPR) into the international trade system. Being part of WTO, it presents a stringent governance regime with enforcement and dispute settlement procedures so that countries can be forced to implement its regulations. TRIPS' primary concern is protection of property rights. It makes no reference to the protection of traditional knowledge. Its basic assumption is simple: if things are not patented, they are not owned; if not owned, they belong to the global commons, in other words they are *terra nullius*, and can be taken by everybody. This position is in conflict with that of the 1992 Convention on Biological Diversity (CBD). The convention is the only major international treaty that acknowledges that indigenous communities have ownership of biodiversity. Its goal is to find a balance between conservation, sustainable use, and sharing of benefits. One way to do that is through benefit-sharing arrangements when bioprospectors are granted access to biodiversity and traditional knowledge. It is clear that TRIPS and CBD have conflicting approaches. [59] The discourse of biopiracy in fact criticizes the priority given to intellectual property rights, and thus patenting. While TRIPS encourages commercialization of biodiversity and monopolization of natural products without addressing benefit sharing, it is in fact promoting biopiracy. [60] In reality, bioprospecting results most often in exclusive monopoly control without any compensation for countries and populations.

Cyclosporin

Dr. Hans Frey, a biologist working for a Swiss pharmaceutical company, spent his 1969 holiday in Norway. The company encouraged employees to bring back soil samples after foreign trips for its screening programs. Frey took many soil samples near Hardangervidda National Park between Oslo and Bergen. One sample contained a fungus that produced cyclosporine A. In 1983, the company released two drugs on the market based on this product. They immediately became blockbusters and are the leading products of the company today. They are patented, so the company has a temporary monopoly for production and sale of the drugs. The question is about fairness. Who owns microorganisms in soil? Should Norway receive royalties on the sales (estimated to be $24.3 million annually)? Should the company contribute to conservation of biodiversity in the National Park? [61]

PATENTING

Many cases of biopiracy, but not all, involve patenting. To obtain a patent, the applicant must show that his or her creation fulfills the criteria of novelty, invention (or non-obviousness), and industrial application (or utility). Scholars have long argued that these criteria are not met for patents on traditional knowledge. In many cases, the products are not new. They are not real inventions, but source countries often cannot demonstrate that the natural products have been used for centuries. In many countries, patenting of plants is not allowed for moral reasons. Patent laws usually grant rights to exclude others from use. Patents for medication were for this reason not allowed in many countries. [62]

Basmati Rice

In 1997, a patent was granted to RiceTec, Inc., in Texas for certain basmati rice lines and grains. The researchers claimed to have developed novel basmati varieties. Basmati rice has been cultivated in India for centuries. India contested the patent because it is not new and not an invention. Documentary papers describing the breeding of the rice had appeared in international journals before the approval of the patent. Nevertheless, only some parts of the patent have been revoked. [63]

Patenting has increased exponentially since the US Supreme Court decided that living matter could be patented. In the famous *Chakrabarty* case (1980), a patent was allowed on a human-made bacterium. The Court famously decided that "anything under the sun that is made by man" can be patented. [64] A genetically engineered microorganism is no longer a product of nature but an invention and therefore patentable. David Conforto argues that it was with this pronouncement that biopiracy truly emerged. [65]

In the recent past, many countries did not have patenting systems. If they did, they were all very different. The United States, for example, did not have strong intellectual property protection. The emphasis was for a long time on the creativity, innovation, competition, and diffusion promoted by patenting. It was not so much related to trade. The global context of the 1980s changed this conception. Businesses argued that intellectual property rights could contribute to the economic competitiveness of the country and that their patents should be recognized abroad.

Susan Sell tells the story of how powerful commercial lobbies first started to push for stricter property regulations within the United States (with the argument that American inventions were pirated by developing countries). [66] They then argued that a trade-based approach to intellectual property was necessary at the global level. The Intellectual Property Committee (IPC) was formed in 1986 by a private sector advocacy group composed of twelve corporations, including Pfizer, Merck, Monsanto, and DuPont. [67] The IPC created a powerful lobby and succeeded in gaining the support of politicians and government officials. It successfully influenced the negotiations on the Agreement on Trade-Related Aspects of Intellectual Property Rights. Almost all of its wishes were reflected in the final document. The agreement not only imposed the American patent system on the rest of the world but also installed a strict enforcement mechanism. [68]

The history of TRIPS demonstrates that the emphasis of the global IPR regime is on trade and economic interest, and that it is applying one set of standards as appropriate for all countries. The first implication of this is that the balance between property rights (regarded as private rights) and public interest is skewed. For this reason, in the 1990s, the current IPR regime began to face more criticism. Opposition to TRIPS (and the exclusive focus on patenting) grew stronger. The expanding AIDS crisis made clear that patents increased drug prices and obstructed access to healthcare. It was also argued that patenting plants endangers food security. Furthermore, patenting life forms is ethically problematic. These arguments produced a reframing of property rights in terms of public health, human rights, and security, rather than trade. This reframing was primarily done by NGOs, social movements, and civil society. [69] These groups reiterated that making profits is evidently more important for multinational companies than access to medication, nutrition, and public health. Patents and profits lead to death. This reframing changed the political and ethical discourse. [70] The main concern was how to rebalance private and public interests. The Doha Declaration on the TRIPS Agreement and Public Health, adopted in 2001, was an effort to introduce flexibility and to give priority to public health issues.

Another implication of TRIPS is that one type of IPR system has been universalized to the exclusion of others. [71] The imposed global regime does not acknowledge indigenous rights perspectives. On this issue, significant changes in critical thinking occurred in the 1990s and early 2000s. Scholars demonstrated the various concepts of ownership that exist in traditional communities. [72] The focus on individual rights (of a specific inventor or author) is biased. Knowledge production is a collective activity. [73] Even if there is no individual owner of traditional knowledge, it is wrong to see it as the property of nobody. The knowledge is shared by the people or the community. There is a strong "sharing ethos" that takes into account that knowledge is inherited by the people and belongs to today's people in order to preserve natural resources and biodiversity for the sake of future generations. [74] The essential error underlying patenting and biopiracy is that they do not consider these long-term efforts in support of sustainability but primarily focus on short-term benefits.

Over the same period of time, efforts were undertaken to articulate the crucial connection between traditional knowledge and practices on the one hand, and biodiversity conservation on the other. Scholars such as Graham Dutfield, Anne Ross, and others claim there is substantial evidence that an ethics of conservation is inherent in indigenous practices. [75] Traditional knowledge indeed contributes to sustainable development. Another significant event was the adoption in 2007 of the Declaration on the Rights of Indigenous Peoples by the United Nations. The document states that these peoples have "the right to their traditional medicines and to maintain their health practices." [76] They also have full ownership, control, and the right to protect their cultural heritage and intellectual property. [77] Respecting traditional knowledge is thus important not only because it has instrumental value for conservation, food, or new drugs. It is important because it has value on its own. It expresses respect for cultural diversity and human rights.

OWNERSHIP

Bioprospecting and biopiracy raise two specific ethical issues. One is the issue of justice, which is related to benefit sharing. The other is ownership. To whom do biodiversity and its natural products belong? [78] The answer depends on how biodiversity is valued. As argued in chapter 2, the economic perspective taken in the CBD regards biodiversity as a collection of resources, an economic asset. Until that time, it was assumed that biodiversity, like air, outer space, and the ocean bed, was the common heritage of humankind. It does not belong to anybody; it cannot be appropriated by individuals or groups as private property. In this view, biodiversity, even if it is merely regarded as a resource, demands international cooperation in order to create common management. Possible benefits should be equitably shared among states, irrespective of their geography. And importantly, it is necessary to preserve biodiversity for future generations. Even when individual nations manage these resources (often spread out over several countries) by considering them common heritage, a shared interest is constructed. Resources need to be preserved in such a way that humanity is benefited rather than the citizens of one or more countries. The resources should also be open for peaceful scientific research that could deliver results for the benefit of humankind. The Antarctic, for example, is important for the study of climate change. Since resources labeled as common heritage are vital for the survival of humankind, global responsibilities are defined: proper management and stewardship transcend any national interests. [79]

This view of biodiversity as a common heritage or global commons was rejected in the CBD [80] because it did not sufficiently protect biodiversity. Regarded as owned by humankind, biodiversity was approached as *res nullius*. It could be appropriated by anybody, as an open access resource (see box, p. 128, top). This leads to exploitation, especially of bio-rich developing countries. The CBD affirms that conservation of biological diversity is "a common concern of humankind." [81] That means that global policy is necessary but does not imply common ownership. Instead of common heritage, the document declares that states have "sovereign rights over their own biological resources." They are the owners of the biodiversity in their territories. This was assumed to address the problems of bioprospecting and biopiracy. States could now control and regulate access to biodiversity resources. With the economic perspective and the change of the notion of ownership, specific rules could be introduced: first, the requirement of prior informed consent for the use of natural resources and traditional knowledge and, second, the need for fair and equitable benefit-sharing arrangements if research results in commercial products. In this way, it was hoped that the objectives of the CBD (conservation of biological diversity, sustainable use of its components, and sharing of benefits) could be better accomplished and ethical problems avoided.

The problems with this approach are that, especially in the developing world, biodiversity is differently valued and other views of ownership prevail. Traditional knowledge and indigenous medicine, for example, are regarded as "medical heritage." [82] It is not merely a package of practical resources but part of a cultural, religious, and spiritual community, developed over time in a specific ecological setting. Because biodiversity sustains human life, it is important to conserve nature and not exhaust its resources. The worth of natural products is not primarily commercial; rather, they are components of a much broader value system. It is also important that, in this value system, traditional knowledge is collective knowledge. It is the legacy of a particular community, tribe, or population. The consequence is that the discourse of private property rights does not apply. In traditional medicine, the emphasis is more on responsibility than ownership. Discoveries and inventions are made by the community over a long period of time. They cannot be private property. However, indigenous populations cannot demonstrate "prior art" if their knowledge and products are patented. This type of knowledge in many cases is not documented or published; it is often based on oral traditions.

Benefit Sharing

Because bioprospecting frequently results in claims of exclusive ownership of natural resources and products used for centuries in traditional and indigenous cultures, it can be regarded as a form of exploitation if these cultures are not respected and compensated. For this reason, bioprospecting has frequently been redefined as biopiracy—the unauthorized and unfair exploitation of biological resources and/or associated traditional knowledge. [83] This highlights another ethical problem of bioprospecting: that of global injustice. To address this problem, the notion of benefit sharing has been introduced as one of the goals of the CBD, specifically in the Nagoya Protocol that established the access and benefit-sharing framework. This approach to the problem is based on the notion of justice as an exchange between two parties. [84] The underlying idea is that countries rich in biodiversity provide access to resources in return for benefits that are derived from the resources. Sharing benefits is thus a compensation mechanism, a form of commutative justice (justice in exchange).

The idea of benefit sharing was promoted by the powerful example of the agreement in 1991 between pharmaceutical company Merck and the Instituto Nacional de Biodiversidad (INBio) in Costa Rica. INBio was the first national institute to regulate access to biodiversity resources. In exchange for biological samples, Merck made significant up-front payments, provided equipment and training, and promised to pay royalties on any commercial products that might be developed from these resources. INBio granted all intellectual property rights to Merck. [85] This agreement became a landmark case of benefit sharing and encouraged high expectations. It illustrated that there are different types of benefits. While many stakeholders focused on monetary benefits, it was pointed out that a benefit is not identical with profit in the economic sense. The Ethics Committee of the Human Genome Organization in its "Statement on Benefit Sharing" has defined a benefit as "a good that contributes to the well-being of an individual and or a given community." [86] Determining what is beneficial will depend on needs, values, priorities, and cultural expectations. This refers to a broad notion of benefits. They can be available even in the absence of profits (such as scientific collaboration, training, joint publications, and capacity building).

However, although the desirability of benefit sharing was mentioned in the Convention on Biological Diversity, it was regarded as a weak statement. Sharing was voluntary and depended on the willingness of users to pay for and assist in biodiversity conservation. [87] But now that a framework has been defined, models for benefit-sharing arrangements could be developed and tested. Also, international negotiations continued, resulting finally in 2010 in the Nagoya Protocol. This supplementary agreement to the CBD provides a legal framework for the implementation of benefit sharing in relation to genetic resources. The protocol took force in October 2016, with 86 countries having ratified it. Prior informed consent and benefit sharing are now required to gain access to genetic resources. [88]

Global Bioethics Policies

The debate on bioprospecting has been ambiguous from the beginning. All parties agree that biodiversity is a rich source of potential drugs (and food). Deterioration of biodiversity means loss of potential benefits for medicine. Indigenous populations are under pressure, and traditional knowledge is disappearing. So, if we want to benefit from biodiversity, we must act.

A number of scholars insist that immediate and wide-scale bioprospecting is necessary before it is too late. We need new leads for drugs, for example, because of growing resistance against antibiotics, which is now regarded as the biggest global health threat. [89] They argue that bioprospecting can be done in an ethical way. The CBD has specified the rules of prior consent and benefit sharing, elaborated in the Nagoya Protocol. Bioprospecting can no longer be regarded as exploitation; biopiracy can be avoided. [90] A stricter framework of global governance is developing, although slowly. [91]

Other scholars are critical. They argue that global governance has so far produced great expectations but minor results. [92] In practice, the economic effects of benefit sharing are limited. There are few successful examples. The deal between Merck and Costa Rica has ended; the company abandoned the search for natural compounds. [93] The development of effective drugs that reach the market is a long and expensive process. [94] Arturo Gomez-Pompa warns that while the potential uses of unknown plants, animals, and microorganisms are valid and real, unrealistic expectations are often created about the riches of biodiversity. [95]

Furthermore, global governance fails to protect biodiversity. [96] Current governance does not prevent biopiracy, [97] largely because the main focus of the governance system remains intellectual property rights. The conflicting perspectives of CBD (conservation) and TRIPS (property rights) often lead to prioritizing the privatization of common resources. The future impact of the Nagoya Protocol is questionable. It has already been criticized for its weak compliance mechanisms. The protocol does not make any provision for the verification of benefit sharing, nor is there any requirement that the source and country of the natural resource be disclosed. [98]

Another critique is that the current framework does not protect indigenous rights. Indigenous communities are often unable to show that the knowledge belongs to them; there is no written evidence, and if there is the evidence, it is often not recognized by courts. Patents privatize knowledge and products; collective rights of local and indigenous communities are not recognized. One remedy is to document traditional knowledge. An example is the Traditional Knowledge Digital Library in India, established in 2001 with now 290,000 medicinal formulations. Patent claims filed on Indian traditional knowledge have declined by 44%. [99] However, traditional knowledge is not always common knowledge. In a number of cases, it is practiced secretly and known only to a few healers and is thus not documentable. In many instances, the rights of indigenous peoples are not recognized, despite the Declaration on the Rights of Indigenous Peoples. This declaration took a very long time to adopt; it is nonbinding. The importance of these rights was shown in the landmark agreement between Costa Rica and Merck. The arrangement allowed collection of natural products on the land of eight indigenous peoples, but they were not consulted or mentioned as beneficiaries. [100] There is no way to reconcile indigenous heritage with the current IPR regime unless indigenous rights are specifically articulated in international treaties and forcefully implemented. [101]

The failures and distortions of the current governance framework have lead Tim Mackey and Bryan Liang to conclude that current global governance has "failed to protect indigenous rights, promote access to life-saving drugs, prevent biopiracy, or provide for responsible biodiversity development." [102] Daniel Robinson goes even further: he does not see any likely changes in practices. [103] One reason is that the United States, as the main creator and supporter of the current system, follows a hard line. [104] It has not ratified the CBD and the Nagoya Protocol; it does not want changes in the IPR regime (e.g., requiring a disclosure of origin of the patentable drug). US companies therefore do not have the same commitments and obligations that constrain companies in other countries.

An Ethical Assessment

Biodiversity is a common concern. Conserving it is essential for the survival of humankind. It is also a source of enormous benefits. These values have been articulated by the international community in the Convention on Biological Diversity. The present framework for governance, however, is dominated by the intellectual property rights regime. This assumes that biodiversity is a deposit of raw materials, unused and ready for discovery; it is a collection of resources that are made valuable by scientific and technological processes, and transformed into commercial commodities. The notion of biopiracy offers a counterdiscourse that criticizes the assumptions of current governance. [105]

Three approaches are available to ethically address this situation. [106] First, rectify the IPR system. For example, it is imperative to improve the examination of patent applications and to revise the criteria for patentability. In this approach, biopiracy is first of all an IPR issue. Ameliorating the system and giving more consideration to the concerns of bio-rich countries and the rights of indigenous people will help to maintain the status quo. The cumbersome history of the Nagoya Protocol and the Declaration on the Rights of Indigenous Peoples shows that this will be a long and complicated road (these two treaties took 18 and 25 years, respectively, to negotiate). Persistent problems remain. The sources of traditional knowledge will be difficult to identify, since communities share the same knowledge, and no individual "owner" is known. Patenting is therefore problematic unless other systems are employed. Also, evidence of discovery can often not be supplied, while the patent process requires facility with technical language, and the procedure is expensive. The basic assumption of the present system is that it provides equal access to all. In practice, implementing the regulations often ignores the power relations that are at work; developing countries are in a weaker position from which to negotiate and bargain. For these reasons, Dutfield concludes that the current governance is unjust: "one type of IPR system is being universalized and prioritized to the exclusion of all others." [107] The main beneficiaries are transnational corporations. [108] As long as other mechanisms for protection of property—such as community or group rights protecting collective knowledge of indigenous peoples or better enforcement measures—are not developed, biopiracy will continue. [109] Model the process on the idea of inverted bio-invasion; instead of working to prevent natural products from invading countries, measures should be taken to prevent those products from being taken outside of their countries of origin.

If the current global system is not just, why continue trying to ameliorate it? Why not acknowledge that the idea of patenting itself is not helpful in promoting global health? Gustavo Ghidini and colleagues, for example, conclude that the current system is no longer acceptable. There should be more focus on human rights. [110]

No Patents

On April 12, 1955, the day that the US government announced that the new polio vaccine was safe and effective, its inventor, Jonas Salk, was asked on television who owned the vaccine. He famously replied, "Well, the people, I would say. There is no patent. Could you patent the sun?" The vaccine was common property; it belonged to the people who had donated money for the public interest. Salk was later pitied for his decision. He could have earned $7 billion if his vaccine had been patented. His attitude to patenting, however, is not unique. Even in the era of excessive property protection, there is a new interest in open sources and access, global goods, and the notion of the commons. Elon Musk, CEO of Tesla Motors, decided in June 2014 to release all of his patents. Technological leadership is not defined by patents. The future of sustainable transport will be better served by openly sharing information and knowledge. "All our patents belong to you." [111]

Second, enhance benefit-sharing mechanisms. New rules for access and benefit sharing have been introduced since the Nagoya Protocol took effect in 2016. The hope is that this will improve practices of bioprospecting and diminish cases of biopiracy. Various benefit-sharing arrangements have been negotiated, implemented, and offered as examples of successful policies. Cooperation with Madagascar, for example, shows that benefit sharing can contribute to conservation. It demonstrates how a range of benefits can be provided—from payments, joint publications, training, and research equipment, to schools and water wells for villages. [112] Positive examples like these may promote a cooperative approach of combining environmental protection and IPR. Compensation for the use of biodiversity is then invested in conservation. Resources that are extracted through, for example, deforestation will be replaced, and thus the focus is on sustainability. However, the problem is that the current context of benefit sharing is still cumbersome and discouraging for developing countries. There are many complicated practical difficulties. It is not clear what the benefits are, how they are distributed, and who should be recipients. Benefits should be determined by the beneficiaries. They are not necessarily benefited by payments. Who exactly are the beneficiaries if traditional knowledge is crossing borders or belongs to several indigenous peoples? Indigenous peoples also have difficulties in demonstrating that knowledge belongs to them. There are furthermore ethical difficulties. Biopiracy is regarded as unethical, since it is an unjust and unequal exchange: benefits go to the receiving party; nothing goes to the giving party. In principle, justice is restored if benefits are shared. The exchange, however, should be fair and equitable. The assumption then is that what is exchanged should be equivalent. Often this is not the case, since the benefits for both parties are often not of the same value. Developing countries and indigenous communities are mostly not in the same bargaining position, or they lack sufficient competence in the field of patent law and international treaties. Furthermore, in global governance, the burden of implementation is most often on bio-rich countries: they have to make or revise laws, create a bureaucracy for compliance, and procedures for access while they already have much weaker local and national governance than user countries and multinational companies. Mechanisms of benefit sharing should therefore be enhanced. The question is whether these improvements will remove the injustice of bioprospecting.

Third, transform global governance itself. The discourse of biopiracy can also have another meaning. Rather than pointing out that the global IPR regime does not work well for indigenous peoples or that benefits from bioprospecting are misappropriated, the notion of biopiracy is used to argue that global governance based on the current IPR regime itself is unjust. Changing the rules of the game will not produce justice if the game itself is unjust. Biopiracy discourse in this approach advocates more than amelioration or revision. It demands a fundamental rethinking. [113] Three ethical concerns play a major role in this critical biopiracy discourse: property, life, and justice.

PROPERTY

Biopiracy discourse is not merely a critique of patenting or (mis)use and (mis) allocation of biodiversity benefits. It points out deeper injustices of the IPR system itself. One issue is the assumption that biological and genetic resources are primarily regarded as tradable commodities. Biodiversity itself has changed into a resource (in the CBD). Framing the debate and governance in terms of intellectual property rights is problematic for indigenous peoples who have other notions of property but also for many governments. This framework is necessarily associated with inequalities. It also means that interests are different. Robinson in his study on biopiracy observes that Western countries are more interested "in ever-higher standards of intellectual property protection, rather than assisting with the protection and promotion of biodiversity and traditional knowledge." [114] The ethical problem is not who owns natural products or who will benefit but what is owned in the first place. [115] More needs to be done than to challenge patents. The patenting system should be scrutinized and discarded. This brings us back to the discussion about common heritage. Should biodiversity as common heritage and global commons be privatized? Similar questions arise concerning the diversity of knowledge, innovation, and creativity. Should one specific form of knowledge from one specific culture be imposed at the global level?

The story of the Intellectual Property Committee and the TRIPS agreement illustrates that global justice considerations did not play a role in setting up the global governance regime. [116] On the one hand, there was a lack of procedural justice, since the balance of power was very unequal and transparency was missing. The establishment of the governance regime was not the result of democratic deliberations and negotiations. The public was not involved, and most countries were under pressure. Like present-day bilateral free trade agreements, they were secret. On the other hand, substantive issues of justice were not taken into account. The dominant value was trade. Considerations of the effect of the governance regime at a global level were absent for a long time. The interests of Northern multinationals prevailed. As a result of the regime, the subject matter that is patentable gradually expanded, while many countries had not previously allowed patenting of plants and animals.

LIFE

In the beginning of her first book on biopiracy, Vandana Shiva explains that "resistance to biopiracy is a resistance to the ultimate colonization of life itself." [117] Renewable resources are transformed into commodities, raw material, and private property. Life has become patentable. It can be owned by somebody. This has ethical implications for the relationship between humans and biodiversity. In fact, life itself has fallen under the dominion of human power as the triumph of Worster's imperialism. The discourse of biopiracy also criticizes another effect of appropriation of natural resources: monopolization. Today biopiracy is no longer primarily associated with traditional and indigenous knowledge, and thus limited to exchanges between the South and the North. [118] A new type of biopiracy is developing under the influence of biotechnology, and it is much more generalized across the globe. Particularly the pharmaceutical industry and agribusinesses are genetically modifying and patenting natural products such as seed varieties and attempting to control and monopolize consumer markets. Traditional medical markets and farming methods are restricted. The difference from the past is that nowadays these practices are no longer limited to the developing world. All farmers everywhere are affected. While farmers traditionally try to increase crop diversity, modern biotechnology promotes monocultures, thereby making crops more vulnerable and reducing biological diversity. Farmers have to buy genetically engineered seeds from companies such as Monsanto and are no longer allowed to select and sell their own seeds. This will lead to soil exhaustion and loss of biodiversity. In healthcare, similar monopolizing efforts are applied by pharmaceutical companies in the developed countries, leading to increasing prices for medication in many countries, even if patents have expired.

The critical discourse of biopiracy therefore has a wider focus than to criticize the system that allows patents or to provide practical examples of misappropriation. The increasing use of biotechnology has introduced a new form of biopiracy. It is new because it is applying sophisticated technologies that are no longer traditional or natural. It is also new because the technologies are no longer restricted to developing countries; they are affecting farmers, healers, and people across the globe. Nonetheless, the practice is the same as before: appropriating knowledge and resources in the public domain, patenting them, and bringing them into the private domain to make profits for a limited number of people. [119] In the long run, however, everybody, including the monopolizing companies, will suffer because biodiversity continues to decline. [120]

The basic problem here is not so much one of power, inequality, and injustice. It is the idea that life itself can be patented and thus controlled. From the indigenous perspective, the idea that life can be bought, owned, sold, and patented is offensive. [121] How can it be assumed that life forms are inventions, the fruits of the human intellect? Again, the biopiracy discourse argues that a fundamental ethical question is stake. Patenting life forms presupposes a specific view of the relationship between human beings and biodiversity that is ethically questionable. It implies a narrow view of biodiversity, whereas everything in nature is actually common property (before it is appropriated). It is not merely raw material to be privatized and exploited; natural resources belong to the human community. This specific view furthermore violates dignity and rights. It is not merely the rights of indigenous peoples that are disrespected but also human rights: for example, farmers' rights are not taken into account. Intellectual property rights may easily conflict with other human rights (e.g., the right to health). It is not obvious that they should have priority. [122]

JUSTICE

A third ethical concern addressed by biopiracy discourse is related to global justice. [123] One of the goals of the Convention on Biological Diversity is the "fair and equitable sharing of the benefits arising out of the utilization of genetic resources." [124] As argued earlier in this chapter, the use of biodiversity is inherently unequal. Bio-rich countries are mostly located in the global South. Brazil, for example, accounts for 20% of global biodiversity. On the other hand, rich and developed countries have the capacity for research and development; pharmaceutical companies and the biotech industry are located there. The overwhelming majority of patents are owned by countries in the global North. From the perspective of critical biopiracy discourse, biodiversity is a resource for medication and food for humanity. That means that the relationship between providers (giving access) and users (making commercial products) should be just for everybody. Fair exchange is a way to accomplish this. The Access and Benefit Framework is therefore based on the concept of exchange (commutative) justice. The idea is that the injustice of biopiracy practices can be undone by compensating the access to biodiversity with benefit-sharing arrangements.

This notion of commutative justice is often criticized as inadequate. Taking into account the entire global context, a tit-for-tat exchange does not seem to do much justice to the inequalities of bioprospecting. The problem is that the exchange includes noncomparables. The future revenues from exchanged plants are rare and hypothetical but can be massive. How can these benefits be compared with providing a school or water well for the community that gave the plants and the associated knowledge? The exchanged products are not equivalent. Commutative justice also assumes that the exchanging parties are in an equal position to make a transaction. As pointed out earlier, indigenous peoples are often not in such position. Viewing benefit sharing as exchange neglects the fact that an unequal global governance system may provide weaker parties with some compensation but does not affect the enormous benefits that will continue to flow to the stronger parties. Justice as exchange does not take into account that traditional knowledge has developed over generations (because it has not been private knowledge, and communities have been concerned about its development and perfection).

Because of these concerns, it is argued that another concept of justice is needed, namely, distributive justice. [125] The benefits emerging from biodiversity should be used to address human basic needs in the long run because resources are scarce. For benefit sharing to be fair, it should lead to an equitable distribution of goods among parties in an unequal position. It is not so important that the goods exchanged are equivalent (if that will be possible at all) and that the transaction is fair. If benefits are available, they should be used to address the needs of populations and communities, whether or not they provide traditional knowledge. From this point of view, it is more important that the weaker parties will benefit more than the stronger ones so that differences in health and well-being will be diminished. Knowing that inequality today is the major source of moral problems, a more fair distribution of goods will effectively contribute to improvement of global health.

Such view of justice connects to the basic motivations of the CBD, especially its highlighting of biodiversity as a common concern of humankind. What does this mean? The later Declaration on the Rights of Indigenous Peoples also refers to the common perspective that "all peoples contribute to the diversity and richness of civilizations and cultures" (which it regards as the common heritage of humankind). [126] Cultural diversity is intimately associated with biological diversity. Biodiversity is crucial for the survival of humankind; it provides essential means for everybody's flourishing. What does this commonality imply for global governance? First of all, we need to assess the moral quality of the current global governance regime based on IPR and acknowledge that it is an unfair system.

An Unfair System

Current global governance is not the result of democratic and transparent consensus at the global level. It is the outcome of the imposition of one particular regulatory system on the rest of the world. Sell has demonstrated that "TRIPS largely reflected the wishes of the CEOs of twelve American companies." [127] The global system naturally gives priority to trade above health. There is no empirical evidence, as Sell argues, that TRIPS works for the benefit of developing countries. On the other hand, there is plenty of evidence that TRIPS is making developed countries much better off, especially in the area of medicines. [128] This does not imply, as argued above, that other considerations—particularly ethical and human rights—do not have an impact on the debate. The priority of trade has been watered down significantly in later agreements (such as the Doha Declaration) which offer the flexibility of giving priority to healthcare. In theory, it means that a better balance can be achieved between private and public interests. [129] Very few countries, however, have used these so-called TRIPS flexibilities. It should be remembered that commerce and trade do not disappear as priorities. We should not assume that global governance has suddenly demonstrated more ethical awareness. Continuous alertness and grassroots activities will be needed to keep attention focused on biodiversity conservation and global health.

This is obvious in the context of bilateral and regional trade agreements. Even though mandates to protect food, health, and the environment are now permitted in WTO arrangements, several governments (especially the European Union and the United States) are introducing requirements in so-called free trade agreements that weaken the WTO requirements. They extend the lifetime of patents, relax some environmental restrictions, and give multinational corporations the ability to sue governments. The negotiations, as usual, are secret; there is not even the possibility of public debate and democratic input. The risk is that standards of public health and environmental protection are lowered. [130]

The major concern is that contemporary trade policies restrict access to healthcare and, particularly, to drugs. Lisa Forman gives examples how avenues to protect public health have been closed. Compulsory licensing rules (countries manufacture or import generic drugs under strict conditions) have become increasingly restrictive. [131] Parallel imports (the importation of cheaper patent drugs) is prohibited. Owing to these global policies, access to generic drugs is getting worse. Forman concludes that the space for producing and importing generic medication is shrinking. Even in developed countries, prices are rising as a result of monopolies. [132]

Developing countries in particular are forced to agree not to use TRIPS flexibilities to provide medication to their populations. Pressure strategies, litigation, and trade sanctions are used to favor commercial interests and profits at the expense of global health. [133] At the same time, generic drugs are widely used. They make up to 80% of all prescriptions in the United States. [134] With many patents for blockbuster drugs expiring, gigantic generic pharmaceutical companies (from India and Brazil) now dominate the market.

This situation is ironic. For decades, it has been pointed out that the governance system for global health is unfair. After long struggles over access to medication and the right to health, especially in connection with HIV/AIDS, the system has been adapted; it can give more room to health considerations. But now this remediation is undone at national and regional levels, where more stringent requirements are imposed to implement the IPR regime than are allowed at the global level. The policies of Western countries in particular continue to prioritize profit over health. Confronted with this situation, the bioethical debate has two options: ignore or repack. Either continue with business as usual, focus on personal autonomy and individual responsibility, and do not touch social, economic, and political issues, or engage with the context in which bioethical issues arise and impact individual health and well-being. The challenge is that many people involved in bioethics regard the IPR regime as an arcane, highly technical, and secret enterprise. They do not appreciate that in fact this regime predetermines the global debate and makes certain ethical considerations impossible to discuss.

Conclusion

Vandana Shiva, the activist from India who was the first to raise the issue of biopiracy, has rightly pointed out that this discussion about TRIPS, patents, and IPR is not just about trade: it is "about the ethics of how we relate to other species . . . about how our biodiversity is used and controlled. . . . It is an issue of justice and human rights." [135] Shiva's work is in fact one long discourse that regards patenting as a form of (neo)colonialism. What used to belong to people is appropriated by commerce and corporations. Globalization is not the sharing of the benefits of open borders among all people to satisfy their basic needs. In reality, borders are only permeable from one side. Elites on this side misappropriate the benefits of biodiversity that can be found in large areas beyond their borders.

Shiva's observation implies two things that have been articulated throughout this book. First, we need a wider ethical view. Second, we need to reflect on what the common concern of humankind implies. Both implications refer to the relationship between human beings and biodiversity. One dimension of this relationship is embeddedness. Humans are incorporated into nature; they are part of biodiversity. As discussed in chapter 2, environmentalists, particularly those from the developing world, have argued that human beings cannot be separated from nature. Conservation policies in the West with their emphasis on wilderness and establishment of parks and reserves are disastrous for the majority of peoples in developing countries. Through an important change of perspective within environmentalism, the role of human beings, and especially indigenous peoples, was redefined. Conserving biodiversity was no longer regarded, in the spirit of John Muir, as protecting nature from human beings and preserving wilderness, thus establishing protected areas and removing people. Protected areas now cover 12% of the land surface of the planet. [136] It is recognized that protective activism neglected concern for human beings, and indigenous peoples in particular. They have been excluded and marginalized into poverty rather than respected as the "custodians of biodiversity." [137] For a long time, these policies overlooked the intimate connections between biological diversity, cultural biodiversity, and human beings.

This view of the relationship between humans, nature, and culture not only created a favorable context for commercial exploitation of biodiversity and the phenomenon of biopiracy. It encouraged an imperialistic worldview. The debate on bioprospecting, biopiracy, and indigenous knowledge critically revised this worldview. It articulates that we are all in the same predicament. The fate of human beings cannot be separated from the future of biodiversity. It also points out that biodiversity (and its possible benefits) is not the product of human beings. They can use, exploit, and enhance natural products, but their activities generally erode biodiversity itself. The consequence is that humankind has a common responsibility to use medicinal products provided and derived from biodiversity for the sake of global health as well as conservation of biodiversity.

The current global governance system, based on private property rights, will no longer suffice against this backdrop. Leonardo Boff, the liberation theologian from Brazil, has explained why: "Being antecedent to humankind, the world does not belong to humanity. Thus, the relationship between human and world is characterized by responsibility; it is "an ethical relationship." [138] In a time when the notion of the Anthropocene is becoming popular, it is important to underline that human beings have profound and dominating influences on the nonhuman world but that there are limitations. Materiality is resisting. We cannot completely control biodiversity. [139] Activities and interventions will not be enough; stewardship, responsibility, and respect are demanded. The idea of an ethical relationship between humans and biodiversity presupposes that human beings share the "common concern of humankind," expressed in the CBD. Boff joined with his colleague Miguel D'Escoto from Nicaragua to promote a Universal Declaration on the Common Good of the Earth and Humanity. [140] The two theologians emphatically make the argument that life is a common and shared project. This is perfectly expressed in the indigenous worldview: "The earth exists not as private property, but as a commons, to be tended with respect and reciprocity for the benefit of all." [141]

These considerations also lead to the conclusion that emphasis on distributive justice to address global inequality is too limited. As argued in the following chapters, a broader approach to justice will be necessary. Social movements for food justice and water justice have articulated that justice requires recognition, participation, and functioning. It is not only the unfair distribution of goods and services that is at stake but the lack of respect and recognition for individuals and groups, particularly indigenous peoples and their traditional knowledge. It is also unjust that individuals and groups are not provided with the capabilities to function and flourish. Furthermore, lack of participation and involvement in decision making is unjust. This chapter has illustrated that many policy decisions regarding trade lack participatory justice. The next chapters show that sovereignty in agriculture and food consumption or water availability has become a major concern. The discourse of justice therefore is broader than distributive justice. For global bioethics to critically analyze the roots of contemporary problems, it is necessary to examine the underlying causes of mal-distribution: the context of power differences, oppression, racism, disrespect, and marginalization in which many people try to survive. [142]

6 | Food

Biodiversity has always been a source of human food. Hunting, fishing, and foraging have provided human beings with nutrition to survive. A variety of species are directly available in the wild for consumption: plants, fruits, mushrooms, nuts, roots, and seeds. A well-known example is the banana. There are almost 1,000 varieties of bananas in the world; half of them can be consumed. From the beginning, humans have domesticated animals and plants to improve and expand food provision. In the wild there are only 13 species of tomatoes, and some of these are poisonous. Now we have cultivated 10,000 varieties. One of the most important food crops in the world is potatoes; there are 4,000 varieties for human consumption, besides 180 wild species that are not edible. [1] With growing human populations, however, wild species were insufficient to feed everyone. To provide food security, food needed to be produced by livestock grazing, aquaculture, and agriculture.

This chapter argues that biodiversity is essential for food and therefore for global health. Food is one of the determinants of health. Most food is now produced through agriculture. The problem is that agriculture often has negative effects on biodiversity. In the long run, the risk is that it undermines its own ecological basis and therefore endangers human health. It is imperative that agriculture sustains biodiversity so that biological and genetic diversity will be preserved. Diversity is crucial for the continuation and further development of agriculture (e.g., to find species that are resistant to diseases). Another issue is that because of the growth of the human population, more food needs to be produced; the question is how this can be done without further loss of biodiversity. How can the human need for food be balanced against the need to conserve biodiversity for future generations?

Biodiversity, as elaborated in the previous chapter, is important for healthcare because it provides source material and ideas for the development of medicines. This is not simply a matter of discovery of natural products or identifying useful plants. Human activities, interpretations, and interventions are necessary in order to deliver a drug. It is the result of collective efforts to transform plants into substances that have therapeutic use. Ethical issues arise when we ask how these products are used to address human needs. Who has the right to claim these products of biodiversity? For what purposes are they developed and distributed? Similar questions arise if we consider the relationship between food, health, and biodiversity.

This chapter explores the ethical issues related to food, which constitutes another product of biodiversity. Again it elaborates the interrelations between biological diversity, health, and healthcare. It shows that these issues go beyond the moral discourse of individual choice and personal responsibility. They are framed in terms of scarcity and abundance, security, rights, and sovereignty and therefore require a broader ethical approach. Ethics is not merely an individual affair but asks questions about the good life as well as the good society. The chapter highlights the generosity of life. Biodiversity nourishes us and therefore requires our respect and protection.

Agriculture and Ethics

The development of agriculture that took place 10,000 years ago was a major step in civilization. [2] It has led to human settlement, founding of cities and civilizations, and growing populations. The fate of empires has been closely related to the success of their food production systems. [3] Throughout history, food has also been associated with declines in health, when the variety of food became narrower and the quality of diets lower. [4] Agriculture furthermore has various impacts on biodiversity. [5] It leads to habitat conversion and destruction; more land is needed to feed more people, so biodiversity is lost. Present-day farming uses pesticides, fertilizers, and drugs. This results in pollution of soil and water, and it affects human health. Modern agriculture tends to grow food in monocultures, thereby reducing genetic diversity and making crops vulnerable to diseases and invasive species. These effects generate a paradox. In order to produce more food and keep human beings alive and healthy, agricultural systems have to expand and be more productive. But in order to do so, they reduce biodiversity and use methods that are adverse to human health. Estimates are that 40% of the earth's land is used for agriculture. [6] Will increasing food production imply more deforestation, or are alternatives available? At the same time, the food supply is already insecure for many people, and the human population is expected to grow from the current 7 billion to 9.15 billion in 2050. [7] Given these challenges, it is clear that the food system is in trouble. [8] This situation also presents an ethical dilemma. It is morally imperative to address hunger and malnutrition, yet it also is a moral duty to protect biodiversity for the survival of humanity.

Examining this ethical puzzle demands critical analysis of the three arguments on which it is built. The first is that in order to prevent hunger and malnutrition, contemporary agriculture should produce more food because the world population is growing. The second is that expansion of modern agriculture will further increase loss of biodiversity. The third holds that biodiversity is necessary for human health. These arguments then lead to the conclusion that, in order to keep people alive and healthy and to feed future generations, modern agriculture is damaging biodiversity as the basis of sustainable production of food and is threatening global health, although in principle it can provide sufficient nutrition for an increasing world population. The following sections scrutinize these contentions.

Food Security

According to the Food and Agriculture Organization (FAO) of the United Nations, in 2015 one in nine people in the world was hungry. A total of 793 million people were undernourished, mostly in developing countries. This is an improvement compared to 1990–92, when 216 million more people were hungry. The reduction in the number of undernourished people took place despite a substantial increase in population. However, the number of undernourished people did not decrease in all areas of the world. In central Africa and western Asia, it became higher. [9] In 2016 in the Central African Republic, 47.7% of the population was hungry every day; in Zambia, 47.8%, and in Haiti, 53.4%. [10] However, even when the total number of hungry people decreased, it remains a moral scandal that millions of people nowadays continue to suffer and die from lack of food.

The heads of states gathered at the World Food Summit in Rome in 1996 concluded that the world can produce sufficient food for every inhabitant. They observed that during the previous 30 years the world population had doubled but food production had increased even faster. Twenty years later, agriculture made available 2,770 calories per person per day (the recommended daily intake is 2,200 calories). FAO studies estimate that in 2050, 3,070 calories will be available. [11] The conclusion is that global resources are sufficient. Increasing agricultural production should be able to provide enough food for a growing human population. [12] Although this is true at the global level, local problems of insufficient production will remain. There are also uncertainties in the prediction, depending on the impact of climate change, disasters, and wars. This optimistic estimate is reflected in the evaluation of the Millennium Development Goals in 2015. It concludes that the target of halving the proportion of people suffering from hunger has been reached. It has fallen by almost half since 1990 (from 23.3% to 12.9% in 2014–16). [13] Nonetheless, despite the progress, one in seven children under age five today is underweight; one in four has stunted growth. [14] Each day 16,000 children die before their fifth birthday, mostly from preventable causes. [15]

The conclusion that modern agriculture will be able to produce enough food for the growing world population is significant for the ethical debate. If food security is possible, how do we explain food shortages and hunger? [16] For a long time, population growth has been regarded as the major cause of hunger and starvation. People multiply faster than food can be produced. But that assumption did not prove true. Despite population growth, availability of food has increased. Countries like the United States and Australia produce much more than their populations need. Australia exports more than half of its agricultural products. [17] In the early 1990s, it was already observed that the majority of malnourished children in the developing world lived in countries with food surpluses. [18] Nowadays shortage of food seems less of a problem than overnutrition. Another issue is that the growth of the human population is decelerating. The increase will be less than in the past, especially beyond 2050, with a peak in 2075 (9.43 billion people) and a decline in 2100 (9.2 billion). The prediction is that after 2050 many countries will enter a phase of population decline. [19] Reduced population growth will diminish the pressure on agriculture.

Current common opinion is that hunger and undernourishment are not the result of food shortages but of poverty and unequal global distribution. For the ethical debate on hunger, this means that the focus moves from food production to better access. The poor are hungry because they cannot afford to buy food. Basic food is often not affordable or available where they live. An example is India. Large-scale agriculture in India produces surpluses, while 250 million rural farmers are malnourished. [20] The first argument—that in order to prevent hunger and malnutrition, contemporary agriculture should produce more and more food because the world population is growing—is therefore questionable. The moral imperative to produce sufficient food for humanity is currently met and can be met for future generations. It is doubtful that ever-greater productivity and expansion of agriculture is really necessary.

The major concerns regarding food are not technical or scientific but ethical. At the same time, it is necessary to consider the environmental impact of agriculture.

Protecting Biodiversity

Biodiversity is a rich reservoir for food. Only a limited number of plants and animals are used in agriculture. Eighty crop plants and 50 animal species provide most of the food for humans. Only 12 plant species provide 75% of our total food supply. Out of the 75,000 edible plants in nature, humans use only 10%. The same for animals; 14 species provide 94% of the total production, with five categories dominating world production: cows, sheep, goats, pigs, and poultry. [21] This means that in biodiversity an enormous range of food sources is potentially available that is currently not exploited. It might be enormously useful in the future. For example, when food crops become infected with diseases, new varieties might be found in biodiversity with resistance against these diseases.

Repeating History

The most commonly eaten banana disappeared from the market in 1965 as a result of Panama disease; a fungus had quickly affected all banana plantations. The alternative was the Cavendish variety of the species, immune to the disease, although of lower quality. There are hundreds of banana varieties in the wild, but for commercial purposes the Cavendish bananas have become a global monoculture. A new strain of the fungal disease is now affecting plantations. It is spreading rapidly because of lack of genetic diversity. The industry hopes to develop a genetically engineered banana with disease-resistant genes. [22]

Referring to the possible future benefits from biodiversity presents the "insurance argument" that is also used with regard to potential drugs and traditional medicine. It is applied to agriculture with the contention that current agricultural methods reduce biodiversity, while diversity is the basis for breeding new varieties of plants—and thus future improvements. Genetic diversity of wild species can protect crops in the future against pests and diseases. It can help to develop species that are resilient to climate change. It can provide biological alternatives to chemical pesticides. Modern agriculture's concentration on a limited number of crops (especially wheat, rice, and corn) and its preference for monocultures makes it vulnerable to pathogens. Genetic homogenization can have disastrous consequences. The classic example is the Irish potato famine in the 1840s. The vulnerable potato monoculture in Ireland was annihilated by blight and led to famine and mass migration. The lesson is to make sure that diversity is not lost. But genetic diversity today is already strongly reduced. Unique genes are lost from the gene pool of species. It is estimated that since 1903, 86% of varieties of apple have become extinct; the same for 88% of varieties of pear. Of the 408 varieties of tomato known to the US Department of Agriculture in 1903, only 79 were available in 1983. [23] Genetic erosion is rapidly reducing the nutritional base for food production. [24] Not long ago, India cultivated 30,000 varieties of rice; now 75% of Indian rice is limited to 1 of 10 varieties. [25] This loss of biodiversity is a great danger for agriculture. If genetic diversity cannot be preserved in the wild and in natural ecosystems, the only other option is to preserve it in special collections such as seed banks and storage facilities. An example is the Svalbard Global Seed Vault in the Arctic region of Norway, which stores 880,000 samples of seeds from almost every country, securing them against global disasters. After four years of civil war, Syria was the first country to withdraw seeds from the vault. [26]

The search for food has an unavoidable impact on biodiversity. Hunting has led to extinction of species. Fishing has depleted fisheries. Agriculture has an even broader impact on biodiversity. Human beings will always give priority to food; they will not go hungry or starve for the sake of conserving biodiversity. However, if a global and long-term perspective is taken, there will not be much difference. In the long run, if we continue our current agricultural methods, biodiversity will be destroyed. From this perspective, there is only one fundamental value at stake, and that is human survival. Agriculture is necessary. It cannot be disconnected from biodiversity, but the way food is produced now is disastrous for biodiversity. It will destroy human possibilities to produce food in the future, thus endangering the sustainability of agriculture itself. Even today it results in serious threats to human life, as documented in previous chapters. The clearing of tropical forests brings human beings in contact with unknown viruses and is associated with emerging infectious diseases. The way human beings deal with animals, especially in factory farming and the poultry industry, endangers human health. Overfishing in West Africa has led to increased demands for bushmeat, risking new zoonotic diseases. Furthermore, modern agricultural methods create risks to human health, through increased use of chemicals such as pesticides and fertilizers, but also drugs, such as antibiotics and growth hormones. Each year more than 80% of antimicrobial agents in the United States are used to feed healthy animals in order to promote their growth (see box, p. 153). This nontherapeutic use contributes to the development of drug resistance. [27] The ubiquitous use of glyphosate as an herbicide has significant toxic effects on human beings (see the discussion in chapter 4). Finally, agriculture is a major source of greenhouse gases. [28] Food production is responsible for 25% of global greenhouse-gas emissions. [29]

Estimates of future agricultural productivity usually pay little attention to the effects on biodiversity. From an economic perspective, food security comes first. The damage, and thus costs, to the environment are not really connected to the primary aim, which is producing food. This is an example of "externalization": the effects of agriculture on biodiversity and health are for others to consider; the costs of these negative consequences cannot be taken into account, since that would make food much more expensive.

The second argument underlying the ethical dilemma between producing food and protecting biodiversity is that expansion of modern agriculture will further increase loss of biodiversity. Apparently, there is sufficient evidence for this connection in the scientific literature. [30] Yet it is not agriculture as such, but rather the way in which it has developed and is practiced that is problematic. The question therefore is: Is it possible to have an agricultural production system that respects the value of biodiversity and does not reduce biological diversity?

Human Health and Food

Biodiversity and health are closely connected. In chapter 3, I argued that without a healthy environment, human beings cannot flourish. Food is an important link between biodiversity and human health. [31] In ethical debates, attention is chiefly focused on consumption. Specific diets are associated with particular diseases. Individual behavior and choice are therefore important. Mainstream bioethics articulates individual responsibility in selecting and consuming food. From a global ethics perspective, it is questionable whether this perspective is adequate, since broader processes are at work. Currently, a global dietary transition is under way. Traditional diets are being replaced by those that contain more refined sugars, fats, oils, and meats. Such diets will lead to more diabetes, coronary heart disease, and—particularly—obesity. In 2014, more than 1.9 billion adults were overweight or obese. This "nutrition transition" that is occurring in both developed and developing countries will seriously impact human health and biodiversity. [32] It is the result not so much of individual consumer choices but rather of decisions and policies at the level of agricultural production. Another highly relevant example is antimicrobial resistance. Neglected for a long time, it is now recognized as an urgent global problem. The World Health Organization appointed a special representative in 2015 to set the agenda for this problem. It is estimated that at least 700,000 people die each year because of antibiotic resistance. It particularly affects the global poor. [33]

Antimicrobial Resistance

Resistance against antibiotics is growing, largely because of overprescription and self-medication. In many developing countries, antibiotics are available over the counter without prescription. China consumes 10 times more antibiotics than the United States. But there is also conflict between health and agriculture. Industrial animal farming in the United States consumes 80% of all antibiotics, usually for growth promotion or disease prevention. Back in 1969, the Swann Report in the United Kingdom recommended that antibiotics not be used for these purposes, but its advice was never implemented. Sweden was the first country to ban nontherapeutic use at the request of Swedish farmers themselves in 1986. The European Union banned nontherapeutic antibiotic use in livestock in 2006 despite the opposition of animal health experts. In the United States, reports and studies have been published since 1972 urging voluntary action. Thirty years later, in 2002, consumer activists put the issue on the social and political agenda. It was only in the fall of 2014 that Barack Obama's administration launched a national strategy to reduce the use of antibiotics in food animals. In that year, only 5% of the meat sold in the United States was free of antibiotics. [34]

If attention is focused on the production of food, is it obvious that the way in which food is produced contributes to the health effects of diets. Individuals usually have little influence on the kind of food produced and even less on how it is produced. Ethical emphasis on personal autonomy and individual responsibility in regard to food production is rather futile nowadays, since food is produced by corporations and rural farmers and families far away. Worldwide there are 570 million farms; more than 90% are managed by an individual or family, producing more than 80% of the world's food, in terms of value. [35] Transnational agriculture corporations control 40% of the world trade in food. [36] In the United States, four companies control 79% of beef packing and 57% of pork packing, while over the past 50 years, four million farms have disappeared. [37]

The current effects of agriculture go beyond the scope of individual decision making and demand collective action. The dietary transition requires more energy (and thus increases gas emissions) and land clearing. [38] Greater use of pesticides and fertilizers will be required. Water and air will be more polluted. This intensification of agriculture will not only lead to biodiversity loss and species extinction but will also be more risky for human health. Intensive production and use of monocultures result in unhealthy consumption. Reduction of crop diversity, for example, has reduced the quality of diets. Micronutrients such as vitamin A and iron are often deficient. [39] It is important not only how much food is produced but what kind of food—not only healthy food but also a sustainable diet. Food production is moreover related to food safety. A series of recent scandals show that food production and processing are vulnerable to manipulation and contamination. Biodiversity loss is a significant factor in emerging viral diseases that infect chicken and pigs, which are major food sources for human beings. Owing to climate change, the burden of foodborne pathogens will increase. Currently, 1.9 million people, mainly children, die each year as a result of diarrheal diseases related to contaminated food. [40]

The third argument connected to the ethical dilemma of food production versus biodiversity protection underlines that biodiversity is necessary for human health. This connection has been illustrated in chapter 3. The food produced in agricultural ecosystems is directly linked to human health. The challenge is not only what food is produced but also how it is produced.

From a global bioethics perspective, it is important to consider the ethical issues of agricultural production. This will require a broad ethical view as well as a long-term perspective. To conserve the health of the world population, it is necessary to look beyond the health of individual consumers. Agriculture will be necessary to feed human beings, now and in the future. It is questionable whether there is a real need to expand agriculture, since there is already more than enough food produced. But if more food needs to be produced for a growing population, can it be done without further loss of biodiversity, and thus without risking and jeopardizing the health of future generations? Can food be produced in a way that promotes health and conserves biodiversity?

Different Ways of Producing Food

Agriculture has been transformed into a modern industry only recently. Mechanization replaced human workforce. Specialization focused on one type of crop or animal. Chemical fertilizers and pesticides increased yields. Science and technology supported plant breeding and genetically engineered crops. This new type of farming delivered enormous productivity gains. It also had a much larger impact on biodiversity and, consequently, on human health than previous methods of producing food. [41] In the current debate about food security and the need for a balance between food, health, and biodiversity, three options are on the table: intensification, ecology, and biotechnology.

INTENSIFICATION

The first, most-advocated option is intensification of the current type of agriculture (i.e., to increase productivity per unit of land). The argument is that more food can be produced with less land, [42] and food security can only be addressed by producing more food. [43] Because this "productionist paradigm" is efficient, there is, according to some experts, little ethical justification to promote any alternatives. [44] Furthermore, intensification has advantages for biodiversity if no new land is cultivated. [45]

However, intensified industrial agriculture is regarded as the source of many of today's problems. [46] It is criticized for not being able to provide solutions to the difficulties it is creating. The productionist paradigm represents an economic system that is driven by efficiency and rationalization. [47] It is primarily occupied with production of food as a commodity and neglects other relevant dimensions of food culture and policy. It is associated with structural problems such as overproduction, disregarding animal welfare, reduction in biodiversity, and risks for diseases. It emphasizes availability rather than access and affordability of food. It furthermore ignores the role of corporate interests. Finally, it is doubtful whether intensification will lead to land sparing. [48] Anyway, it is projected that land suitable for agriculture will be exhausted by 2050 or earlier. [49] The system will destroy itself in the longer run. The point is that these problems have been recognized for a long time. But in addressing them, industrial agriculture itself is not questioned. What we need is an ethical reorientation toward a new type of food production. [50]

ECOLOGY

The second option is an "agroecological approach" that combines food production and biodiversity protection in a sustainable way. [51] This is what Tim Lang
and Michael Heasman call the "ecologically integrated paradigm." [52] This approach balances the interests of humans and ecosystems. It combines a number
of alternative agricultural methods: traditional practices, ecological farming, and
organic food production. The best solution to the current problems is widespread
conversion to alternative agriculture. In many countries, the relation between
society and the food industry is being transformed, and farmers are willing to
change their production methods. [53]

Organic Food

Organic farming as an alternative form of agriculture is one of the fastest-
growing sectors of world agriculture (although it represents only 1% of the
world's agricultural land). Organic food is grown in harmony with nature.
Only organic materials are used to fertilize the soil. Chemical pesticides and
fertilizers are avoided. Genetically modified cultivars, hormones, and
antibiotics are prohibited. While it is clear that organic farming has advantages for health and is less harmful to biodiversity than industrial agriculture,
it can also produce sufficient yields to feed future generations. Nonetheless,
it is often neglected in policy discussions. Consumers, on the other hand,
are increasingly interested in food without chemicals that is "natural" and
contributes to preservation of biodiversity. They purchase and consume
organic food because it is healthier, tastier, and safer than conventionally
grown food. [54]

One argument in favor of this approach is that it can increase global food production without further damaging biodiversity and increasing land use. [55] By
drawing on the ecological idea of connectedness and interdependence, [56] it is
possible to find a balance between optimizing food production and minimizing
damage to the ecosystem. [57] Wildlife-friendly farming, for example, is focused
not only on the crop but also on the management of the land, the water, and the
ecosystem as a whole. Searching for another type of agriculture revives the idea

of the commons. If food is regarded as a commodity that can be produced and marketed, it is separated from community, culture, and context. Traditionally, it has been connected with humans and land, being part of biodiversity as one of its gifts, like medicinal products discussed in the previous chapter. [58]

Another argument is that, as Michael Chappell and Liliana LaValle explain, "the problem of food insecurity is a matter not of total availability but indeed one of access, political power, and equity." [59] Improving access to food is more important than producing more food. That means that policies should aim not at producing more but at producing differently. [60] About 2.3 billion people reside in rural areas dominated by smallholder agriculture. In these places, traditional agricultural methods are commonly used.

A third argument is that an ecological approach will lead to sustainable agriculture. Sustainability here means that humans preserve their capacity to produce food that satisfies human needs for an indefinite period of time. [61] Sustainability is a balance between production of food and the needs of future generations, depending on how fast the world population is growing. If we don't make agriculture sustainable, we are harming not only existing people but also future people. Therefore, it is argued that we have a moral obligation to promote this type of agriculture. [62]

BIOTECHNOLOGY

The third option is that science and technology are the best way to deliver more food. Science, and in particular biotechnology, has contributed much to the development of agriculture. A famous example is Louis Pasteur, who in the nineteenth century at the request of the French wine and beer industry, developed the process of pasteurization. Nowadays, genetically modified crops promise to feed growing populations through their higher yields while promoting sustainable practices. [63] Genetic modification can produce crops that are resistant to insects and diseases and that also have higher nutrient quality. This can reduce the level of pesticide and herbicide use and thus lead to less exposure to toxic chemicals and improvement of human health. Because of these potential advantages, biotechnology is increasingly used in agriculture. About 11% of the world's cultivated land is covered by genetically modified (GM) crops, particularly in the United States, Canada, Argentina, Brazil, and India. [64] On the other hand, in the European Union, 57% of the population is not willing to support GM food. They consider it unsafe, inequitable, and worrying; 19 European countries have banned GM crops. [65]

Super Bananas

In February 2016, protesters gathered at Iowa State University to prevent human testing of genetically modified bananas. The bananas were modified in Australia with a gene producing beta carotene, a precursor to vitamin A. The idea is that the "biofortified" banana could help to overcome vitamin A deficiency. This is a serious problem, particularly in Africa, where it is a major cause of blindness. [66]

Many scientists argue that biotechnology is the best option we have for food security. Nobel laureate Norman Borlaug, regarded as "the father of the Green Revolution" in the 1950s and 1960s, declared that biotechnology is the only way to increase food production and prevent future starvation. [67] Nevertheless, since the 1990s, debates have raged about the benefits and risks of GM crops. [68] The claim that biotechnology can secure food production is contested. There is evidence for increased yields in some crops, but this higher productivity is often associated with reduced profits for the farmers because the costs have increased. [69] GM seeds cost twice as much as non-GM seeds. [70] Major concerns exist regarding human health and biodiversity. Research findings differ, and the evidence is often ambiguous. Data are mostly provided by the industry; it is not clear how objective and reliable this information is. Most academic and international organizations, however, assume that GM food poses no risk to health. Similar controversies exist in regard to biodiversity. Studies show that crop biotechnology reduces the use of agrochemicals, but the findings are not significant and convincing. [71] One of the potential risks is invasion of natural habitats. GM seeds can contaminate native genetic resources. GM organisms can also spread to other fields and jeopardize the organic certification of farmers who do not want to grow GM crops. Contamination is almost unavoidable. Past experiences show that transgenes will escape and spread rapidly and widely. The decision to introduce transgenic crops may be irreversible. [72]

The Schmeiser Case

Monsanto has patented a canola seed that is resistant to glyphosate. Farmers have to purchase the seed every year and sign a licence agreement in order to use the patented product. The seed was introduced in Canada in 1996. In 1997, Monsanto found that its herbicide-resistant product was growing on the farm of Percy Schmeiser while he had not paid royalties. Schmeiser argued that his crop had been contaminated by genes from neighboring farms. Arguing that he owns the seeds he is harvesting, Schmeiser grew the GM crop the following year. Monsanto then sued him in 1998. The Canadian Federal Court in 2004 judged that Schmeiser had infringed Monsanto's patent rights.

This long debate demonstrates that many people remain unconvinced that biotechnology can provide sufficient food for the future without damaging human health and biodiversity. There is no evidence that biotech food is better for human health and that it will enhance biodiversity. On the other hand, there is still no convincing evidence that it is risky for health and the environment. A further complication is that the assessment of risk is not uniform. As discussed in chapter 4, the usual risk governance is based on science, and it operates with a utilitarian framework; it is outcome-oriented, data-driven, and comparative. Such focus on expected costs and benefits does not take into account the socio-economic consequences of biotechnology. [73] Another risk assessment approach is based on the precautionary principle. This approach is advocated by United Nations organizations and is used by many European countries. It emphasizes caution where there are threats of serious damage and insufficient scientific certainty. It also integrates ethical concerns and broad public participation in decision making. The difference in risk assessment partially explains the variety of perceptions about genetically modified food across countries.

Another major concern in regard to biotechnology is related to manipulation of life. Genetic modification implies transfer of genes from one species to another. It also involves the creation of new or mixed species by human beings. In many worldviews and religions, this is morally objectionable. Life is not a simple commodity that can be managed as a resource to be changed and enhanced by biotechnology. Animals and plants are not machines that can be manipulated and patented. [74] Considering the gene as the core of living beings does not do justice to the dignity of life. Similar ethical considerations have also been raised in the debate about ownership of life forms. [75]

A third ethical concern is also acknowledged by supporters of biotechnology. It addresses the socioeconomic consequences of this technology. The emphasis on science and technology has articulated the inequalities in global food production. The benefits and harms of bioengineered food are not equally distributed; transnational corporations have benefited most, while farmers and consumers have received little direct benefit. [76] A limited number of very large companies control the world food market. Patent rights confer upon them almost monopolistic power. [77] One of the effects is that farmers have become more and more dependent on agribusiness. [78] Another is that policies are not necessarily formulated to address fundamental problems of farmers or consumers but rather are primarily concerned with the profitability of crops for large companies. [79] Not much research is done on crops that are important in Africa, such as sorghum or yams. [80] GM food is often presented as a panacea that can address food scarcity and hunger. But in practice it seems a mere commercial instrument in a market that is already overproducing food and that reinforces the injustice of the present system. What is the impact of bioengineered food on the quality of life of the poor in developing countries when most food in those countries is produced for export? What are biotechnology's benefits and to whom are they directed if after almost four decades of research only two new features (resistance to herbicides and to insects) have been added to just four species? [81] The ethical issues generated in biotechnical agriculture are associated with those in medical research and development of new medication, discussed in the previous chapter. The natural products of biodiversity belong to humanity; they are used as drugs and food. But today they engender injustices and inequalities. Indigenous people and local farmers who for ages have bred and improved crops and preserved biodiversity are not respected and compensated. Economic interests determine how we live in the world. Against this backdrop, the ethical question is whether biotechnology in agriculture is the best option available for the future.

Food Regimes

The foregoing discussion illustrates how food, globalization, and politics have become intimately linked. While biodiversity has always provided food and human beings have been responsible for growing, harvesting, processing, and preparing it, we now live within what has been called a "food regime" or "food system." These are complicated networks that produce, process, transport, market, and sell the food that we consume. [82] It includes farmers, corporations, retailers, supermarkets, consumers, and many others. The current food regime has two characteristics: it is globalized and corporate. Food we consume is often produced far away and in multiple countries. Recent studies show that 69% of all food supplied originated in other regions; in the United States nearly 90% was "foreign." [83] Hardly anyone eats a native or traditional diet. Because consumers and producers (mostly in the global South) are separated, we have no idea how our food is produced, what technologies and chemicals are used, and under what circumstances it is grown and harvested. Efficient global systems bring it into our supermarkets and restaurants. At the same time, similar food is available almost everywhere.

This globalized regime is also a corporate one. Over the past 30 years, an industrial type of agriculture has developed based on neoliberal models. Food supply chains and food policymaking are controlled by a few actors, mostly a limited number of multinational companies. The main concern of this regime is production, efficiency, and quantity of food. Since agriculture has been included in the arrangements of the World Trade Organization, it has been deregulated and privatized while international standards have been adopted, trade liberalized, and import restrictions removed. Critics argue that this regime, though it has increased food production, has serious negative consequences all over the world. Free trade agreements are destructive for farmers everywhere, even though they are the main producers of food. Millions of small farms have folded. [84] Since the 1990s, safety and nutritional quality standards have been lowered and multiple food scares have occurred, because the preoccupation of policymakers has been promoting trade. The power of agribusinesses has enormously expanded not only in food production but especially in food consumption. These consequences have led to a critical counter-discourse that emphasizes food sovereignty and many alternative approaches to current agriculture; agroecology, organic farming, fair trade, slow food, and local food are well-known examples. [85] But there are also important consequences for the bioethical debate about food. This should promote a rethinking of the way in which the food we buy and use is produced as well as consumed.

Food Consumption

With one in nine people hungry and with a child dying of hunger-related diseases in developing countries every 10 seconds, it is not a remarkable observation that the current food system is in trouble. [86] This conclusion has become even more evident given that, at the same time, more than two billion people suffer from overweight and obesity. The system obviously does not promote health. This is not due to general lack of food. Enough calories are produced to reduce hunger. But the qualities of those calories are insufficient for a nutritious diet. [87] This observation refers to two ethical concerns. First is the issue of access, availability, and distribution; it is unjust that an abundance of food exists while people die of hunger. Second is the issue of the common good; food should be produced not only for individual survival and satisfaction but should also promote health, protect biodiversity, and thus be available for future generations. Over the past few decades, we have seen that more intensive food production alone is not sufficient to reverse malnutrition. A broader perspective is necessary. [88] This conclusion relates not only to the production of food but also to its consumption. Food consumption has significantly changed. This "nutrition transition" means that Western diets heavy in fat, sugar, and salt have become more popular in countries such as China, India, and Brazil. Paradoxically, there is also a growing interest in quality and local food. This tension between global and local is explored later in this chapter.

Food choices are ethical choices. [89] Whether we prefer to eat organic food, meat, or fast food has consequences for the food system. It is often the expression of various values. We may be concerned with our own health or simply the taste of food. But our preferences also have an impact on biodiversity or on global justice in terms of the availability, distribution, and affordability of food for other people, often far away. In the ethical debate about what we eat, there is a growing emphasis on values, particularly health and sustainability. At the same time, the debate is characterized by a particular ethical focus on individual responsibility. There is an enormous supply of food products, and individuals have to make choices. The kinds of food that they buy and eat are ultimately their responsibility. The assumption is that human beings are consumers who follow their personal values. This view reflects the neoliberal model that people are rational agents who make decisions based on their own preferences. The market only provides products. There are no good or bad foods. Another assumption is that foods are commodities like other products in the marketplace. However, this is a serious reduction of the meaning of food. It is primarily nourishment, derived from an ancient French word that refers to sustenance, fostering, nursing. It is not simply alimentation, substance, fodder, or fuel to stay alive. Of course, food is a basic requirement for our existence, but it also shows that human beings are part of nature (we need its materiality because of our corporeality) and at the same time connected to others, present and future humans and nonhumans (in the choices of what we eat and how we use food). Eating therefore is an ethical act, bridging biology and culture, private and social conduct. [90]

This broader meaning of food is often lost in current debates. It is assumed that food has no exceptional status. These assumptions are specifically articulated by food companies and in food policies. Regarding obesity, for example, emphasis is placed on weight and physical activity. Individuals should take care of themselves; display moderation, self-control, and temperance; and observe the balance between calories and exercise. They should be adequately informed and educated. [91]

This focus on individual choices neglects how choices are strongly influenced by the "food environment." [92] Individual choices are also determined by ease of access to food and what kind of food is available. Some people live in areas where healthy food is scarce (so-called food deserts). [93] Supermarkets offer an abundance of choices, but it is difficult to find out what is healthy and sustainable. It is argued that an "obesogenic" environment contributes to and promotes obesity. [94] There is more at play than individual behavior.

It is important to recognize the role of the food industry in shifting attention away from the environment to individuals. A lot of evidence has been presented to demonstrate that the food industry is not really interested in producing and selling food that is healthy and environmentally sustainable. Its interest is profit. Marion Nestle gives many well-documented examples of how food companies undermine dietary advice provided by governments and experts. [95] Companies argue that governments should not restrict the choices of individual consumers; they are free to choose their own food, and they are responsible for their choices. The industry provides generous financial support to experts, researchers, journals, and professional associations to promote the idea that there are no bad or good foods (giving rise to multiple conflicts of interest). Enormous amounts of money are spent to influence efforts to regulate food, for example, the availability of soft drinks and fast food in schools. Companies shield themselves against accountability and do not assume any responsibility for harms and errors. They argue that external regulation to protect the consumer is not necessary. They collude with governments that are often serving business interests rather than the public interest. According to the industry, self-regulation will be sufficient, but studies show the contrary. [96] Nestle concludes that diet is a political, not only a personal issue. [97]

The shift toward blaming individual consumers for any unhealthy effect of food is accompanied by extended efforts to shape the choices and preferences of these individuals. Food choices are not made in a vacuum, so food producers continuously try to influence the environment of choice, particularly at home, work, and school. Enormous budgets are spent on advertising. The marketing of junk food especially targeting children and young people has long been controversial. [98] As vulnerable populations, they have limited ability to make rational choices and often have difficulty distinguishing between advice and advertising. A global and rising problem such as childhood obesity can only effectively be addressed at the level of populations, not by blaming and educating individual persons. [99]

An ethical concern in this context is trust. Public confidence in food producers is low. Most Americans do not know how their food is produced. Only 17% of consumers trust food producers and manufacturers. [100] Trust has diminished, particularly in regard to larger food companies, because of debates about genetically modified food and animal welfare but also because of lack of transparency. In fact, consumers have different types of moral concerns. [101] One type is relevant for all consumers. Individuals cannot assess whether food is safe; they have to rely on others, specifically the government, for that assurance. A second type of concern is specific to particular groups of consumers. They want certain foods that accommodate their values. Jewish people want kosher food. Vegetarians do not want meat. They need to be sure that the food they buy and consume fits into their value system. A final group of consumers has social concerns. As global citizens, they want to contribute to a better world. They do not want to be involved in practices that are unjust or contribute to inequality. For example, they do not want to buy food that was produced in exploitative conditions, so they buy primarily fair trade products. However, a long series of food scandals have diminished all these types of trust about food. [102]

Horses and Babies

In December 2012, food inspectors in Ireland detected that meat sold as beef contained horse meat. Similar results were found in various European countries. Horse meat is regularly eaten in France and Italy but not in other countries. Consumers were horrified. Hamburger sales plummeted. Millions of beef burgers and beef ready meals were recalled. It was eventually discovered that the horse meat originated in Romanian abattoirs and was then distributed as real beef by French and Dutch companies. The scandal revealed that it is difficult to control the food chain. An appropriate European watchdog is lacking. Horse meat in itself is not unsafe. Rather, the scandal raises the issue of food authenticity. This makes it very different from the baby milk scandal in 2008. Chinese producers of infant formula mixed it with the chemical melamine, used in the production of plastics. More than 50,000 babies were hospitalized, and several died. [103]

There are diverse reasons why it is difficult to develop food policies. [104] Nutrition is a matter of health as well as agriculture. There is no natural institutional home for policymakers. Conflicts can arise between institutions representing health and farming (at the global level between WHO and FAO) so that policy development can be incoherent. Also, the nutrition community is fragmented, so there is a disconnection between agriculture, nutrition, and health. There is furthermore a multiplicity of stakeholders. The role of the private sector is strong, so commercial interests frequently prevail. The contribution of ethics is the final piece in the puzzle. Food ethics is a relatively new field of interest that has recently emerged along with the notion of global bioethics. [105] Studies that integrate issues in food ethics into the framework of global bioethics are scarce. Global bioethics, as is argued later in this chapter, can focus attention away from the ethics of individual responsibility toward the broader neoliberal context and the underlying roots of many problems. It can provide a different framing of problems and solutions in terms of human rights, poverty, and public health. As Nestle reminds us, tobacco use could only be regulated by focusing on environmental issues rather than by merely educating and informing individuals. [106]

The insufficiency of an individual perspective is illustrated in the debate on food labeling. Labels provide information and thus support consumer autonomy. They also indicate that the products have an alternative value (e.g., green labels and fair trade labels). Labeling of GM foods has become controversial and in 2003 led to a trade dispute at the WTO between the United States and Canada versus the European Union. Labeling of such foods has been mandatory in the EU since 1997. Especially in the United States, producers have strongly resisted such labeling and argue that it should be voluntary. They contend that because in terms of health and safety GM and non-GM foods are equivalent, labeling would unfairly stigmatize bioengineered food. [107] More than 70% of food in US supermarkets contains genetically modified organisms. [108] Producers were rightly concerned that labels would reduce sales. In 2015, polls showed that 89% of Americans are in favor of mandatory labeling. [109] The protracted, acrimonious debate over GM labeling makes it clear that the food industry is emphasizing the notion of personal responsibility for consumer choices but has for a long time denied consumer autonomy, the right to know what is in the food. One cannot be responsible for what one does not know. Producers regard food simply as a commodity, ignoring its cultural and social value. For many people, it is not just material to fill the stomach. Of course, consumers are concerned about safety and health. But food is also related to identity and values. [110]

GM Food Labeling

In July 2016, the US Congress approved legislation requiring the labeling of genetically modified ingredients in food products. It is questionable, however, whether major food crops such as GM corn will fall under the definition of bioengineered food as stipulated by the law. In fact, it is not a real label; consumers have to scan a code on their smartphone in order to access the information. Or they have to dial a phone number. The information is not on the package of the product.

Ethics and Food

Ethics is not just about making individual choices but is also a way of life. Since human beings are not isolated decision makers but are always connected to others, the focus should not merely be on how to live a good life but also on how to build a good society and promote human flourishing. This implies that the perspective of mainstream bioethics is often too narrow. Nevertheless, for a long time, the ethical reflection on food has primarily been concerned with individual choices. Only more recently has the moral perspective shifted toward the social dimension of food production, distribution, and consumption. [111]

Human life has always been threatened by hunger, disaster, poverty, and disease. [112] From an ethical perspective, hunger is often regarded as the worst evil. [113] Without food, people die; it is a necessity of life. Hunger can also easily be relieved with food aid. However, hunger has often been regarded as a failure of morality. It is seen as evidence of ignorance and inadequate choices; it is not due to poverty but to laziness, moral weakness, and lack of discipline; the hungry are "authors of their own misery." [114] Historian James Vernon shows how this attitude changed with the "humanitarian discovery of hunger" in the nineteenth century, when human rights were also rediscovered. The hungry and poor became the object of compassion; people became aware of the suffering of distant strangers. People noticed that the hungry were the victims of forces beyond their control, regardless of moral failure. The Irish famine, for example, produced general sympathy. It was recognized that hunger hurts individual people but that it is also a collective social problem. [115]

Before the 1970s, hunger was primarily framed as a technical issue. Conceived of as a general scarcity of food, hunger was the domain of agricultural experts, whose main challenge was to grow more food. This framing changed in the 1970s, a time of major famines in the Sahel, Ethiopia, Bangladesh, and Cambodia. Hunger became more clearly a moral problem. This was due to two influential publications. Peter Singer argued in 1972 that especially people in developed countries have a moral obligation to relieve hunger. [116] Amartya Sen showed in 1981 that hunger was not the result of scarcity; it is not a lack of availability but of accessibility that creates this problem. Global inequality is at its roots. Lack of access and unequal distribution specifically harm the most vulnerable populations. Sen also argued that individuals are entitled to food. [117] This moral framing had two consequences. [118]

RIGHT TO FOOD

First, it reactivated the discourse on the right to food, formulated in the Universal Declaration of Human Rights as part of the right to a standard of living adequate for health and well-being. The right to be free from hunger as well as the right to adequate food was explicitly recognized in the International Covenant on Economic, Social and Cultural Rights in 1966. [119] The impact of this right nonetheless has been limited. One reason is that the US government is adamantly opposed to any right to food approach (the United States is one of the few countries that have not ratified the covenant). Another reason is that FAO policies have for a long time articulated the neoliberal approach of the World Bank. Only in 2009 did the organization advocate for a right to food approach. [120] Finally, some countries used to make a stringent distinction between negative and positive rights. Civil and political rights such as freedom of speech require noninterference, whereas those like the right to food call for some positive action. This distinction is now rejected. Even negative rights demand positive action—for example, setting up institutions to guarantee free speech—while all rights have correlative duties. [121] It is also more accepted nowadays that states should implement the right to food. They have a responsibility to make sure that activities of transnational food companies do not have a negative impact on this right. One major difference since the time when the right was first formulated is that there is now a global community that is assuming moral responsibilities. Especially in the face of hunger and famine, it is often the global community that is supposed to respond, whether it does so or not. [122] Human rights obligations do not stop at the border.

INSECURITY

Second, the moral framing of hunger also encouraged the political framing of "food insecurity." In the late 1970s, the FAO introduced the notion of food security to emphasize national self-reliance. It encouraged national and regional food reserves for vulnerable developing countries. However, this strategy was quickly reversed in the 1980s, when the emphasis shifted toward market dependency. Food security was redefined, articulating access by all people at all times to enough food and focusing on individuals rather than nations and regions. It is lack of purchasing power of individual consumers that is determinative, not supply, social context, economic inequality, disease, or land ownership. Associating hunger and poverty, the neoliberal redefinition of food security emphasized individual income and the functioning of the food market. Most important in this view is economic growth; the assumption is that this will lead to eradication of hunger. [123] This political framing sees food insecurity primarily as an economic, not social or moral problem. Policies based on this framing had disastrous effects. Most developing countries became dependent on imports from the global North. Nigeria, for example, was self-sufficient in the 1960s. After that, its importation of wheat increased, and its indigenous production declined. The result was that in 1983 nearly one-quarter of the country's earnings were spent on wheat imports. [124] Another example is Zimbabwe. The World Bank encouraged the country, known as the breadbasket of southern Africa, to sell its food stocks and buy grain on the world market. Following devastating droughts, malnourishment and infant mortality rapidly increased in the 1980s. [125]

The World Food Summit in 1996 in Rome reinforced this approach to food security. [126] The policies of FAO focus on agriculture as the main driver, the market as the main producer and distributor, and individuals and households as consumers responsible for feeding themselves. In her study on definitions of food security, Lucy Jarosz concludes that "food security is indistinguishable from neoliberal development discourse." [127] The 1990s was a time of food surpluses. Fears of food shortages have diminished as a result of the progress in agricultural production. The concept of food security recognized that famines and hunger are not only a matter of food scarcity. It was obvious that hunger is also occurring within developed countries. The aim of food security should not simply be survival but should enable human beings to have a good life, that is, a healthy and active life. The notion of food security therefore is not only concerned with the alleviation of hunger but also with accessibility and distribution of food, as has been advocated by Sen. At the same time, hunger is primarily regarded as a technical problem with scientific and biotechnological solutions. Hunger and famines are caused by war, drought, or natural disasters and require humanitarian aid. Usually, little attention is paid, at least in official policies, to the underlying structural inequalities in the global food regime. This has changed recently.

The notion of food security assumes that the best way to guarantee availability and access to food is the market. Open borders and minimal trade regulations will guarantee that sufficient food is produced and distributed. [128] But there are serious doubts about whether trade liberalization can take care of future global food security. Food patterns are changing; richer people eat differently; they eat more animal-based protein such as meat and cheese. There is increasing use of land to produce biofuels. A "land grab" is going on in developing countries, where foreign companies and countries buy up large areas of fertile land to grow food for their own populations or to sell it on the market. [129] Climate change has serious adverse effects on agriculture. And food prices have been rising since 2008. [130] It is doubtful whether these challenges can be controlled by market forces, especially in regard to biodiversity losses. Such control of diverse challenges will require that the market is embedded in ethical values such as health and respect for biodiversity. However, such subordination and embeddedness of food markets is very weak at the moment. Related to this is the predominant view of food as commodity. The idea of a right to food assumes that food is an exceptional necessity of life—one that cannot be treated as another market issue. It relates to human dignity. Food cannot be denied to people because profits cannot be generated. Furthermore, the value of food is not limited to nutrition and health; it also has a symbolic value. [131] It is a matter of identity. Food determines differences between social classes. It signifies exclusion (difference) and inclusion (acceptance). As anthropologists have shown, food is not just material substance but also a symbol of our social status and identity. [132] In addition, framing food as a commodity deflects our view from the social context and human interactions within which food is normally produced. Small farmers are responsible for most of food produced in the world. In the context of trade, they are regarded as "food entrepreneurs," not as agrarian citizens, connected to a particular place and having a collective responsibility as citizens toward their fellow human beings. [133]

These criticisms of the notion of food security affect basic elements of the global food regime. The current system does not provide security. It is not certain that it guarantees the sustainability of agriculture. It cannot promote global health, since it does not deliver healthy diets. It cannot meet social needs, since the focus is on maximum output for sale on the global market. It forces countries to export food rather than to provide sufficient resources for their own populations. It promotes privatization and deregulation, and does not defend peasants, small farmers, and rural communities. Finally, promoting food security has individualized the problem of hunger. It has transformed it into a technical and economic concern, and effectively removed it from the discourse of ethics. Food security as a narrow market concept is therefore not able to merge the values of health and biodiversity into a sustainable food system. Against this backdrop, many alternative approaches to the dominant agricultural model have developed, as discussed earlier in this chapter. But it also gave rise to a critical ethical discourse in connection with the current global food regime.

FOOD SOVEREIGNTY

A new critical discourse has emerged around the notion of "food sovereignty." It is a reaction against neoliberal globalization and the industrial food system but also a project for the democratization of the global food regime. This new discourse emphasizes public control, agency, autonomy, and democratic participation. [134] Food sovereignty is formulated and advocated by La Via Campesina, a grassroots organization that was formed after rural farmers from Latin America, Europe, and the United States met in Nicaragua to adopt the Managua Declaration in 1992. The next year, they founded La Via Campesina in Belgium. The organization represents over 200 million farmers from 73 countries. This was the time of international trade negotiations, resulting in the establishment of the World Trade Organization, which defined new trade rules for agriculture. The new, transnational network gave voice to small-scale farmers, rural women, peasants, indigenous people, and agricultural workers. [135] But La Via Campesina is more than a farmers' movement. It has a much broader agenda that is relevant for global bioethics. [136] The current debate introduces two bioethical considerations: power and justice.

La Via Campesina: Managua Declaration (April 1992)

"For us, the maintenance of our families and a vital rural life is a fundamental objective. . . . Current agricultural policies, imposed and exploitive of the environment, do not benefit either producers nor the society as a whole. . . . Trade and international exchange should have as their fundamental goal, justice and cooperation rather than competition and the survival of the fittest. . . . We reject policies which promote low prices, liberalized markets, the export of surpluses, dumping and export subsidies. . . . We demand real participation in the formulation of policies which affect the fundamental condition of our sector in order to overcome the injustices which we bear. . . . We draw your attention to the lack of respect which has been shown to our productive culture." [137]

IMBALANCE OF POWER

As a countermovement, food sovereignty claims that people have the right to determine their own food and agriculture. The current governance system is primarily controlled by transnational food companies with their allied governments. These corporations first of all have instrumental power: they can influence policies and change decision outcomes with lobbying and political campaigns. They also have structural power, setting agendas and making proposals that limit the range of choices that can be made—for example, threatening to relocate investments to other countries or locations. Finally, they have discursive power, determining the standards set for themselves and establishing the rules that are intended to govern their activities—for example, private certification standards and voluntary codes of conduct in order to avoid state regulation. [138] The consequence of this domination of power is that the present food system leads to dispossession (land, seeds, and knowledge are no longer owned by small-scale farmers and indigenous populations), displacement (closure of farms, mass migration to cities, and creation of slums), insecurity, and general vulnerability. The notion of food sovereignty underlines the fact that food security has to do with resources but also with control of resources. Security means that we have adequate food now but also in the future. We should therefore pay special attention to the social and institutional arrangements that guarantee we will be able to produce sufficient food in the longer run. Rather than focusing on productivity and creating precarious conditions and vulnerability, care should be directed toward the people who produce food. It should also be recognized that the human right to food is not only an individual right. Human beings are social beings. The right to food can only be realized within a social context, when humans are connected in a community. People should be free to shape the conditions in which they live, work, and farm. George Kent states the conclusion succinctly: "The fulfilment of human rights requires a democratic social order." [139] La Via Campesina therefore argues that the global food system must be democratized. Since hunger is an affront to human dignity, everybody is a stakeholder and should have the opportunity to participate in decision making.

The inequality in power requires, on the one hand, resistance to the current regime. Particularly the neoliberal ideology underlying the regime should be criticized and rejected. In fact, it is the same system that dominates healthcare by controlling patents and intellectual property rights and by limiting access to care in developing countries. Ethical considerations concerning drugs and food should direct attention to the root causes of problems that are often not located at the level of individuals but within the dominating governance system. This system is a globalized localism. A specific value system favored by the global North has been introduced, and sometimes imposed, on the global South. Usually, this neoliberal regime does not take into account the localized values of the South. But as the example of La Via Campesina shows, the South is now demanding governance from below. On the other hand, it is important to present alternatives to neoliberal models of agricultural production and consumption. Food sovereignty is connected with the agroecological approaches analyzed earlier in this chapter. It champions production of food that is compatible with concerns about health and biodiversity. Different visions for agriculture are emerging. Transnational social movements are linking local action and global norms. Local farming initiatives are expanding. Consumer activists are influencing food companies and supermarkets. [140]

Food Justice

The second bioethical issue articulated in the discourse on food sovereignty is justice. The fair trade movement has clearly established that the global food system based on free trade is not equitable. Fifty years ago when it started, this movement revealed to many people that international trade was characterized by numerous injustices. The coffee trade, for example, is an oligopoly of four major companies making huge profits while paying farmers only a negligible proportion of the final price. [141] Many small-scale producers live in poverty and are exploited because they are not in a position to negotiate. The coffee trade is also associated with degradation of the environment. For these reasons, alternative trade networks were created with direct connections between producers (often farm cooperatives in the global South) and consumers (mostly in the global North) so that fair prices could be paid for coffee. Fair trade importers purchase directly from organizations of producers. The first fair trade labeling initiative was undertaken in 1988. Since then, fair trade sales have boomed. [142] It is putting pressure on corporations to change their policies. It has created a market for "ethical" coffee, appealing to consumers who are concerned about biodiversity and social justice. The aim is to provide an alternative to the logic of neoliberalism.

The desire for greater equity in the international food trade has promoted interest in food justice. [143] The global trade regime is not fair but also not free. Massive subsidies are provided to farmers in the United States, the European Union, and Japan, undermining domestic agriculture in other countries. Countries in the Organisation for Economic Co-operation and Development gave $52 billion in foreign aid to the poorest countries but subsidized their own agricultural products with $311 billion. [144] In 2002, each European cow received a subsidy of 1.40 euros each day. [145] This is not open competition on the world market. Exporting cheap food to developing countries undermines the ability of farmers in those countries to fairly compete. Their governments have been forced by the WTO to deregulate, that is, no import restrictions, elimination of tariffs, and no production subsidies. Their farmers are not protected and must export most of their agricultural products, while the US Department of Agriculture spends $25 billion or more a year on subsidies for farm businesses. The European Union provides farm subsidies of 59 billion euros annually. [146] Developed countries are protecting their own markets while forcing "free trade" on other countries. Competition is encouraged between unequal parties. [147] In fact, a net transfer of financial means has occurred from developing countries to developed ones under the neoliberal regime. [148] The global food regime, with its emphasis on markets and competition, has only allocated more power to food companies and reinforced inequality and injustice. [149]

It is now recognized that food justice is a significant ethical concern throughout the food system. Food is not only unequally distributed; it is also difficult to obtain access to a healthy and sustainable diet. Food suppliers receive unfair returns. Food is primarily harvested by low-wage immigrant labor. Human trafficking may be involved. Resources such as land and water are unequally distributed. More than half of the workers in the food system of the United States live below the poverty line. [150] Working conditions in factory farms are horrible, and animal welfare is not a priority. Seeds for small-scale farmers are difficult to obtain and expensive. People living in food deserts have no access to healthy food. Schools are the target of intensive advertising campaigns for soda and junk food. Walmart, the largest employer in the United States, is also the leader in low wages and abusive and discriminatory working conditions. [151] Most farmers in developing countries are women, but they are mostly ignored in FAO policies. [152] In 2007 and 2008, the silent violence of hunger was broken with food riots in 25 countries after significant increases of food prices. [153]

Global Bioethics: Going to the Roots

Biodiversity provides food to humanity. It sustains human survival, health, and flourishing. Production and provision of food now require continuous human intervention and transformation. Intensive agriculture, science, and technology are necessary to grow, harvest, process, transport, and distribute food products around the world. These intensive activities, however, also generate ethical concerns. More than enough food is produced to feed the world, but hundreds of millions of people still suffer from hunger. Other people die from diseases that are the result of too much unhealthy food. Can adequate food continue to be produced for a growing world population, or are we overexploiting biodiversity and making food production unsustainable for future generations? Can we produce our food in such a way that health and biodiversity are no longer compromised?

Ethical questions like these are usually discussed at two different levels of moral discourse. Most of the time, the ethical analysis and discussion takes place at the level of individual persons. Food is an issue of personal taste and choice. Obese patients are held accountable for what they eat; if dietary information and education are sufficiently provided, they are responsible for the consequences. The same argument applies to hunger. If this is primarily a matter of poverty, it is lack of individual income and control of resources that is important; the remedy is to provide humanitarian assistance to individuals in need. Until recently, ethical issues were mainly connected with consumption of food, rather than its production. That has significantly changed now that individual consumers are increasingly concerned with values such as health, biodiversity, and social justice.

Ethical analysis and discussion has also targeted the collective level, that is, the food regime that operates around the globe. The stable and significant number of malnourished people, the increasing need for food, the loss of biodiversity, the shortage of productive cropland, deforestation, degradation of soil, diminishing water resources, water pollution, and scarcity of fossil energy are not the result of individual decisions and actions but are associated with the way in which production, distribution, and consumption are currently structured. [154] The global food regime is an important component of the global governance system. It expresses the values of neoliberal globalization. It demonstrates the various forms of inequality that are produced by this type of globalization. At this level of ethical analysis, it is therefore important to engage these neoliberal values. Food security today is not a question of resources. It is first of all an issue of social ethics. [155] The global food regime manifests the moral symptoms of a neoliberal value system that dominates global interactions. This conclusion has important implications for bioethics. It should no longer be limited to analyzing the individual level of discourse but also focus on the social, political, and economic context of bioethical problems. In other words, it should no longer regard itself as an externality. It should introduce ethical, social, and political criticism into the marketplace of medicine and healthcare itself. [156]

Many citizens, social movements, grassroots organizations, nongovernmental organizations, and government policies and regulations attempt to change the food regime. They argue that food is exceptional. It cannot be treated as another commodity on the market. It is not just an economic asset. People have a right to food. It should be fairly available to all human beings. Food is furthermore related to health and the environment so that not merely quantity but quality and sustainability are essential. Against this backdrop, alternative food networks have been created. Consumers and producers are connected in fair trade. Unhealthy food campaigns have been boycotted. Citizens and cities have been mobilized to prevent food waste. Concerns about how to grow food in harmony with nature and health have promoted the fast growth of organic farming and farmers' markets. The focus of many of these activities is now on institutional and social change. Corporations nowadays are more vulnerable to social movements.

The basic assumption of the current food regime is that food security is best delivered by the market. [157] The market, however, is driven by specific values. The underlying neoliberal ideology emphasizes efficiency, productivity, profitability, and private property. The current food regime is agribusiness. It favors monocultures and deforestation, reshaping the landscape, and use of genetically modified organisms. The market also relies on externalization of the costs and harms. Food is relatively cheap because it is subsidized but also because the deleterious effects on health and the environment are not taken into account. Jules Pretty has argued that we pay three times for our food: in the shop, through taxes (subsidies), and to address the environmental and health effects. [158] The negative impacts of food production are shifted to the public domain. It is exactly against these mechanisms of the neoliberal market model that social movements such as La Via Campesina protest and resist. Agriculture is not simply business, just as food is not merely a commodity. It has a broader meaning. The land that we use to grow food has more than merely economic value. It is, repeating the message of Aldo Leopold, a community. Agriculture should protect the diversity of life; it is engaged with the landscape, rural areas, conservation of biodiversity, local communities, local culture, and employment. [159] The neoliberal food regime has reinforced inequalities. It is no longer driven by farmers and smallholders. Food conglomerates, processors, traders, retailers, and supermarkets are now in power. Corporations, states, and civil society battle for control. [160]

The neoliberal values underlying the current food regime present a challenge to global bioethics. Why should the market be imbued with values that lead to inequality, injustice, hunger, diseases, and biodiversity loss? Economic systems have always been embedded in broader political and social contexts in which non-economic principles such as reciprocity, integrity, redistribution, and social obligations had a key role. [161] The neoliberal market model has displaced social obligations with purely economic relations. Instead of a market embedded within society, society is now determined by market values itself. The current criticisms of the food regime as well as many alternative movements argue otherwise. The focus should be on human rights, sustainability, responsibility for nature, and the common good of humanity. What is needed is an ethical reorientation.

Our World Is Not for Sale

OWINFS is a transnational association of 251 organizations, social movements, and activists opposed to corporate globalization and in particular the policies of the World Trade Organization. It critiques the antidemocratic practices of global governance (with often secret negotiations and deals), the reduction of public services in the name of free trade, increasing market control by corporations, patenting of life-forms, and the intellectual property rights regime. [162]

BIOETHICAL ARMAMENT

As argued elsewhere, global bioethics provides a broader conceptual and normative vocabulary to address global problems than mainstream bioethics. [163] It highlights a set of ethical principles that goes beyond individual interest and articulates the common good and the public interest of humanity as a whole. Global bioethics therefore widens the scope of ethical discourse. There is a need for such wider discourse because many of today's ethical problems are associated with globalization. This is clear in the area of food. Most of our food comes from other countries. Food diseases can emerge in some countries and quickly travel to others. Drought and civil war can create hunger in specific parts of the world, creating a moral obligation to help in other parts. But it is not only globalization that requires a broader scope of ethical inquiry. It is also the type of globalization. Individuals face many ethical problems today, but often these problems are not produced by individuals. They are the result of underlying processes or root causes that continue to exist at the level of neoliberal globalization. This particular type of globalization can only be addressed by a wider ethical discourse that includes principles that transcend the respect of autonomy that dominates mainstream bioethics. As this chapter has demonstrated, such principles are already inherent in the criticisms of the global food regime.

HUMAN DIGNITY AND HUMAN RIGHTS

One of the notions mentioned in food ethics discussions is dignity. Hunger is regarded as an affront to human dignity. The same applies to obese people; they may be blamed for a deficit of self-control, but they still should be respected because they have the same dignity as anybody else. The idea of food sovereignty militates against the loss of dignity of farmers and their diminished self-reliance in a powerful food regime. They have no agency, no voice, and limited control over what they eat and can produce. Human dignity is the basis for the right to food. This is not simply a normative claim of the hungry and powerless. It is a moral obligation of the powerful agents and agencies that are able to transform the structural conditions within which the hungry are suffering. Only they can change the rules of the game. [164] The problem of hunger is not solved by feeding people. It is only addressed if people can live with dignity.

VULNERABILITY

A notion that more often arises in food ethics discourse is vulnerability. [165] In global bioethics this is a crucial concept as well as an ethical principle. [166] The global food regime itself is vulnerable, since it is a complex and extended network of multiple actors. This is expressed in terms of security. There are many inside and outside threats. Climate change, environmental change, and natural disasters are examples. But intensification and specialization of agriculture can also create more vulnerability. Monocultures and genetic erosion can increase vulnerability of crops. Intensive pig and poultry farming lead to vulnerability to diseases. Vulnerability has three components: exposure, sensitivity, and coping capacity. This allows specific analysis and assessment, so that vulnerable regions and groups can be identified. Usually, the emphasis is not on individual persons, as it is in mainstream bioethics discourse, which often understands the notion of vulnerability as lack of individual autonomy. Even when consumer vulnerability is examined, attention is focused on powerlessness, lack of control, and dependency on food corporations. Advertising junk food and soda to children, for example, is criticized because they are a vulnerable group. Also hungry people are regarded as vulnerable, not because as individuals they lack autonomy but because of the conditions in which they live. Poverty is one of the most important determinants of vulnerability. The context of living without social security, safety nets, or government protection creates general vulnerability. This is the lesson taught by decades of neoliberal governance: it produces or exacerbates human vulnerability. [167]

JUSTICE

Concerns about justice play a major role in relation to the global food regime, as discussed above. The case is not simply that resources are not equally distributed. There is a more fundamental issue. Biodiversity is a rich source of plants and animals. Human beings can flourish because they benefit from these products of biodiversity. For centuries, indigenous peoples in particular have developed traditional knowledge associated with sustainable practices that have conserved biodiversity. Currently, these natural products are transformed into drugs; natural food is transformed into processed food. The point is that medication and food provided by biodiversity is converted from common to private property that is no longer available to all humankind as global commons. Rural farmers regard themselves as guardians of the commons, but they are losing. The disappearance of the notion of common heritage in regard to biodiversity has not only led to depreciation of indigenous and traditional knowledge but also to such injustices as patenting, biopiracy, and land grabs. Today, transnational land acquisitions remove local land users from commons in many developing countries and eradicate traditional systems of food production. [168] Privatization of commons has furthermore deflected attention away from food as a common good. Since biodiversity provides food, it does not belong to anyone. Every person has a right to food. Now that food is the product of labor and human manipulation, it has become privatized; it has been transformed into property and commodity that can be exchanged. Finally, the loss of perspective about commons has transformed citizens into producers and consumers. For many generations people have cultivated plants and animals within specific places. Farmers are connected to communities, land, and particular places. They assume a collective responsibility for tending and maintaining the commons. They regard themselves as citizens contributing to the welfare of communities and societies by using and developing traditional ecological knowledge not only in producing food but also in sustainable stewardship of biodiversity. [169] Industrial agriculture has blown up this idea of the "citizen farmer" who takes care of ecological as well as social and cultural commons. [170]

DIVERSITY

The importance of diversity for the global food regime cannot be overemphasized. Nowadays food is shared from all over the globe. Food has become an expression of cultural diversity; it also reflects the connectedness of countries and regions. [171] At the same time, it is argued that industrial agriculture has led to a reduction in dietary diversity. Global foods, especially fast foods, are homogenized and simplified. It is also obvious that the future of agriculture depends on biological diversity, yet only a small number of crops species are used to produce food. [172] The role of diversity is often underestimated. Genetic erosion can jeopardize our future ability to grow food. Genetic diversity should be regarded and protected as a global public good. The food regime should be more flexible and protect the rights of farmers to choose what crop varieties they want to use, exactly as advocated by the food sovereignty movement. [173]

SOLIDARITY

The creation of the World Trade Organization has established a global structure of governance. The political project of neoliberal globalization has not been the result of global agreement but has in fact been imposed on unwilling member-states, particularly in the developing world. The food regime that is part of this structure is driven by enormously powerful, large, transnational companies. Such a strong regime cannot easily be resisted, let alone amended or changed. Nonetheless, it is challenged and criticized. Alternative food networks have been created. Consumers are changing habits and food patterns. This is only possible through solidarity. Individuals may not be able to change the system, but they can take initiatives as the first step toward collective action and projects. La Via Campesina is an example of a grassroots movement that mobilizes farmers from different countries embedded in local places. They realize that they share common goals and are in similar situations. They see that the only way to address exploitative practices and structural violence is to overcome individual perspectives. They unite in solidarity and engage in collective action. Acknowledging solidarity makes them move from private to public interests. This type of solidarity focuses on policymakers, international organizations such as FAO, and producers of food. Creating a transnational network and sharing experiences has made agents aware that collective action is the best way to defend rights, articulate specific needs, and prioritize local concerns in a global context. [174] Solidarity can also be the underlying ethical motive for consumer organizations and actions. Consumers identify with values such as health, biodiversity, or social justice. They perceive shared concerns and see that common action is a more effective method through which to influence food companies than individual consumer choices.

SOCIAL RESPONSIBILITY

Social responsibility for health is a recently formulated principle of bioethics since its inclusion in the UNESCO Universal Declaration on Bioethics and Human Rights of 2005. Promotion of health and social development is first all a concern for governments but also for other actors in society such as private corporations. The principle points out that health is essential to life itself and thus a social and human good. That implies that advances should be focused on access to quality healthcare, adequate nutrition, and clean water; improvement of living conditions and the environment; elimination of marginalization and exclusion; and reduction of poverty. [175] Although the pharmaceutical industry is often regarded as the prime target of this principle, it can also be argued that it should equally engage the food industry. Food is intrinsically related to health, environment, and social justice. In fact, consumers and social movements increasingly associate food with such values; first and foremost, they want healthy and sustainable diets. In response, food companies are pressured to show that they also adhere to these values and are not merely driven to maximize profits. Showing that they share broader consumer concerns, many food companies nowadays have corporate social responsibility programs. Some have also joined the Global Compact, the UN initiative of 2000 to encourage business corporations to work responsibly and in particular to promote 10 principles in connection with human rights, labor, environment, and anticorruption. Among the 9,000 companies and 4,000 nonbusiness organizations that have signed up, 409 are active in food production. [176] These efforts to focus on social responsibility are often critically examined. The Global Compact is voluntary and nonbinding. There is no global monitoring or check on compliance. Is the food industry really interested in restoring societal trust, or is this a relatively easy public relations effort without any real changes in practice? Given food companies' long history of legal convictions and fines for pollution, misconduct, bribery, and corruption, as well as their vigorous and continuing efforts to shape the formulation and implementation of policies, prevent and undermine regulation, and influence children's consumption and school meals, there is sufficient reason to be skeptical.

FUTURE GENERATIONS

Although the principle of protecting future generations is not often explicitly mentioned in food ethics discourse, it is clearly implied in the notion of sustainability. This notion focuses on the needs of future generations as well as the needs of the present. The major worries about the current approach of intensive agriculture are related to the question of whether it can continue to provide food for humanity in the longer run without seriously damaging biodiversity and undermining its own biological foundations. Feeding future generations is therefore a major argument to make agriculture more sustainable and for developing alternative agricultural approaches. Concerns for future generations are also high on the agenda in debates about genetically modified crops and genetic diversity. Similar interests arise when the nature of a sustainable diet is discussed. Usually, the notion of sustainability is related to resource sufficiency. Given the pervasive global inequality, policies and practices that are sustainable are regarded as a matter of how resources are distributed, not so much within generations but also between generations. It is argued, however, that sustainability is more than a resource issue. Paul Thompson, for example, emphasizes that agriculture has a special significance in human societies and civilizations. It is not just another sector of the economy. How we produce and consume our food determines our way of life. If this is integrated into the social and ecological context, the food regime will have a better chance to survive than when it is separated and isolated as commercial production of commodities. Sustainability then means functional integrity. [177]

PROTECTION OF THE ENVIRONMENT AND BIODIVERSITY

This principle reflects one of the innovations in the Universal Declaration on Bioethics and Human Rights. It clearly introduces the broader perspective of global bioethics, as earlier reflected in Potter's concept of bioethics. This chapter reiterates that environmental concerns can no longer be banished from bioethics. The significance of food for human health cannot be disconnected from our interactions with the environment and biodiversity.

Now that a range of ethical principles are available in global bioethics discourse, and in fact already occur in many debates on food ethics, the question is how they can be employed to ameliorate the ethical challenges in relation to food. It is important to acknowledge that FAO is working within a broad framework of ethical principles such as vulnerability, solidarity, justice, and social responsibility. FAO also refers to food and agriculture as global commons. [178] However, in practice this discourse has little impact; FAO has been busy in implementing, not criticizing the neoliberal food regime. [179] Instead of trying to imbue market approaches with values other than competition, profitability, and self-interest, it promoted technical solutions. Super bananas, for example, are considered to solve the problem of vitamin A deficiency. But what is the context? People in developing countries are so poor that they can only eat rice. They cannot afford green leafy vegetables that contain vitamin A. Producing GM bananas can ameliorate the deficiency, but why are the original sources of vitamin A no longer available? It seems that the roots of problems are not addressed.

Conclusion

This chapter has given many examples of resistance against the global food system, which is being contested at various levels and in many countries. There are numerous alternative approaches. It is obvious that this resistance and contestation is focused on the underlying neoliberal values of the system. Ultimately, protest and action are motivated by an ethical controversy. The moral imagination trusts that a more humane and just world is feasible. [180] Basically similar struggles are playing out in the field of pharmaceuticals. Private interests are pitted against public interests; the rational, self-interested *homo economicus* contrasts with the vulnerable social human being embedded in society and community; conflicts emerge between individual versus collective moral responsibility or trade and profit versus health and human well-being. The difference between the ethical debates surrounding drugs and food is that the latter more clearly shows how contestation can be successful. Confronted with powerful and global actors, individuals can feel powerless and desperate. They pessimistically argue that there is nothing to expect from corporations: their main purpose is to make money, and they will never focus on promoting the public interest. [181] They also have almost unlimited power without any accountability. There is an obvious lack of global regulatory authority. Transnational companies have and will continue to resist all efforts to regulate their activities. They simply have no interest in the common good. [182]

These observations may be true, but that does not mean there cannot be change. The critical ethical discourse on food is playing out on various levels. According to neoliberalism, the major actor in ethics is the individual person. Precisely here significant changes are observable. It is argued that consumers are moralized and that their preferences are turned into values. [183] Many people no longer want to be approached as mere consumers. They want to act as citizens concerned with global health, biodiversity, and social justice. [184] They are changing perceptions and conduct, and they create and support alternative food systems. More important, it is argued that it is not only individuals agents who can ethically deliberate about food problems. Institutions and collectivities also have agency and responsibility. One such actor is the state. From the neoliberal perspective, its role should be reduced. From the perspective of public health, it is recognized that ultimately only regulation by states can solve problems and protect population health. An example is the widespread use of antibiotics as growth promoters in the food industry. Since the 1950s, this has been permitted without prescription. It has become clear that this practice contributes to developing resistant bacteria. Many efforts to change the practice have failed. The problem will affect the whole of humanity. Finally, the European Union has prohibited the nontherapeutic use of antibiotics. The United States, however, is still struggling and hopes that voluntary measures will be sufficient. [185]

Between states and individuals, a broad spectrum of social activities and organizations exists. From the neoliberal perspective, this intermediary field is completely taken over by the market. The qualities of education, healthcare, transportation, recreation, and sports are all determined by trade. Interactions among human beings are simply transactions in the market. Interestingly, in the field of food ethics, most critical activities are happening in this intermediary field. However, it is not only the market but civil society that seems to flourish. Evidently there is a public sphere of social life, a public arena for activities of citizens that is not incorporated in the state, the market, or the private sphere of individuals and families. It is in this shared lifeworld that networks, associations, organizations, and movements emerge. [186] It is here that individual action can inspire and motivate collective action. It is also here that neoliberalism is "domesticated" and social life remade on the basis of different ethical values. [187]

In this intermediary field of civil society it is furthermore argued that institutions and corporations have moral agency. They can ethically deliberate and accept moral responsibilities. They can be held accountable for the harms and wrongs they have produced. They can learn and can be redesigned on the basis of moral experience. This is exactly what is happening today. A new ethical environment for food is emerging. For the first time in decades, Americans drink more water than soda. Consumers massively prefer fresh, natural, local, and organic food. The annual volume of packaged food sold in the United States is decreasing. [188] Food companies are seriously rethinking how they operate. The ethical concerns with present-day food will induce them to give moral weight to considerations other than narrow economic interests.

7 | Water

Water is essential for human health and survival. It is perhaps the most fundamental substance in biodiversity. Human beings cannot survive without it. They themselves consist mainly of water, which constitutes 70%–80% of human body weight. Two-thirds of the surface of the earth is covered with water. It is understandable that water is frequently equated with life itself. The ancient Greek philosopher Thales of Miletus argued that water is the first principle in nature. It is not simply a substance or resource that human beings continuously need. In fact, the language of resource is too limited for water. Water is not produced by human beings, like most of the food and drugs discussed earlier. Water is not made but rather is found in biodiversity. Human beings are very much dependent on rain. [1] Unlike other resources, water circulates naturally throughout the world within hydrologic cycles. It is locally stored and available, but water inherently is a global matter. Furthermore, water is more than a physical substance. It is endowed with a wide range of meanings. [2] In religious discourse it is often associated with rebirth and purification (e.g., baptism in Christianity and the Ganges as a sacred river for Hindus). Water has also been regarded as a source of rejuvenation (the idea of the fountain of eternal youth) or of miracle cures (e.g., spa towns in Germany and France). Water contributes significantly to recreation and tourism. On the one hand, there are different types of water, such as holy water and healing water. Hydrologists distinguish between blue water (fresh, surface water and groundwater), green water (rainwater stored in the soil), and gray water (polluted water). [3] On the other hand, water is unique. While there are various types of food, water cannot be replaced. There are no substitutes. Drinks and juices are all based on water. Water can also be a source of risk. It can be dangerous and destructive when it causes flooding and inundation. It can be contaminated and lead to diseases. Water, like food, has played a crucial role in the rise and decline of empires and civilizations. [4]

Human beings use water for various purposes, notably for drinking. Most water is saltwater in oceans (97.2%). Potable freshwater accounts for only 3% of all water, but 2.5% of this water is not available because it is frozen in Antarctica, the Arctic, and glaciers. [5] The available drinking water is stored in underground aquifers, in natural lakes and rivers, or in human-made reservoirs. Water is furthermore used for domestic purposes such as cooking, cleaning, and bathing. Most of the water (70%) is used for growing food and fibers. Furthermore, water is used for transportation and producing energy. Biodiversity provides abundant water for human beings and sustains the survival of humanity. But water is also essential for biodiversity itself. Ecosystems cannot survive without water. It is necessary for aquatic systems to support fish. In fact, all life in biodiversity depends on water. Therefore, water is not only a health but also an environmental concern. Without healthy biodiversity, human health cannot flourish. This interconnection of human and environmental health has created growing concerns about the ethics of water. [6]

Until recently, water has not been regarded as a problematic issue. Water was abundantly provided by biodiversity. Human beings in general had easy access to water. Most countries could manage to provide water to their populations. Nowadays, however, water has become a global problem. Pollution, contamination, scarcity, and unequal access and distribution have emerged as ethical problems. As discussed in the previous chapter, biodiversity provides food for human beings, but agriculture is threatening biodiversity. Water is similarly provided, but its use by human beings has a negative impact on biodiversity. This tension generates a question similar to the one examined in the previous chapter. How can the use of water, which is necessary for human health and survival, be balanced with the sustainability of biodiversity?

Water therefore provides another example of the need to reconnect environmental and biomedical ethics discourse. It demonstrates that a broad notion of bioethics is necessary, since the ethical concerns about water extend beyond national borders. This chapter first discusses how water became a global ethical concern. Water ethics has become a burgeoning new field of research and debate. National and international institutions are engaged in it, and global governance is drawing up ethical principles to guide policies at various levels. This chapter examines different frames in which the ethical debate on water can be developed. Water can be regarded as basic human need, as a human right, as a commodity, and as common property. But why did the debate emerge in the first place?

Water Crisis

The World Health Organization estimated in 2015 that 663 million people lacked improved sources of drinking water. At the global level, at least 1.8 billion people used a source of drinking water that was contaminated with feces. [7] Contaminated water transmits diseases. It was estimated that unclean water, along with lack of sanitation and poor hygiene, was responsible for 842,000 deaths from diarrheal disease in 2012. [8] Many of these victims are children under five. Dirty and unsafe water is a source of diseases because worms and parasites can infest water and indirectly transmit, for example, yellow fever, dengue, and schistosomiasis. The total burden of disease worldwide could be lowered by 10% through better water resource management and improvement of drinking water, sanitation, and hygiene. [9] Water is therefore a primary example of the connection between health and environmental factors. Death and diseases can be prevented through healthier environments. WHO argues that 23% of global deaths are due to environmental factors that can be modified. Diarrheal morbidity could be reduced by 45% with improved drinking water, by 28% with upgraded sanitation, and by 23% with better hygiene. [10]

Thirsting for a Future

According to the United Nations Children's Fund (UNICEF), "Every day, more than 800 children under 5 die from diarrhea linked to inadequate water, sanitation and hygiene. Unsafe water and sanitation are also linked to stunted growth. Around 156 million children worldwide suffer from stunting, which causes irreversible physical and cognitive damage and impacts children's performance in school." [11]

Water has not been a serious global concern until recently. In the 1970s, when the United Nations started to organize global mega-conferences (e.g., on the environment, food, and population), a UN Conference on Water was convened in

Mar del Plata, Argentina. The forum was mainly concerned with water management. It focused on specific concerns such as water for food production, water pollution, desertification, and floods and droughts. Data on water resources and water consumption were insufficient, and water use was inefficient. However, the conference did relate water use to health and environmental consequences, and therefore argued that water management was not merely a technical issue but needed to integrate other perspectives. It also pointed out that access to water supplies was highly unequal. [12] It noted that water is heritage of humankind. [13] In its resolutions, the conference articulated for the first time the right to water: "All peoples, whatever their stage of development and their social and economic conditions, have the right to have access to drinking water in quantities and of a quality equal to their basic needs." [14] The normative consequence of this position has been pointed out. The aim of policies should be that: "water . . . is attainable and is justly and equitable distributed among the people within the respective countries." [15]

After the conference there was no follow-up. Water simply was not on the international agenda. [16] It was hardly mentioned in the report of the World Commission on Environment and Development (1987); it was not considered an important environmental concern during the UN Conference on Environment and Development in Rio (1992) that adopted the Convention on Biological Diversity. Only in the late 1990s did political interest in water start to grow. New organizations were established outside of the UN system by scientists, water companies, and nongovernmental organizations like the World Water Council and the Global Water Partnership. The council is organizing World Water Forums every three years (the first one in Marrakech in 1997; the upcoming one in Brasilia in 2018). Its mission is to place greater priority on water in the political agenda. The council has also established the World Commission on Water for the 21st Century. In 2000, this group published *World Water Vision*, an influential report about how to deal with future and long-term water issues. [17] The commission, chaired by Ismail Seragaldin, vice president of the World Bank, presented a double message. First, humanity is facing a major problem: a water crisis. Second, the solution is privatization. More than before, water has become an urgent global problem. Growth of population, agriculture, industry, and energy needs imply that much more water will be required to meet demands. At the same time, aquatic ecosystems are degraded, and water quality is deteriorating and pollution increasing. Aquifers are depleted rapidly. Water diversion for irrigation has devastating consequences.

Aral Sea

The Aral Sea used to be the fourth-largest freshwater lake in the world. Now it is a notorious case of environmental catastrophe. In the early 1960s, Soviet authorities decided to use two rivers that emptied into the sea for irrigation of neighboring areas to grow cotton, fruit, and vegetables. Over the years, most of the water was drained away, and the sea shrank considerably. The salinity of the water increased. All 24 native fish species disappeared. The water was contaminated with pesticides. Only 10% of the water now remains. Former fishing ports are located in parched wasteland. Most of the area is now a dusty salt desert whose particles fill the air and cause respiratory diseases. People's health has been seriously affected in other ways as well; the prevalence of anemia, cancer, tuberculosis, and allergies has increased. The mortality rate in villages surrounding the former sea is higher than elsewhere in the region. [18]

Some rivers (such as the Yellow River in China) are overexploited and do not run into the sea anymore. Freshwater fish species are in the process of extinction. Many parts of the world are now undergoing "water stress," that is, there is not enough water for all uses (domestic, agricultural, and industrial). Currently, 36 countries experience very high levels of water stress. The latest research shows that water scarcity has been seriously underestimated. Some four billion people experience water shortages for at least one month of the year. Half of those people live in India and China. [19] These shortages will increase. It is estimated that by 2040, nearly 600 million children will live in areas with extremely high water stress. [20] Climate change will not help. The future effects of climate change are not precisely known, but they will certainly make the global water infrastructure more vulnerable and influence water use and availability, thus affecting food security and global health. [21]

According to the World Commission on Water, water is now a scarce commodity. Unless we implement drastic measures, we will face a combination of water shortages, environmental degradation, and threats to health. By emphasizing scarcity, the commission defines the main problem as an economic one. This also determines the solution. The commission sees only two solutions: market arrangements and technological innovation. The most important one is full-cost pricing of water, a system under which users have to pay for the water they consume. That means the creation of water markets. Water should not be made available for free or at low cost. Following typical neoliberal discourse, the commission's report argues that free water leads to waste and inefficiency. [22] Governments, it claims, particularly in developing countries, are not sufficiently investing in public goods such as water; they are unaccountable, corrupt, and inefficient. On the other hand, users should be offered choices. Private companies will offer such choices while they invest in and use new technologies and will provide better services for lower costs. This solution to the water crisis therefore requires a reduction of the role of governments. They should no longer be providers of water, but they should have an "enabling and regulatory role," facilitating the introduction of the private sector as a service provider. [23]

Water Ethics

The report of the World Commission on Water put the issue of water on the global agenda by articulating the notion that there is a water crisis and by advocating a neoliberal approach to the problem. This created a new dynamic and more intensive debate about water. Policymakers became more conscious of impending global scarcity. Access to water is no longer a given. Humankind needs to take care to avoid a global thirst. It should not be forgotten, of course, that at the present moment hundreds of millions of people already lack access to reliable drinking water. But the situation will become worse in the near future because demand for water will rise.

By 2025, half of the world's population will live in areas that have water stress. The water crisis is associated with other crises such as the environmental crisis, the food crisis, the agricultural crisis, and, more generally, the crisis of biodiversity. [24] The report also frames the debate in a specific way. Using "crisis" language, it articulates a sense of urgency. At the same time, it promotes privatization as an immediate solution. What it does not mention are concerns about inequality of access, water as a common good, and rights to water—concerns that were important in the UN Conference on Water, decades earlier. The responses to the commission's "vision" all highlighted the need for an ethical discourse on water.

One response criticizes the notion of scarcity. [25] Is there really a lack of water, or rather a lack of access to water? This discussion mirrors the one in the previous chapter about lack of food: scarcity is the result of a disturbed balance between supply and demand. [26] It is not clear that there will be an absolute scarcity of water in the future. The problem is more that water will be available but will be used unwisely, thereby increasing water stress. Growing populations will need more water. Freshwater may be abundant but not safe to drink because of contamination. We should focus then on evaluating water use for various purposes, which will be an ethical assessment. As mentioned earlier, 70% of water is used for agriculture. If water is becoming scarce, it does not mean that it is scarce for domestic purposes. The notion of a "water footprint" has been introduced to measure the amount of water used to produce goods and services. [27] Currently, 1%–4% of a person's water footprint is in the home. It takes 15,000 liters of water, however, to produce 1 kilogram of beef. In order to increase food production for a growing population, more water will be needed for irrigation, while the environment will also require more water in order to remain sustainable. It is argued that these challenges are not technical but ethical. Instead of identifying new supplies as a response to scarcity of water, it is necessary to focus on better use of available water.

There also is a global distribution problem. Most of the available freshwater is possessed by only a limited number of countries (Brazil, Canada, Colombia, Russia, China, the United States, India, Peru, and Indonesia). Water use, however, is very unequal. In the United States, each person uses 700 liters per day for domestic purposes. In Senegal, the average use is 29 liters. [28] The water footprint of the global average consumer in the period 1996–2005 was 1,385 cubic meters per year. Consumption for the production of agricultural products takes about 92% of the water footprint, while 4.4% goes to consumption for the manufacture of industrial products and 3.6% to domestic water use. Differences between countries are substantial. The water footprint for the average consumer in the United States is 2,842 cubic meters per year; in China, 1,071 cubic meters per year; and in India, 1,089 cubic meters per year. [29] Two-thirds of the global population will be confronted with water scarcity in the near future. Water will be especially scarce in areas with low rainfall and high population density (such as North Africa and central and western Asia). [30]

Another response to the crisis vocabulary argues that this is not simply an issue of quantity. Perhaps more important is water quality. Water should be safe so that it does not transfer diseases. It should not be toxified by chemicals. This means that the water supply should be monitored and also that agriculture (e.g., animal farming) should be practiced in ways that reduce pollution of water resources. Another critical factor is accessibility of water, not only physical (through water services) but also economic (so that people can afford to pay for water services).

The discourse of water crisis is often linked to conflict. [31] If freshwater is scarce, it will lead to growing competition among users but also to increasing stress within regions. Many rivers are shared by multiple countries, and disagreements can arise about who is entitled to use the water. Downstream users are usually in a weaker position than upstream users. In the nineteenth century, the Rhine had many productive salmon fisheries; by 1958, the salmon had disappeared. The river had been channeled, and many chemical plants along its banks polluted the water. Only in 1999 did the neighboring countries agree to cooperate, when they signed the Convention on the Protection of the Rhine. While it is sometimes assumed that future wars will not be over oil but over water, it is remarkable that conflicts are actually rather scarce. In practice, people collaborate and find agreements to conserve and share water, finding solutions for the benefit of all. John Fleck, for example, shows that the idea that water scarcity inevitably leads to conflict is a myth. Water shortages in the Colorado River basin could be managed collectively by the seven US states affected by them through collaborative agreements and collective problem solving. Cooperation, reciprocity, and sharing have been more common than self-interest, competition, and fights. [32] Pessimism and visions of impending doom may be overstated; humans will be able to find solutions. Vandana Shiva has rightly pointed out that water leads to "paradigm wars" rather than real wars. Different moral and cultural perceptions of water are at stake. The present-day neoliberal culture that sees water as commodity is at war with the cultures of sharing, which regard water as a free gift from biodiversity. [33]

Finally, a major concern about the crisis narrative is that it is often associated with specific neoliberal solutions. The primary solution for water scarcity is the market. Scarcity creates market opportunities, especially for big water corporations. However, the experiences with the private sector in many countries are contested. Water companies often increase water rates while water quality deteriorates. [34] Privatization of water resources often reinforces social inequality and further marginalizes vulnerable groups. The underlying ethical question refers to the moral status of water: is it a resource or commodity that can be owned and exploited?

The critique of the characteristics of the water crisis discourse (scarcity, distribution, quantity, conflict and privatization) has produced a change in focus. Rather than emphasizing a pessimistic future and a sense of urgency, it has led to a call for "water ethics." Instead of pursuing technical efforts to increase water supplies, it will be more helpful to address issues of equity, social justice, and ethics of care. How can we make sure that not only human needs are fulfilled but also the needs of the environment? How can we ensure that justice is done to present and future generations? Where should the available water go? How do we deal with the current politics of inequality? [35] These questions are primarily ethical. Management of water always includes values and decisions about values. This is the consequence of the fact that water is not simply a resource. It is not only necessary for survival but also indispensable for leading a life with dignity. [36] Every human being needs a minimum amount of water for drinking, hygiene, bathing, and cleaning. Yet water is not merely a personal good; it is a common good, a community resource with a variety of meanings that people value beyond the fact that it sustains human life. Regarding water as a resource reduces its broader meanings. This language is based on the dichotomy between humans and nature. It reflects the ecological tradition of imperialism criticized by Donald Worster, emphasizing the dominion of humans over nature and the view that biodiversity exists to provide for human needs. [37] It assumes that human beings are in control of nature. It does not accept that water belongs to all people as common heritage. [38]

These considerations lead to the conclusion that ethical debate and analysis will be necessary before we rush to the conclusion that the market will provide solutions. The global water crisis is ultimately a governance crisis and even a moral crisis. [39] To address this, we need to employ an ethical framework in analysis and policy. [40] Water and water management first of all raise ethical issues. And ethical principles are available to direct the analysis toward a broader set of issues. [41]

The rise of ethical discourse about water and water policy over the past two decades has been greatly stimulated by UNESCO. Its Commission on the Ethics of Scientific Knowledge and Technology (COMEST), established in 1998, selected the ethics of freshwater use as one of its priorities. [42] The mission of COMEST is to study issues "beyond the purely economic and scientific." [43] The commission undertook a wide-ranging study and published a series of reports. These studies were not only widely disseminated (in different languages), but they also provided information to governments and policy bodies, and in some sense alerted them to relevant issues in water governance. [44] It is obvious that COMEST as a global platform of experts from different countries immediately focused on the controversy of water as an economic or ethical value. Several experts, especially from Islamic countries, opposed the view that water is a commodity. They argued that water belongs to everybody; it is a free gift of God. [45] COMEST defines water as a common good. It is a shared resource, and therefore concerns regarding human dignity, social equity, and solidarity come first. In fact, the commission places water issues within the context of social ethics. [46] It poses critical questions about efforts to develop a water market. It notes that there is enough water if human beings and countries cooperate. It criticizes the power of international corporations and stresses that they "must be held accountable and bound by ethical guidelines." [47] It observes that the debate in Europe is moving from the dichotomy between private and public to one that focuses on public regulation and common property–based governance. More than 90% of domestic water and wastewater services worldwide is provided by the public sector. [48] In practice, there are myriad options between privatization and public administration. The commission emphasizes that state intervention is "vital to ensure equity among users." [49] Furthermore, it points out that there are different implications depending on whether we consider ourselves "water citizens" or "water consumers." Finally, the commission proposes a set of universal ethical principles to guide the ethics of water.

Guiding Principles for the Ethics of Freshwater (UNESCO)
- The principle of human dignity: there is no life without water
- The principle of participation: all individuals, especially the poor, must be involved in water planning and management
- The principle of solidarity
- The principle of human equality
- The principle of the common good
- The principle of stewardship: finding an ethical balance among using, changing, and preserving our water resources and land [50]

The UNESCO initiatives in the area of water ethics have provided a broader framework for the global discussion of water issues. The proposed principles are global in nature and go beyond the well-known tenets of mainstream bioethics. They were incorporated some years later in the principles stated in the Universal Declaration on Bioethics and Human Rights. They also go beyond the neoliberal view of water as primarily an economic good. The major ethical concerns about water discussed in this chapter much reflect the ones related to food: security, common good, human rights, justice, and governance.

Water Security

The notion of water security arose in the 1990s within the context of concerns about biosecurity (see chapter 4). Initially, scarcity was considered the main threat. [51] Policies were aimed primarily at the quantity and availability of water for human use. As with the notion of food security, however, the perspective gradually transformed. It was recognized that water is important for drinking but also for health (e.g., sanitation and cleanliness) as well as for agriculture and industry, and thus economic growth. The concept of security, therefore, should include the quality of water as well as wise use of water resources for other than immediate human needs. Less attention was given to the environmental needs for water. It was assumed that the main solution would be provided by the market. Water governance was thus fundamentally a technical rather than ethical issue. It was analogous to food security as an example of neoliberal policy. The main challenge was to secure physical access to water to as many people as possible. This interpretation of water security expanded in the new millennium. Security issues became connected to discussions about the right to water. Not only access to water should be provided, but water should actually be provided and be affordable for everyone. Providing water should also be sustainable. That means that governance should not only aim to meet human needs but also to protect biodiversity. Finally, water management should be based on equity and social justice. Water resources should be shared and equally distributed. The broader notion of water security now integrates different perspectives and promotes the idea that water governance is no longer a technical but an ethical challenge. This is demonstrated in the concerns that are prominent today: safety, scarcity, and conflicts.

SAFETY

Water and health are strongly connected in positive and negative ways. [52] Water safety is therefore an important consideration for global health. That health and water are linked was evident from the beginning of medicine. Remember the discourse of Hippocrates about the significance of water for health as well as disease (see chapter 3). Notice the rise of the sanitarian movement in the nineteenth century (see chapter 4). Contaminated water is a threat to all forms of life. Contamination can be the result of human activity, such as pollution by chemicals or agricultural and industrial waste. It can also occur naturally, as in the cases of arsenic and microorganisms. Water can furthermore operate as a medium to transmit diseases like malaria and schistosomiasis. Guaranteeing the safety of drinking water requires three types of activity: protection of water sources against contamination, treatment of water to remove contamination (e.g., filtration and disinfection with chlorine), and protection of water during distribution. These are complicated tasks. The ethical question is how much effort is required and how much safety is demanded. James Salzman has observed in his historical study on drinking water that "for most of human history, safe drinking water has been the exception, not the norm." [53] Another problem concerns the water provision infrastructure. In Western countries, these water-delivery systems were largely built at the end of the nineteenth century as a result of the sanitarian movement. In many countries, they have not been well maintained or renovated and have now become very vulnerable to damage and leaking. In the United States, a major water line bursts every two minutes. [54]

Lead Poisoning from City Water

The city of Flint, Michigan, is one of the poorest in the United States. Because of its fiscal difficulties, the state's governor appointed an emergency manager. Public services were drastically reduced. In order to lower costs, this manager ordered municipal water services to be redirected to the Flint River. In April 2014, the water supply was switched to the river. Soon people started to have illnesses and complaints. Tests showed that water in many homes contained very high levels of lead; an increasing number of children showed elevated levels of lead in their blood. It turned out that municipal services had not treated the water to prevent corrosion of the old lead pipes, so lead particles had entered the water. State and federal regulatory agencies denied any problems, did not listen to the affected population, and manipulated test data. Federal agencies were aware of the problem but stayed silent. It was not until October 2015 that the water was switched back to its initial source, and the governor waited until January 2016 to declare a state of emergency or take any action. The Flint case is now regarded as a clear example of governmental failure in the richest country in the world. It is the result of neglect and underfunding of aging water distribution systems, but also a lack of concern for decaying cities whose populations are mostly poor and black. It has led to thousands of lead-poisoned children who will suffer for the rest of their lives. [55]

SCARCITY

Insecurity regarding water is most often associated with the threat of insufficient supplies. Water stresses are expected to increase because of the uncertain impact of global climate change. Throughout history, there have always been fears of drought and thus food insecurity and hunger. Water shortages have led to the demise of empires and civilizations that once flourished as a result of their wise management of rivers and other water resources. [56] However, scarcity is a relative concept. It requires an ethical reflection on the proper balance between demand and supply. Studies show how demand can be significantly lowered by deliberate policies and voluntary reductions of water use. [57] Moreover, water is increasingly reused nowadays. Water stress encourages reflection on the various uses of water. Different types of water can be utilized for different purposes. The challenge of scarcity is ultimately connected to the debate over the right to water. That means it is necessary to determine the minimum level of water that must be provided in order to sustain human life. This development has also encouraged a new discourse of sharing and solidarity. [58]

CONFLICTS

It is a common idea that scarcity provokes competition and thus will lead to conflict, violence, and war. There is much talk about future water wars. In fact, evidence is weak. Water can and has been used as a military and political weapon, but the association between water scarcity and violence is not evident. [59] Rather than lack of water, it is injustice and inequality that lead to conflicts and violence. When the available water (e.g., in rivers) is not fairly allocated among neighboring riparian countries, disputes arise. That means that conflicts are provoked by ethical disagreement. Contrary to neoliberal assumptions, research has highlighted that in times of scarcity cooperation is more likely than violent conflict. [60]

The security framing of water issues has consequences similar to those of the biosecurity framing discussed in chapter 4. It accentuates threats and the need for protection against diseases, contagion, and contamination. Securitizing water has the same importance as prevention of war and terrorism. Managing water resources is the business of experts; it requires monitoring, control, and containment. The military metaphor may therefore distract from more important concerns, such as cooperative management to alleviate water stress and to provide fair and affordable access to safe water. In the practical reality of water governance, however, the security frame competes with other powerful frameworks. One of the most pervasive ones concerns water ownership.

Water as Commons

The most contentious issue in water ethics is the status of water. From the dominant neoliberal perspective, water is an economic good because it is scarce and the demand is unlimited. This implies there is a market for water. This perspective is more and more contested nowadays. The view that water is a commercial product is rejected by two related arguments. The first is that water is a commons. [61] Water is provided by nature. It can be used but not owned. It is the ecological basis of all life. It is indispensable for the survival of humankind. The second argument is that water is the heritage of humankind. It is therefore common property. Water is not the result of human activity, creativity, or labor. Unlike other goods, it is naturally occurring. Water can only be found, not produced. It cannot be claimed by anyone as his or her property. The arguments of commons and common heritage are connected. [62] This is the underlying idea of European water governance: "Water is not a commercial product like any other but, rather, a heritage which must be protected, defended and treated as such." [63]

COMMON HERITAGE

Water has been the main reason why the discourse of common heritage of humankind was introduced in international law in the late 1960s. Concerned about the possible exploitation of common resources such as the ocean floor and outer space, the international community discussed how to regulate these resources. Maltese ambassador Arvid Pardo argued in a famous speech to the General Assembly of the United Nations in November 1967 that the seabed and the ocean floor are a common heritage of humankind and therefore should be used and exploited for peaceful purposes and to benefit humankind as a whole. The new concept in fact went back to the older tradition of Roman law that distinguished the separate category of *res extra commercium* (things outside of commerce; property that cannot be exchanged). This category defines objects of law that cannot be economically exchanged, thus not sold or owned by anybody. The distinction between private and common property was famously used in 1609 by Hugo Grotius in his book *Mare liberum*, where he argued that the sea is *res omnium communes* (thing belonging to everybody; common property) like the air and the sun: these are objects that are used and enjoyed by everyone, but they cannot be owned by anyone. [64]

COMMON SPACE AREAS

The idea that there are objects and spaces outside of the usual frame of private and public property was initially used for areas with material resources that are outside the borders of nation-states, such as the ocean floor but also the moon, outer space, and the Antarctic. [65] However, the use of the concept gradually expanded to vital resources within the territory of individual nations, such as tropical rain forest. The basic elements of the notion of common heritage are maintained. Common areas cannot be appropriated. They demand international cooperation in order to create common management. Possible benefits should be equitably shared among states, irrespective of their geography. They should only be used for peaceful purposes. And finally, it is necessary to preserve them for future generations. Even when individual countries manage these resources (often a group of several countries), by considering them as common heritage, a shared interest is constructed. These resources need to be preserved in such a way that humanity is aided rather than the citizens of one or more countries. They should also be open for peaceful scientific research that could deliver results benefiting humankind. The Antarctic, for example, is important for the study of climate change. Since resources labeled as common heritage are vital for the survival of humankind, global responsibilities are defined: proper management and stewardship transcend any national interests.

The Common Heritage of Humankind: Basic Characteristics
- Nonappropriation
- Management by all people
- International sharing of benefits
- Peaceful use and freedom of scientific research for the benefit of all people
- Preservation for future generations

In the 1970s, the concept came to include culture and cultural heritage. The member states of UNESCO in 1972 adopted the World Heritage Convention, which asserts that humankind has a specific legacy in the cultural as well as the material realm that it has received from previous generations and a duty to preserve it for future generations. This heritage is important for the sustainability and quality of life on earth.

Common Heritage and Global Community

The use of the concept of common heritage has significantly contributed to articulating the idea of a "global moral community." Just as specific communities can only thrive through commonly shared ideas, symbols, and values, a global community is created through identifying particular material and cultural objects as world heritage and constructing a new global geography of symbols indicating that humanity itself can be regarded as a community. In other words, a global grammar is produced in which diverse and local phenomena receive a universal significance and require global management. Common heritage is the expression of human identity at a global level. It is part of the quest of citizens of the world. It becomes an indicator of global culture. The concept of common heritage therefore motivates a global civilization project that seeks to create a new global community that can represent humanity as a whole and enable the identification of world citizens and evoke a sense of global solidarity and responsibility. On the one hand, it places high regard on diversity but demonstrates, on the other hand, that it can only be preserved within a global perspective underlining its importance for all human beings.

Global Commons

Another relevant and related notion is "commons." While common heritage refers to the historical traditions and the past, commons are orientated toward the future. Both notions go beyond the perspective of the individual; they emphasize what all human beings share and what they need to respect and preserve. They assume inclusion rather than exclusion and emphasize the central role of transmission: we need to conserve what we have received from previous generations, and we need to safeguard what we can transfer to future generations. Both notions evoke the context of mortality, the finitude of humankind, and concerns about survival. The emphasis on commons, however, shifts the attention more clearly from shared past to future. Commons raise the question of how we can be good ancestors. Commons inspire concerns for planetary resources, future generations, and social transformation.

COMMONS IN HISTORY AND TODAY

In the history of humankind, commons were the rule. [66] They can be shared domains, materials, products, resources, or services. A classic example is land available for public use that is managed by the surrounding community. Irrigation systems managed by local communities are another example. The commons are not owned by somebody. They represent a shared interest and require cooperation and collective action. They are ruled by self-governing institutions set up by the stakeholders themselves. In the Netherlands, one of the oldest forms of local government is regional water boards established back in the thirteenth century to control dikes, water levels, and water quality, usually in small districts and representing various types of users. In the United Kingdom, the common use of pastureland, forest, or fields was officially recognized in the Charter of the Forest, connected to the Magna Carta in 1215. In most countries, natural commons have been tribal and communal property. Many indigenous populations did not have a conception of private property. Colonial regimes in particular have dismantled local and self-governing systems of communities that took care of commons such as forests and pastures. The interest of these regimes was on extraction rather than conservation of resources.

Recently, attention has been focused on the so-called global commons: Antarctica, oceans, deep seabed, atmosphere, and outer space. These are areas that are un-owned and, in principle, accessible to everyone. They are indispensable for the survival of humankind, so the aim of management is sustainability; they should not simply be used as resources that can be exhausted. At the same time, they are only manageable through international cooperation.

The commons are marked by a long history of destruction and dispossession, especially since the eighteenth century. They were expropriated, based on the legal fiction that land without private owners was empty—just as newly discovered territories were considered uninhabited before colonization. The first person occupying the territory is considered the owner. The keyword here is *enclosure*: blocking access, sometimes literally by fencing of land and constructing barriers to common use, at other times by legal restrictions and privatization of collective property. Commons have disappeared because public space has been restructured as private space, excluding others from use. In recent times, the development of countries in the global South has often been the story of transforming commons into private property. On the other hand, while traditional commons are disappearing, new commons have been created, through new technologies and through the interaction between technologies and nature. Examples are digital commons based on the Internet, along with airspace commons (for international air transport), ether space commons (for radio and television broadcasting), weather forecasting systems, and pools of crop genetic diversity. These new commons are cosmopolitan; they are necessitated by globalization and facilitated by new technologies. The notion of "commons" has an evolution similar to that of "common heritage": starting with material, tangible, and often localized objects and resources that were common by nature, the notion evolved into covering objects and resources that were constructed as shared and common for purposes that override mere individual and localized interests. These so-called cosmopolitan commons transcend national boundaries; they are shared through technologies and user communities; they require collective action and negotiation; they are valuable for human existence and nature. At the same time, they are always vulnerable to deterioration, overuse, and hazards; they therefore require protection in order to sustain them. For example, when radio technology was first developed, effective use required the creation and management of commons for the electromagnetic spectrum.

CHARACTERISTICS OF COMMONS

Commons are of several types and are distinguished by their different qualities: natural commons (such as fishing grounds, lands, forests, and water supplies); social commons (e.g., caretaking arrangements, public spaces, waste removal, and irrigation systems); intellectual and cultural commons (knowledge and information, as well as cultural products); and digital commons (notably the Internet and the World Wide Web). They can be depletable (natural commons) or renewable (social and cultural commons). They can refer to rivalrous or nonrivalrous goods: use by one person diminishes the amount of goods left for others, as is the case in many natural commons; but using knowledge, ideas, and information does not engender competition that decreases opportunities for other people—in fact, it increases them. For rival goods, it is therefore important to restrict access, whereas open access should be required for nonrival goods. Global commons are nonexcludable (they are open to everybody, since access cannot be restricted) and nonrivalrous (use by one individual does not affect use by others). Cosmopolitan commons usually have regulations governing accessibility.

Despite heterogeneity, commons share several basic features. First, they do not fit into the usual distinction between private and public ownership; they refer to collective property, jointly owned by a group of people (as in local commons) or belonging to all persons on earth (as in global commons). Public goods are also collective property but are managed by the state for its citizens. Common goods are managed by users who negotiate rules based on traditions and collaborative practices. Water is a good example. It is usually managed for drinking and irrigation by communities; it is a "heritage resource" that cannot be appropriated. [67] Second, common goods are essential for human subsistence and long-term survival (since they provide water, food, and shelter but also health and knowledge). Third, commons need to be protected for future generations, so sustainability is important. This is the reason that they are often connected with mechanisms of inclusion and exclusion, regulating and monitoring access and use to prevent overexploitation. Commons thus imply collective action and cooperation. They are often connected with identifiable communities of users that are interdependent and that engage in local self-governance. Since commons are not static, they require continual revision. Commons, finally, are not simply resources; they are social practices (some advocate the use of the verb *to common*). They express a discourse of sociality, reciprocity, sharing, and social harmony. Commons refer to togetherness (as in the word *commonwealth*), not only with people but with nature, environment, and land. Human beings use commons, but they are also part of them. Commons are social arrangements within human society based on common interests and ideals.

Characteristics of Commons

- They are collective property: what is held in common.
- They are essential for long-term survival.
- They are managed in behalf of future generations.
- They require collective action and cooperation.
- They are not simply resources to be managed but are also expressions of sociality: reciprocity, community, regulation of conflict, and maintenance of social harmony.

THE PROBLEM OF THE COMMONS

Until recently, commons were viewed as problematic and a major cause of environmental degradation. The drawback of collective property is that nobody feels responsible. It invites a free ride. If, for example, aquifers are regarded as common property, everybody with access will use as much groundwater as they need, depleting the water in the long run. Individuals will maximize their benefits without sharing the costs, with the result that individual self-interest will ruin everybody's long-term interest. This conventional theory of the commons has dominated since an influential publication by Garrett Hardin in 1968, where he argues that what is rational for a single individual may be detrimental for all other individuals.

The Tragedy of the Commons

Picturing the commons as pasture open to all, Garrett Hardin argues that each herdsman, as a rational being, will seek to maximize his gains by adding more animals to his herd: "Each man is locked into a system that compels him to increase his herd without limit—in a world that is limited. Ruin is the destination towards which all men rush, each pursuing his own best interest in a society that believes in the freedom of the commons. Freedom in a commons brings ruin to all." [68]

In the 1990s, it became clear that Hardin had created a misconception about the commons. They are not merely resources that are ready to be appropriated. Not being privately owned does not mean that they are *terra nullius*; there is a community to which they belong. This community has developed institutions and practices that govern the use of commons with norms and sanctions. Overuse and free riding have been prevented by collective action. Commons are not the same as open access systems. Empirical research has demonstrated the inaccurate presentation of commons. Common property is often well maintained. The problem is not free use but inadequate management and the lack of well-designed rules for access and creative incentives for maintaining the resources. A pessimistic view of the commons is not warranted, given the many examples of their sustainable use. [69] Especially local and regional commons, such as fisheries and irrigation systems, can be, and have been, well managed with a view to sustainability and protection against overexploitation and deterioration.

Hardin's stance has also been criticized for its narrow view of ethics. In order to save the commons, he suggests two strategies: either privatization or collectivization, thus turning the commons into either private or state property. In his view, community regulation and reciprocal cooperation will not be feasible. The only choice is between the market and government. The assumption is that all rational individuals are selfish and that social sharing is impossible. However, there is ample reason to believe that individuals are equally capable of learning that global problems cannot be addressed with self-interest. They require cooperation rather than competition.

Regarding water as commons and common heritage has two consequences. First, it rejects the view that water is a commodity. It is not simply a substance or resource that exists to serve human needs. Water has various values that go beyond the instrumental and consumptive needs of human beings. It is interesting that even economists admit that water is not like any other economic good. It is special because it is essential for life and nonsubstitutable. That means that in fact there is no choice; we need water. Economic theory therefore is not applicable. [70] Second, water is not private property. As a significant social good and community resource, it transcends an individual perspective and should be provided as a public service. It also implies equitable availability; water as a common good should benefit everybody. Neoliberal policies of privatization should therefore be rejected. They result in dispossession and increase inequality. An infamous example is the water war in the Bolivian city of Cochabamba in 2000. The first thing the water company did after privatization was raise the price of water. For weeks, the population protested. After sometimes violent clashes, the government cancelled the privatization project. [71] The idea of commons demonstrates that there are alternatives between state and market. Civil society and communities can provide effective collective management, focused not merely on human needs but also on ecological requirements. Water management should not be driven by the pessimistic view that self-interest will prevail.

Right to Water

While the right to food is explicitly formulated in founding human rights documents, the right to water is not. Neither the Universal Declaration of Human Rights nor the International Covenant on Economic, Social and Cultural Rights refers to the right to water. Framing it as a human right has become an increasingly popular means to articulate ethical concerns about water. It is used as an argument to criticize the view that water is an economic good and to counter neoliberal water policies. Emphasizing water as a basic human right aims to eliminate the inequalities in the supply of and access to water that exist today for hundreds of millions of people across the globe. However, using the framework of the right to water proves more difficult than the right to food. One challenge is that the right to water is contested; the other is that it is problematic to implement.

That water is not mentioned in early human rights documents (as opposed to food, clothing, housing, and medical care) is not really surprising, since it is generally assumed that water is a necessary precondition to food. Without water, two explicitly stated rights in the Universal Declaration of Human Rights cannot be realized: the right to life (article 3) and the right to a standard of living adequate for the health and well-being of a person and his or her family (article 25). The general interpretation is that water, though not included, is implicitly assumed. It is a core element of health, just like air, which is also not mentioned. For the rights discourse, it is important to note that the right to water evolved gradually and has now been explicitly articulated. The UN Conference on Water in Mar del Plata in 1977 was the first global platform that advocated the right to water. Explicit recognition of this right was included in the Convention on the Elimination of All Forms of Discrimination against Women, adopted in 1979, and in subsequent conventions related to children and disabled persons. [72] A significant development was that the UN Committee on Economic, Social and Cultural Rights in 2002 formally stated the right to water, stipulating that this right is indispensable "for leading a life in human dignity." The following year, WHO endorsed the right to water in a publication. [73]

The Right to Water

As proclaimed by the UN Committee on Economic, Social and Cultural Rights, "The human right to water entitles everyone to sufficient, safe, acceptable, physically accessible and affordable water for personal and domestic uses. An adequate amount of safe water is necessary to prevent death from dehydration, to reduce the risk of water-related disease and to provide for consumption, cooking, personal and domestic hygienic requirements." [74]

Finally, in 2010, the UN General Assembly recognized the human right to water and sanitation, asserting that water is "essential for the full enjoyment of life and all human rights." [75] The right to water has now clearly become an independent human right. Later in the same year, the Human Rights Council adopted a resolution affirming that the right to water is implied in the right to an adequate standard of living as well as the right to life and human dignity. It also reaffirmed that "states have the primary responsibility to ensure the full realization of all human rights." The resolution explicitly underlines that when the delivery of water services is delegated to third parties, states are not exempted from their human rights obligations. [76]

These steps in the process of articulation of the right to water have been important. Strengthening the right provided and reinforced arguments to counter the neoliberal discourse. It shifted the frame from water as commodity toward water as legal entitlement. Water is not optional but a right. Even when it is lacking, states have an obligation to provide it. Framing it as a right also implies that water should be available to everybody, even those who cannot afford it. Inequities in the supply of water and access to water are not acceptable from this perspective. Privatization of water resources and particularly the principle of full cost recovery of all investments related to the provision of water may violate the right to water. The price may be too high for some populations. Endorsing the right to water therefore means that poor, marginalized, and vulnerable groups should receive special attention. [77] Articulating the right to water has furthermore encouraged efforts to determine what is the basic amount of water necessary to meet the essential needs of human beings. Water should satisfy basic domestic needs: drinking, bathing, cooking, hygiene, and sanitation. It is estimated that a human being needs 50 liters of water per day (5 liters for drinking, 20 for sanitation, 15 for bathing, and 10 for food preparation). [78] In many countries, people currently lack these basic amounts. Water is of course also needed for agriculture and industry, as well as a healthy and viable ecosystem. It is uncertain what the minimum amount of water is to meet these needs, but the right to water first of all focuses on human needs.

Making the right to water more explicit has given greater impetus to the ethical debate. It has encouraged the international community to be more active and to set priorities. It has put some pressure on states and international organizations. Water issues now receive greater public attention. [79] Countries such as South Africa and Uruguay have included the right to water in their constitutions. Policymakers are more aware that water management includes issues other than mere quantity (and thus scarcity). Availability implies more than quantity; it also includes equitable distribution, quality of water, and accessibility for vulnerable sections of society. A human rights approach to water makes a difference. It provides stronger safeguards compared with commercial approaches. It makes clear that the ability to pay should not determine access to water. It argues that water services should not be disconnected if people are unable to pay. The focus on respect for human dignity and the need to protect vulnerable groups will suggest different policies. [80]

Despite the advantages of rights framing, its effectiveness has been called into question. Compared to the right to food, the right to water is weak. Its legal status is uncertain. A resolution of the UN General Assembly does not create binding international law. It is also important to remember that although no nation opposed the decree, 41 countries abstained for various reasons. It is furthermore pointed out that although mentioning the right in conventions does create binding obligations for the parties, the right is recognized only in connection with women, children, and disabled persons. It is not a universal right. [81]

There also is, as with all human rights, the challenge of implementation. If the right to water is recognized, who has the obligation to do what? Given that water is essential for human life, non-enjoyment of the right to water precludes enjoyment of any other right. In principle, it is the obligation of the state to make sure that its citizens have access to water. But the right to water can only be implemented if sufficient water is available. Several countries currently experience severe water stress. If there is no water, it cannot be provided. The obligations of such states are therefore physically limited. In what sense does the ethical responsibility to provide water transcend borders? When it is argued that water is common heritage, there is some kind of collective ownership. A particular state cannot claim that it owns the water to implement the right to water. In this commons frame, there is a global responsibility to implement this right. [82] A further question regarding the implementation of the human right to water concerns what exactly should be provided. A narrow interpretation argues that the right refers to a quantity of water necessary to meet individual needs. But water should also be safe and affordable, and its possible uses should be prioritized. This will require a collective decision-making process, because water is a common social and cultural good. The right to water is therefore broader than the right to a quantity of water. [83]

Another, more fundamental critique of the right to water is that rights discourse is not an effective instrument against neoliberal policies of privatization and commodification. The right to water does not exclude private sector management. It articulates that water should be available and accessible to everyone, but water can be provided in various ways. Even private companies have started to talk about water as a human right, assuming that the emphasis on this right can promote privatization as efficient services. [84] The right to water therefore does not imply that the public sector must be the provider of services. This discourse does not address property rights and the question who owns the water. If one wants to question neoliberal privatization policies, the frame of water as commons will be more useful. [85]

Water Justice

Finally, the dominant conception of water as a tradable commodity is critically examined in terms of justice. As discussed in previous chapters, justice is one of the main ethical concerns of global bioethics. Major disparities in health exist across the world. Poor and vulnerable populations are more exposed to environmental harm. They also have reduced or no access to medication. At the same time, corporations appropriate traditional knowledge and natural products, usually available in low-income countries, while the same countries can never afford to buy the new drugs developed. The same inequalities exist in regard to food. It is therefore no surprise that a global water justice movement has emerged. It is part of similar movements related to health justice, environmental justice, and food justice.

In the area of water and water governance, several inequalities are evident. There are major disparities in water use. The average person in North America is using much more per day than persons elsewhere. [86] The availability of water differs substantially. Millions of people lack improved water resources. In many countries, people experience water shortages, while in other areas water is abundant and used for golf courses, lawns, and swimming pools. With climate change, these experiences will get worse—but not everywhere. Water is used for agriculture and industry. But water footprints are not equally distributed. Will we continue to use 15,000 liters of water to produce 1 kilogram of beef? Furthermore, water quality and safety are very different across the globe. Many children die every day because of contaminated water. Access to water differs significantly between the rich and the poor. It is also clear that the benefits of water privatization go almost exclusively to the wealthy sectors of society. Furthermore, water also is not only for human use but should also be available for nonhumans and the environment, but the balancing of these needs is widely different among countries and regions. Finally, there are many concerns about future generations. Will they indeed have adequate water resources if the current trend of aquifer depletion continues?

These concerns have encouraged looking at water issues in the framework of water justice. [87] A just society is one in which all citizens receive sufficient water to lead a healthy and dignified life. The focus of this frame is not so much on water as substance but on the context in which water is provided. The global water crisis is not the result of scarcity. The roots of the crisis are in poverty, inequality, and unequal power relations. [88] The focus on justice and power goes beyond the rights frame. Rights are individualizing and do not generally address injustices. Water companies can argue that the emphasis on the right to water can help to increase access (so that they can sell more water), but this does not provide a focus on fair access. [89] Advocating the frame of water justice has several implications. It implies first of all a rejection of the commodity frame. Water is a social good that should not be privatized. It should be available to everyone and not depend on the ability to pay. Because water is an exceptional good (without water human beings die) everybody should have it. The market will not work, because there is nothing to choose. It also means that water should not be disconnected if people are not able to pay. [90] The water justice frame thus implies opposition to privatization of water. All human beings should have universal access to water. Another implication is that special attention should be given to poor and vulnerable groups. Two groups are especially hurt by inequity: women and indigenous people. Another group that deserves special attention is future generations. The water justice framework in the end implies a different governance perspective. Addressing inequalities does not only demand a focus on access. It also requires addressing power differences. Water is a public good. As part of the commons, it requires democratic control. Equitable distribution of water means a participatory process focused on the needs of communities and civil society. In most cases, however, decisions are imposed by international institutions such as the International Monetary Fund (IMF), which require privatization as a prerequisite for debt relief or loans. Private water companies are often introduced without any public debate, and their decisions to increase water prices are not democratically contestable. [91]

The Case of Plachimada

Coca-Cola started a bottling plant in the small village of Plachimada in the state of Kerala, India, in 2000 to produce mineral water and soft drinks. It relied on groundwater for its manufacturing processes and extracted approximately 1.5 million liters per day. After six months, the villagers, mostly indigenous people occupied in farming and agriculture, noticed that the water was no longer acceptable for drinking and cooking. They also developed medical complaints. The water wells they had always used were emptying, and crops yields were decreasing. The wastewater from the plant was dumped in neighboring fields or "donated" to villagers to be used as "fertilizer." Tests showed that the water contained cadmium and magnesium. Particularly the women complained that they now had to walk five kilometers twice a day to fetch water. The villagers started protests and demanded closure of the plant. After a long legal battle, the plant closed in 2004. The inhabitants were never compensated. In June 2016, a criminal case against Coca-Cola was filed in Kerala. [92]

A particularly problematic issue is bottled water. In 2015, US consumption was equivalent to five bottles of water every week by every inhabitant. A year later for the first time ever, Americans drank more bottled water than soda. Bottled water sales in 2016 grew 9% (totaling 12.8 billion gallons or 48 billion liters). [93] The bottled water industry is severely criticized. Several colleges and US national parks have banned bottled water. One argument is that water belongs to everyone. If water is common heritage, it cannot be appropriated and sold. Also, the claim that it is safer and healthier than tap water is false. In the United States, 55% of bottled water is spring water. The other 45% comes from the municipal water supply, from the tap. The water industry uses the same advertising tricks as the food industry (in the United States, bottled water is regarded as "food"): it deceptively claims all kinds of health benefits, yet it does not provide information about sources and contents. One of the main problems of bottled water is that it takes much more energy to produce and deliver than tap water. In many countries, tap water has superior quality while bottled water is mostly unregulated and not monitored, so quality can differ widely. [94] This raises the question of why people are drinking bottled water in the first place, when tap water is in most circumstances easily accessible, affordable, and safe. The environmental costs of bottled water are considerable: enormous plastic waste, extraction of groundwater, and reduction of water for vulnerable populations. Bottled water therefore is, as Peter Gleick concludes, "the result of a failure to provide satisfactory public water systems and services for everyone." [95] Although bottled water is now available all over the world and seems to provide water for all, the industry actually reinforces water injustice. Water has become a commodity, depending on ability to pay. Obtaining the water to fill the bottles flows from policies of dispossession. Vulnerable populations are robbed of the water they have used for centuries. While the benefits go to transnational water companies, the environmental pollution caused by plastic bottles is left for others to address (see box, p. 217).

Water Governance

The Sustainable Development Goals (SDGs) adopted by the United Nations in 2015 present the priorities and actions that according to all countries are absolutely necessary to overcome global problems in the next fifteen years and to protect the planet and its people. Goal 6 (of 17 goals) is: "Ensure availability and sustainable management of water and sanitation for all."

UN Sustainable Development Goals (2015)

Goal 6: Clean water and sanitation

6.1. By 2030, achieve universal and equitable access to safe and affordable drinking water for all.

6.2. By 2030, achieve access to adequate and equitable sanitation and hygiene for all and end open defecation, paying special attention to the needs of women and girls and those in vulnerable situations.

6.3. By 2030, improve water quality by reducing pollution, eliminating dumping and minimizing release of hazardous chemicals and materials, halving the proportion of untreated wastewater and substantially increasing recycling and safe reuse globally. [96]

Meeting these targets will be difficult. It will require $114 billion per year, which is three times more money than currently available. Government spending is increasing, but 80% of the countries provide insufficient financing to meet their national targets. Half of the countries cannot recover costs. Aid commitments for water and sanitation have declined since 2012. Policies in most countries are insufficiently focused on vulnerable groups. [97]

Water, like medication and food, illustrates the interconnections between biodiversity, health, and human well-being. It is another example of how the discourse of mainstream bioethics needs to be extended to include concerns about biodiversity and how global ethical problems are demanding more analytic consideration and a more important place on the agenda of bioethical reflection. The recognition of the vital role of water for human flourishing, survival, and health, as well as the associated ethical concerns, is recent. Because it was so common and naturally provided by biodiversity, water has in general never been a problem, except in extraordinary circumstances of drought, flooding, poisoning, and pollution. But water has become a problem during the past few decades. Today we are concerned about a water crisis, water security, water pollution, and water wars. As argued earlier in this chapter, the challenges in regard to water are primarily ethical rather than technical. This is reflected in the emerging discourse of water ethics. Water now raises challenges similar to those presented in previous chapters in regard to other products of biodiversity such as drugs and food. The issue of biopiracy, for example, has called into question the value of biodiversity and the status of the natural products that it provides to human beings. Similarly, the status of water implies fundamental questions regarding whether it can be owned and by whom, and how water should be used. Is biodiversity first of all a resource providing drugs and water that can be transformed into commodities? Such a view is at odds with one of the basic lessons of biodiversity: that is, that human beings are embedded in biodiversity. If humans dominate or exploit biodiversity, they are in fact ruining their own survival in the long run. Similar ethical discussions are going on in regard to food, which, like water, is different from other commodities. It is an exceptional substance. Human beings do not have much choice; they cannot live without water and food. The discourse of human rights stipulates that humans not only need them but are entitled to them. It furthermore articulates that water and food are not individual matters but necessarily engage the community, not only in providing water and food but also in making sure that everybody receives these essential, life-sustaining supports. The right to food and water are implied in the right to life. Water and food should therefore receive at least as much ethical attention as abortion and euthanasia. They are inevitable themes in a social bioethics.

It is important to note that the ethical debate on water takes place within the neoliberal context of globalization. This context directs the debate in specific ways. It assumes certain values and posits particular power relations that determine the ethical analysis from the start. Like drugs and food, water has been subjected to neoliberal policies of international organizations such as the World Bank and the IMF, the power of transnational corporations, and governments too weak to protect vulnerable populations. But in contrast to the regimes surrounding other products of biodiversity, resistance against such water policies has been stronger and more successful.

Most of the water infrastructure in Western countries was built in the later part of the nineteenth century for the purpose of providing clean and safe drinking water as well as reliable sewerage systems. Initially, the majority of water supplies started as private enterprises, especially in major European cities. These private approaches, however, illustrated clear market failures. Because major public health and economic issues were at stake, water provision rapidly became a public service. Private practices showed that concerns about public health, and particularly the risk of infectious diseases, could not be left to private companies. Safe water management and an efficient water supply were regarded as a public good and thus a public responsibility. [98] Since water is essential for health and hygiene, and also for national development, mismanagement could not be tolerated. It is also important that water be provided to all citizens so that poverty is not an obstacle to safe drinking water. These considerations were used to justify control by the state. Water as commons means universal provision. This approach to water changed in the 1990s. Under the influence of neoliberal ideology, there was a return to privatization. [99] Especially many developing countries were pressured to shift control of water resources from the public to the private sector and to transfer ownership of water supply infrastructure to international water businesses. Such transfer of ownership was often a condition for loans from the World Bank and the IMF. This time, the idea of state failure was used as an argument. It was assumed that many countries were not able to provide adequate and efficient state services. The language of water crisis and water scarcity supported the idea that immediate action was necessary. It was generally assumed that the market would be an efficient solution. If it aims at full recovery of costs, the infrastructure can be maintained and renovated. It would also make people aware that water has a price, so conservation and reduced use would be promoted. From this perspective, it is better to regard water provision as a business rather than a public service and users as customers rather than citizens. Privatization of water services, particularly in developing countries, increased in the 1990s, peaked in 1997, and then started to decline. [100]

As predicted by earlier experiences, the shift from state to market did not improve water supplies in most developing countries. Private providers were first of all concerned with profits. Water rates sharply increased. Economic efficiency was more important than social equity. There was often no public participation in decision making; there was even less transparency and accountability than during times of public provision; and poor and vulnerable populations were denied access, especially in developing countries. An example of the adverse effects of privatization is South Africa. In the 1990s, the World Bank and IMF pressured the South African government to adopt the full cost recovery principle as a condition for loans. The country's waterworks were commercialized. Since the right to water is included in the nation's constitution, all citizens have a right to a basic amount of water. But beyond that amount, consumers were forced to pay the full costs of drinking water. Many were not able to do so. Early in 2000, 10 million households were disconnected. Many people took water from polluted rivers, ponds, and lakes. This had a serious public health consequence: cholera in August 2000. It illustrated how diseases are linked to water policy. [101] Protests and actions against private water companies broke out in many other places: Casablanca, Jakarta, Johannesburg, and Manila. The "water war" of Cochabamba in early 2000 was a turning point. The Bolivian government had to reverse the privatization. In the following years, many transnational water companies withdrew from developing countries. In most cases, water services returned to public ownership. [102]

These experiences have produced a more nuanced view of the issue of privatization. Despite the promotion of private sector participation in water supply, it is evident that this will not help to achieve the SDGs' water and sanitation targets. Private companies raise prices and invest little in infrastructure; they do not consult the public; they focus on the wealthiest parts of the population; poor and vulnerable households are neglected and often disconnected. [103] Karen Bakker analyzed the example of privatization of water in England and Wales in 1989. The government promised that private ownership and management would be more efficient and lead to lower bills for consumers. In practice, bills rose steadily while the salaries of the top managers of the companies skyrocketed. Numerous direct interventions by the government were necessary to redress inequity and negative public opinion. [104] Privatization not only generates conflicts but in fact leads to decreased public access as a result of increased prices. This has real health consequences. Reduced access to safe water not only diminishes hygiene and sanitation but also disseminates infectious diseases. [105] On the other hand, public provision is often no alternative. Although there are cases of effective, efficient, and equal provision of water, there are also many examples of corrupt and incompetent water management. [106] The fact that millions of people have no access to water or do not have safe water demonstrates that in many countries public provision of water is seriously insufficient. The example of Flint, Michigan, makes clear that this is not only the case in developing countries.

Two conclusions can be drawn. The first is that the ethical frameworks of water as commons, as human right, and as social justice have provided strong motivations for people and civic organizations to be engaged in debates and confrontations about water at all levels—local, national, regional, and global. Especially in the case of water, the basic idea of commons has apparently been stronger and more inspirational than in regard to drugs and food. The second conclusion is that in the critique of neoliberal policies, the opposition of private versus public is not satisfactory. In practice, private provision of water and private ownership of water infrastructure can be as bad, unsafe, inefficient, and incompetent as public provision and ownership. From an ethical perspective, there are relevant differences. If water is regarded as a commons and if the desired policy goals are social justice, universal provision of water as a right, and protection of public health and the environment, then the implication is that governments need to provide water and must undertake regulation. But many governments do not currently fulfill these ethical expectations and obligations. Between market failure and state failure there should be another way. The challenge is to overcome the dichotomy between state and market. [107]

The crucial role of water in global health points to the need for an effective and fair global governance system. Such a system should be based on a broad ethical perspective. For water, the same bioethical principles are at stake as for food: human dignity, human rights, vulnerability, justice, diversity, solidarity, social responsibility, future generations, and protection of the environment and biodiversity. Current discussions specifically target three ethical concerns: an integrated approach to environmental impacts, recognition of different values, and participation of citizens in decision making.

INTEGRATION

The paradigm of water management is "integrated water resources management." This approach has been promoted since the UN Conference on Water in 1977. It argues that good water governance is a balance between human and environmental needs. It incorporates values such as social justice, health, and protection of biodiversity. This paradigm, however, has been difficult to implement. Water management traditionally is regarded as a technical issue. It is the business of engineers and hydrological experts. The focus has also for a long time been on the supply side. To sufficiently provide quality water to citizens requires practical engineering skills, control, monitoring, testing, and rational planning. In most cases, environmental and social issues did not have priority. [108] The emphasis shifted with the discourse of water crisis and water ethics. If water is scarce, it becomes essential to conserve and reuse water, not only for humans but also for biodiversity. Instead of water management, what is needed is water governance:

a broader set of expertise, cooperative action, and participation of all stakeholders, with emphasis not only on instrumental values such as efficiency and efficacy but also ethical values such as justice, solidarity, and protection of future generations. [109] An example of this broader approach is the notion of "hydrosolidarity." Introduced in 1998, it includes notions of equity and fairness in order to counter "hydro-egoism." [110] Water users upstream and downstream on the same river should use an ethical framework of equity to negotiate on the basis of dialogue rather than potential conflict. Equitable water allocation has become a more serious concern than water scarcity. But it can only be achieved if water solidarity is revived.

More Plastics than Fish

In April 2017, researchers discovered that the Barents and Greenland Seas harbor astonishing amounts of plastic debris. Ocean currents drive this waste from distant sources into these once-pristine Arctic waters. Plastics are not biodegradable but instead break into small pieces (microplastics). Estimates are that by 2050 there will be more plastics in the oceans than fish. Plastics not only release toxic chemicals, but fish mistakenly eat them. One in every four fish on the market in Indonesia and the United States in 2014 was found to have plastics in its guts. [111]

Another illustration that water governance is nowadays much more focused on social and ecological concerns as well as human health is the criticism of water that is bottled in plastic. This rapidly growing industry is associated with enormous ecological problems. The production of plastics is rapidly growing (up by 20 times over the past 50 years). The amount of plastics produced in 2014 was equivalent to more than 900 Empire State Buildings. It is expected to double again in the next 20 years. [112] Most plastic packaging is used just once; only 14% is recycled. Plastic bags are used for only 12 minutes on average. Each day, 500 million straws are used in the United States. [113] Producing plastics accounts for 6% of global oil consumption. A major concern is the enormous waste. Globally, 72% of plastic packaging is not recovered at all. That means that we dump the equivalent of one garbage truck per minute in the ocean. Americans used 50 billion plastic bottles in 2016; only 23% were recycled.

DIVERSITY

Another lesson from the debate on water ethics is that water presents more than physical and biological challenges. For human beings, water has a range of meanings beyond meeting basic needs and ensuring survival. Humans also employ water for spiritual and social purposes. The implication is that water governance should take into account economic, medical, and ecological values but should also be attentive to other value contexts. Indigenous people with long and deep traditional knowledge about water management (e.g., in ancient irrigation systems) are usually forgotten in debates about water management. They are not included in policies and activities. One model, privatization, is imposed. [114] World Water Forums are designed and run by the water industry. The idea that human beings can learn from traditional knowledge based on the notion of water as commons and that wise management of water is a social effort rather than an economic one is not even examined. [115]

Water ethics has been practiced by indigenous populations for centuries. Traditional First Nations, for example, assume a relationship of mutual responsibility between humans and water. Rather than rights, there are responsibilities. Human beings take care of water while water takes care of them. They also assume that the earth is a living entity; water is a physical and spiritual partner in life. Finally, it is assumed that women have a special responsibility; they are charged with protecting and providing water. [116]

PARTICIPATION

A final ethical concern is participation of all stakeholders in water governance. This is particularly important since water management used to be an issue of power and control; it is often a top-down and technical approach. Power politics have in fact dominated the agenda of water governance for a long time. During the 1980s and early 1990s, water was not high on the political agenda; the initiative was taken by organizations that promoted privatization of water. In the United Nations several agencies were engaged in water issues, but there was no leading organization with a mandate in this area. Only in 2003 was United Nations Water established to coordinate the efforts of over 30 UN organizations with water and sanitation programs. But this is an interagency coordination mechanism, not a specialized organization. [117] The void in water policy was taken over by the World Bank and private water companies. They established the World Water Council and the Global Water Partnership (both in 1996). Through their World Water Forums (since 1997), they promoted a major shift in water policy, pushing the idea of water privatization and shaping a consensus that water is an economic good. [118] Private water corporations dominate the membership. Corporate strategies command the discourse. Increasing criticism of corporate approaches, along with the growing interest in water ethics in the later 1990s, made this situation no longer acceptable. One of the ethical principles that UNESCO formulated in 2000 is the principle of participation: all stakeholders must be involved in planning and management. [119]

Participation was justified from various points of view. Because water is a commons and therefore common property, shared governance is a logical consequence. [120] Whether or not one agrees with this view, water is a public good that everybody needs, so responsibility for it should be shared. Water crises can only be solved by collective action. [121] Water is the object of power and appropriation. This power should be shared in order to promote the benefit of all. [122] The notion of hydrosolidarity implies emphasis on participation, integration, and cooperation. It emphasizes water sharing because it is a common good. Sharing implies a reordering of economic goals and priorities. A wider set of values needs to be taken into account, and this requires broader participation in decision making. [123] One additional argument is that sustainability requires cooperation, especially at the international level. People will not cooperate if they have no trust that arrangements will be just and if they have not been involved in the process. They will demand democratic participation to guarantee that the decision-making procedure is just, and not merely the outcomes. [124]

The necessity of participation is articulated in the water ethics discourse for two historic reasons. In the past, during the 1950s and 1960s, water policies had focused on mega-projects such as huge irrigation systems and large dams in rivers. The purpose was almost exclusively economic growth. The human and environmental costs were neglected, but they were significant. In India, for example, between 16 million and 38 million people were displaced by dams. [125] The mega-projects were not examples of democratic decision making; the population was not involved. In the 1980s, policy emphasis shifted toward neoliberal approaches. These policies, however, did not find a proper balance between economic efficiency, equitable access, and environmental integrity. On the contrary, human welfare was not a priority, and generally, participation of stakeholders was limited. So-called participation was often a cover-up for top-down marketing. Decisions were already made or preestablished by loan conditions. [126] Another reason for the emphasis on participation is that until now vulnerable groups and communities have been infrequently involved. Indigenous peoples have been ignored. People living near large water projects such as dams have not been consulted but have simply been displaced.

Conclusion

Biodiversity, as well as human beings, needs water. It is essential for the survival of living beings. Once abundant, water resources are now threatened. Freshwater ecosystems are far more affected by biodiversity loss than terrestrial ecosystems. Many species of fish and amphibians are confined to freshwater. They are especially vulnerable to human activities. [127] Freshwater biodiversity is threatened by overexploitation, pollution, degradation of habitat, and invasion of alien species. Water is also essential for human health. Many diseases are associated with water and affect human populations across the world, especially children. Improvement of water provision and water quality, and enhancement of sanitation and hygiene will generate significant health benefits. [128] Water therefore is an important public health concern. Finally, water is essential for food production. Most of the water in the world is used for agriculture. Many food shortages and famines are the result of drought. [129]

Water security is thus connected to health, environmental, and food security. Water cannot be regarded as an abstraction. It cannot be separated from the environmental, cultural, religious, political, and social context in which it is embedded and valued. However, this is exactly what happens in the contemporary setting. Water is more and more regarded and treated as a resource. As discussed in previous chapters in regard to medicinal products and food, water is likewise considered a product of biodiversity that can be commodified and traded. Categorizing water as a resource is driving the global narrative about water management. If water has now become scarce, it is an issue of security and thus requires drastic and new approaches. [130] Presenting water as a commodity or natural resource isolated from its social context also interprets water scarcity as a problem of nature not attributable to human beings. It is primarily a technical problem of supply, not demand. In this view, access rather than distribution is the real problem. [131]

The implication of this focus on water as commodity is that economic approaches have become dominant. If a resource is scarce, it must be efficiently allocated in order to provide access to all who can afford it. For many policymakers and international institutions such as the World Bank and IMF, the most efficient allocation mechanism is the market. If water is a commodity, it should be distributed on water markets and therefore be privatized. Privatization of water should follow the same trend in regard to ownership and control that has been applied to natural products, indigenous knowledge, and gene plasm. [132]

However, this neoliberal project of commodification and privatization has run into serious difficulties. It has encouraged the development of different ethical frameworks: water as a human right, water as an issue of social justice, and water as commons. All are based on the basic conviction that water is a matter of life and death, not simply a resource. Transforming water into a simple commodity is for this reason morally problematic. [133] Water, as a human right, should be universally accessible; it should not be dependent on economic power or ability to pay. Privatization of water is furthermore associated with problems of distributive justice. Its concern is with recovering costs and making a profit. It also favors human use of water rather than taking into account the needs of biodiversity. [134] Finally, restricted access to water can increase disease risks, as has happened in several countries after privatizing water services.

Even though neoliberal water policies have been imposed with only limited success, particularly in developing countries, such policies are still undertaken and enforced, especially now that austerity measures are being applied in developed countries. This means that ethical debate about the status of water will continue. A recent example is Ireland. One of the agreements with the IMF included full cost recovery in the provision of water services. The government imposed user fees for household water consumption. It also created a national utility (Irish Water) as a profit-making entity that later could be privatized. The population was massively opposed, and there were serious public protests in the autumn of 2014. [135]

Perhaps the strongest argument against neoliberal water policies is that water is one of the best examples of global commons. The planet is not owned by anybody and claims to the contrary are absurd. Similarly, biodiversity and water resources are not the property of a particular person or group of persons. Claims of ownership are not appropriate. Human beings share the earth. Water is a commons that is collectively owned by humanity. That means that human beings have equal status in regard to water. They can all claim similar benefits. Biodiversity provides equal opportunities to satisfy human needs. The conclusion is that privatizing water services and restricting access to water on the basis of inability to pay is ethically unacceptable.

The problem in practice is that policies are based on neoliberal globalization. Magdy Hefni exactly describes the situation: public space, the commons, and common heritage "have been hijacked by the forces of private greed." [136] What is going on is a power struggle to control water resources. Articulating the importance of water for global health and human survival, global bioethics argues that, as in many other areas of ethical debate, it is necessary to consider water from the broad perspectives of human rights, social justice, and common good.

8 | Global Bioethics in Practice

This book started with the assumption that the ethical debate about healthcare is too narrow. Usually bioethics means that we examine the ethical questions of treatment and care for individual patients. We weigh the pros and cons of a medical intervention, asking what kind of benefits and potential side effects can be expected. The important issue is that we can make sure the patient is properly informed so that he or she can make the best decision possible. Ethical considerations may also focus on groups and categories of patients. In that case, guidelines and practice standards may be developed for offering treatment to patient populations on the basis of scientific research so that evidence-based interventions can be proposed to patients. In both approaches, the individual patient is at the center of ethical concern. What is much less common is that the focus of healthcare goes beyond this individual perspective. What does it mean, for example, that patients are poor or uninsured and cannot afford suggested medicines? How do we care about the air and water pollution that are producing a patient's illness? How do we respond to patients complaining about the safety and quality of drinking water in their neighborhood?

Nobody will deny that the individual lifeworld is determined by the physical, sociocultural, and economic environments in which it is situated. However, in everyday healthcare such "environing" concerns are not frequently addressed. There may be several reasons. On the one hand, it is argued that the goal of healthcare is individual well-being and health, rather than social engineering or improvement of societies. Medicine as a scientific discipline has limited possibilities; it is first of all directed at the human body. Healthcare therefore should restrict itself to what it knows best. On the other hand, the contexts in which human life is unfolding are pervasive and complicated; they are hard to influence. Even when we know that the context is relevant for the complaints of patients, it is not up to health professionals to try to change it. That should be the task of politicians and policymakers.

The dispute about the proper focus of healthcare is not a new one. In the nineteenth century, when the adverse effects of industrial development became clear, the sanitarian movement emerged with the purpose to promote the idea of public health. The health of individuals was threatened by harmful influences from the physical and social environment. A focus on individual healthcare was therefore insufficient. It was necessary to develop public health at a level beyond the individual perspective. This change in focus was further articulated by Rudolf Virchow, who argued that healthcare was essentially "social medicine." Rather than being concerned with clinical issues and thus focusing on the individual hospital bed, health professionals should primarily be engaged in addressing the settings in which such beds are located. Virchow argued that it is the task of health professionals to explore why and how people became ill, and therefore to examine the conditions that produced illness in the first place. However, the subsequent development of modern healthcare has diverged significantly from Virchow's recommendations.

The aim of this book is to demonstrate that it is time to reverse course. Today's healthcare should transcend the individual patient perspective. Earlier chapters have presented many examples showing that in practice medicine and healthcare increasingly take into account the contexts and conditions in which individual health is produced. What has changed is that humanity is now confronted with a critical environmental crisis. It is no longer possible to deny that biodiversity loss, climate change, and environmental degradation do have an impact on health. This is not only the case for groups of patients. Also, individual patients have to consider how the quality and safety of their living environment contributes to their individual flourishing and health. Given the enormous transformation of the ecology of human life, the usual distinction between clinical and contextual approaches to healthcare is more and more difficult to maintain.

The emphasis of this book is on bioethics. It poses the question of the implications of this environmental crisis for the ethical debate on healthcare. Bioethical discussions today cannot be imagined without consideration of environmental impacts. It is time to realize Van Rensselaer Potter's aspirations when he proposed the new discipline of bioethics: a broad approach to ethics that covers medical, social, and environmental health concerns. While biomedical and environmental ethics have been developing separately since the 1970s, it has now become inevitable that environmental concerns are incorporated into mainstream bioethics. The impact of biodiversity loss and climate change on health and healthcare can no longer be ignored. Also, the global nature of contemporary bioethics problems connects biomedical challenges with environmental ones. Pollution in one region, for example, can generate diseases in another region. Food contamination in one country can affect consumers in another country. Deterioration of biodiversity seriously threatens global health, since human beings are dependent on healthy environments as well as the products of biodiversity, such as medicinal substances, food, and water. Health threats and diseases are often the result of how we deal with biodiversity. In order to properly analyze such problems, a global bioethics is needed that concentrates on individual perspectives but even more on cultural, social, economic, and political ones.

It is one thing to argue that a broad perspective on bioethics is needed. Biodiversity is necessary to sustain human health, life, and survival. It provides for basic human needs, particularly food and water, as well as helpful drugs. Increasing degradation of biodiversity is a major threat to human health, as the concerns about emerging infectious diseases make clear to everybody. But it is something else to consider what such a broad perspective on bioethics might mean in practice.

Global bioethics is first of all a theoretical discourse. But, like other approaches in ethics, it should also yield practical implications. It is concerned with providing moral reasons for action. The ethical principles of mainstream bioethics have produced established practices of clinical consultation. Many hospitals have engaged professional bioethics consultants. In research ethics, institutional approaches have been developed with the help of research ethics committees and institutional review boards. These practices demonstrate the value of ethical discourse in everyday healthcare. The particular competence required for this task is making helpful contributions to daily care. How does this play out in the field of global bioethics? As a new and emerging discipline, it has not yet established the experiences and practices associated with the 50-year history of mainstream bioethics.

This chapter explores how global bioethics can be applied in practical settings around the world. What are the opportunities and possibilities to translate the theoretical examinations of the previous chapters into concrete and specific activities in global bioethics?

Individual and Social Ethics

Years ago, a well-known expert in neonatology expressed his frustration during a bioethics conference. In his pediatric intensive care unit, he had managed to treat a child with a complicated medical condition. Sophisticated technologies were efficiently used so that the child could be discharged within a few weeks. However, one month later the child was readmitted, undernourished and with a severe infection. Living in the poorest neighborhood of town, his parents had not been able to take good care of him. Poverty had undone the high quality care of the academic department. The pediatrician felt powerless. His medical duty is to the patient, and he therefore had focused on providing the best possible care. He did not see how he could address poverty. Like many others at that time, he could acknowledge that this is a serious social, economic, and political problem, but not an ethical one. This is reflected in the common list of relevant ethical issues in pediatrics textbooks. The list includes genetic testing and screening, confidentiality, end-of-life issues, pain management, brain death, organ transplantation, research, immunization, and enhancement. [1] These are all familiar ethical issues addressed in mainstream bioethics. The focus is on individual patient care as well as on the benefits and limitations of sophisticated interventions. The social context of such care is not often considered in ethical analysis and debate, and if it arises incidentally, it is not regarded as an ethical responsibility of physicians. Nonetheless, this approach is currently changing. The American Academy of Pediatrics is now asking its members to screen for poverty. During a patient visit, they should not only ask whether patients regularly have enough to eat but also whether they have difficulty in making ends meet at the end of the month. Poverty has severe and long-lasting effects on children's health in all countries across the globe. In the United States, more than one in five children live below the federal poverty level. In cities such as Cleveland and Detroit, the child poverty rate is 54% and 59%, respectively. Problems are growing, since the possibility of economic improvement for the poor is declining. Comprehensive care will include much more than treatment facilities; it will need to enhance the social determinants of care. Significantly, the academy calls upon the professional responsibility of its members. Child poverty is unacceptable. Pediatricians should advocate its elimination. The academy's policy statement provides many examples of how physicians can influence and improve the social context of child care. [2]

A major threat to children's health is lead poisoning. Every year, 310,000 children between one and five years old in the United States are at risk of exposure to unsafe levels of lead in their blood. [3] This is associated with many serious health problems, and importantly, possible damage to the nervous system and the development of the brain. Toys and paint can contain lead, but it also can leach out of water pipes. The Flint, Michigan, water crisis discussed in the previous chapter is the most recent example of lead poisoning. After the city's water supply source was changed in April 2014, it was found that lead levels in the drinking water were dangerously high. City and state authorities did nothing. Dr. Mona Hanna-Attisha, director of the pediatric residency program at Hurley Medical Center in Flint, found that the number of children with high lead levels in their blood had doubled. She held a press conference to alert the community. The state authorities again did nothing but attacked her, denying all data. Of course, the scandal was unavoidable. The state had to admit that it had been wrong and dysfunctional. The important point here is that an individual physician took a broad ethical responsibility. In her view, advocating for her patients meant a responsibility for the health of the environment. Chemicals that are causing diseases should be the concern of physicians. In this case, 27,000 children exposed to lead in drinking water will need treatment for the rest of their lives. This health damage would have been entirely preventable if considerations of cost saving had not prevailed. [4]

These two examples of poverty and water poisoning illustrate that healthcare can be administered at different levels, namely, that of the individual patient and that of the social and environmental context in which patients are living. Both levels are interconnected. Human beings, however autonomous they are, are always embedded in a context. This provides them with the means and opportunities to flourish but may also damage them, temporarily or permanently. In medical practice, most activities are focused on individual and clinical interventions. This is associated with the traditional role of healthcare providers. They should focus on the patient; his or her interests should have priority. Healthcare nonetheless is not solely an individual enterprise. It also involves the creation and provision of special contexts to assist patients: care arrangements and institutions aiming at improving living conditions for groups of patients such as children, the elderly, the abused, and the disabled. In some cases, medical professionals undertake advocacy for groups of patients such as the children in Flint. But efforts to influence and critique contextual conditions that impact health and disease (e.g., food production and drug development) are exceptional; many health professionals will not regard such efforts as within their medical competency. Their expertise is clinical, not political. The same tendency exists in mainstream bioethics. As in pediatric ethics, the standard ethical concerns in medical disciplines are related to medical-technical issues (e.g., whether or not to treat) and individual decision making (e.g., is the patient competent to decide about surgery?). Mainstream bioethics most often addresses individual welfare rather than the contextual conditions on which it depends. For many, air pollution, water quality, and food safety are not taken into consideration in medical practice. They are the business of public health as a separate area of medicine or other specialized disciplines. For others, even if these conditions are relevant, it does not imply that they are acted upon. One can agree that the scope of bioethics should be expanded to include social and environmental issues yet at the same time argue that issues such as poverty, social injustice, and inequality are too difficult or too remote to influence. When bioethicists address these issues, they "will waste their time . . . and anger politicians and the public." [5]

Although healthcare and thus also bioethics are applied at various levels, there is nevertheless no antagonism between individual and social perspectives. Approaches at different levels are complementary and contribute to a more integrated and complete picture of what is happening to an individual patient. Global bioethics therefore is not an alternative to mainstream bioethics. It provides a much broader field of explanation, interpretation, and intervention for individual cases. Why is this person suffering from Ebola? How can we treat the lead intoxication of this child? Complementariness means that mainstream bioethics as an individually orientated ethics and global ethics as social ethics are not opposed. They represent a different focus and scope. The ethical principles of mainstream bioethics, and specifically the principle of respect for personal autonomy, are included within the theoretical framework of global bioethics. Global bioethics introduces a wider perspective that places considerations of mainstream bioethics within a specific context, showing the influences of social, economic, cultural, and political conditions on the origin and evolution of the specific illness of an individual. This move will also locate the principles of mainstream bioethics within a framework of ethical principles beyond the emphasis on individual autonomy.

In order to clarify how global bioethics as social ethics can be applied in practical settings, it is helpful to examine first in what sense global bioethics is similar to and different from mainstream bioethics.

Global Bioethics

When Potter introduced the notion of global bioethics, he did not present a systematic agenda or plan of action for a new discipline. His basic argument is that bioethics should be a broad discipline, including medical, social, and environmental dimensions; it should also not be limited to individual perspectives. The term *global* for Potter has two senses: one is encompassing, the other worldwide. [6] In the first sense, bioethics is comprehensive and all-inclusive. Bioethical discourse should include all types of knowledge and practices that are relevant for health and healthcare. As explained in chapter 1, this sense of bioethics includes, in Potter's view, three basic ideas for a new bioethics that is global. The first is interconnectedness. Human beings are not alone and isolated. They are embedded in communities, societies, cultures, and nature. They are part of biodiversity. This means that ethics transcends the level of individual concerns. Ethics is at the same time individual, social, and environmental ethics. The second basic idea, according to Potter, is orientation toward the future. Global bioethics should take into account not merely well-being and survival of individual persons as they currently exist (as the major concern of mainstream bioethics) but also the survival of humanity. This requires a concern for future generations, intergenerational solidarity, and sustainability. The third idea is emphasis on an evolutionary perspective. For Potter, progress (what nowadays would be called development or growth) is not a given but is dependent on our efforts to adapt to and transform our environment. We cannot be sure that humanity will survive, especially when we consider how we continue to damage biodiversity. Several of the previous chapters have demonstrated how human policies and practices seriously affect biodiversity and therefore undermine the basis of our existence. Potter argues that human survival requires a cultural evolution primarily based on values. Although he did not produce a synthetic response, his ideas have inspired the agenda of global bioethics. One of the fundamental concerns in bioethics today is the question: In what kind of world do we want to live, and what kind of world do we want to leave for future generations? Global bioethics will need to encompass all available resources in order to provide comprehensive normative answers to these questions.

The second sense of *global* is worldwide or planetary. Because the ultimate goal of bioethics for Potter is human health and the survival of humanity, its perspective is necessarily global. The issues at stake not only require a combination of biomedical and ecological bioethics, but they also go beyond local and domestic concerns. Even more important, the goal of bioethics can only be reached by forging compromises between individual interest and social good, and between the quality of the environment and the "sanctity of the dollar." [7] The underlying problem is the "dollar dilemma," but unfortunately Potter does not elaborate this. [8] One example that he discusses concerns future water supplies. These supplies are contaminated, degraded, and depleted everywhere. This is all driven by economic purposes. In such cases, present gain has priority over future health. The same is true for hazardous waste and air pollution. His point is that human health and survival should have priority over economic growth.

When Potter published his book *Global Bioethics*, processes of globalization were in full swing. In the 1980s, a set of economic policies (trade liberalization, privatization, and deregulation) was constructed and applied by agencies such as the World Bank, International Monetary Fund, and the US Treasury Department that resulted in neoliberal globalization—which in 1989 was labeled the "Washington Consensus." From the beginning, the emergence of global bioethics has therefore been associated with the expansion of this specific type of globalization.

While global bioethics and mainstream bioethics differ in scope and focus, they are similar in theoretical structure. Both provide principles as the starting point of analysis and reflection, and they identify power differences as the source of ethical problems. Global bioethics uses principles that direct attention to the context of populations, culture, society, and community rather than to the individuals involved. In this way, it introduces another frame to look at the ethical issues. For example, using the principle of vulnerability to examine differences in health orientates research and analysis toward the social determinants of health. Particular persons are affected by diseases because they live in an environment with pollution and toxic waste. There is nothing wrong with their individual autonomy, but what kind of decisions concerning health can they make? They are vulnerable not because of individual fragility or weakness, but because they belong to populations that are affected by their living conditions, gender, age, or race. Such vulnerability is the result of underlying structures and power constellations that negatively influence health. The ethical reply should have a different focus. Protecting the vulnerable individual will not be sufficient; it is more urgent to change the conditions of vulnerability. Another example is the principle of justice. Confronted with cases of patients in developing countries without adequate medication, the main ethical challenge is not at the individual level of patient care. It is clear that patient X or Y needs medication; it is our ethical obligation to provide it, and we should make every effort to improve the health of this patient. International cooperation and charity may help to solve these individual cases.

But the real challenge is at another level. It reflects the lack of justice of the current drug regime. By articulating the principle of justice, global bioethics introduces another perspective. It focuses attention on structures and systems that determine the health circumstances for many patients.

Mainstream bioethics and global bioethics use different principles, but basically, they have a similar theoretical approach. Ethical principles start a theoretical discourse that analyzes and explains the ethical problems people are facing. On the basis of elaborating and applying principles, new perspectives may be offered that can promote practices to address, influence, and transform the problems at hand. A second similarity between the two forms of bioethics is that both have an explanatory theory for the origin of ethical problems. Mainstream bioethics developed in the 1970s because of the growth of medical power. Many new opportunities for medical intervention emerged, but medical decisions were generally made by a paternalistic profession. Citizens and patients, however, claimed the power to make their own decisions. Out of this tension of power differences, the field of bioethics produced a new discourse of health, emphasizing patients' rights and respect for personal autonomy. Global bioethics emerged from a similar power differential. Processes of globalization transformed the context of human existence. They created structures that are not controllable for most individuals. At the same time, they had a deep impact on the lives of most people on the planet. Biodiversity loss, for example, is influencing daily existence but cannot be attributed to actions of individuals. People therefore feel powerless and subjected to anonymous forces. At the same time, globalization is not a collection of impersonal, natural, or unchangeable processes. It is driven by political decisions. It is the outcome of policy arrangements. It is promoted and implemented by specific organizations and structures. It is furthermore the application of a particular ideology of globalization that endorses some values at the expense of others. That means that globalization can be scrutinized and contested as the subject of ethical debates. Global bioethics regards the power differences in globalization as the source of today's global ethical problems. It therefore aims at rethinking the values that underlie the processes of globalization.

Differences between mainstream and global bioethics are more visible in the arena of practical application. Both are examples of applied ethics. As argued earlier, the practical use of mainstream bioethics is more evident for many health professionals, while the implications of the new discipline of global bioethics are not obvious. Especially when biodiversity is taken into account, the question becomes how a broad concept of bioethics will produce specific practices of bioethical analysis at the global level. To answer this question, we need to explore the specific characteristics of global bioethics that determine its possibilities for change.

Globalization

A major determinant of possibilities for social change is the nature of globalization itself. How are processes of globalization interpreted? Currently, globalization is more criticized than ever before. It offers a "neoliberal fantasy" that imagines everybody as an entrepreneur making him- or herself into a successful and wealthy global citizen. In practice, there is a long narrative of humiliation, disrespect, envy, rivalry, resentment, and failure. People feel abandoned by their governments and politicians. Globalization preaches formal equality while it is associated with enormous differences in power, ownership, status, and health. It imposes an impersonal economic order that primarily benefits a fortunate elite minority. For many, global processes only produce a "sense of being humiliated by arrogant and deceptive elites." [9]

For the application of global bioethics, three dimensions of globalization are especially relevant: the interactions between global and local, the rise of global consciousness, and the primacy of economics.

The Global and the Local

Globalization is often perceived as increasing homogeneity. Historically, it first emerged in Western culture and is now spreading its products and ideas over the world. Globalization therefore is the domination of a particular culture across the globe, and it is particularly applicable to ethics. Moral views are the expression of ideals of good life and good society that are associated with specific cultures, religions, and communities. They articulate cultural identities that are not easily changeable and that can differ substantially from culture to culture. In this view, global values usually are related to Western cultures. For other cultures, they come from outside; they are experienced as unavoidable, leaving no choice, and thus as "imposed"—just as in former colonial systems, where the values of the colonizers were hard to reject and gradually marginalized and replaced indigenous values. This view of globalization "from above" is no longer accepted. Processes of globalization are dynamic and interactive. The global cannot exist without the local; the local is continually influencing the global. [10] An example of this dialectic is a global food regime that makes the same food available in all corners of the world. This has encouraged local food production with specific and culturally sensitive foods based on tradition and locality. Rather than homogenizing what people eat, globalization has led to a growing diversity of food. While global processes may lead to destruction of local communities and traditions, they may also inspire and motivate their rearticulation and revival. Usually there is a dynamic interaction within a context of hybridization and creative mixture. The same dialectic view applies to global bioethics. [11] Presenting a theoretical framework as a set of common ethical principles will not have sufficient impact in various practical settings. That framework must be "translated" and "domesticated" within such settings so that relevant applications emerge. Referring to global values is not enough. Within the local lifeworld, patients, policymakers, and other stakeholders engage in activities to interpret and apply those values. People are not victims of global processes; they do not merely imitate and uncritically accept global values, as if they are passive consumers of global culture. This is also clear in the role of human rights. Formulating and advocating the right to food or water is a first step. It clarifies and promulgates the idea that all human beings, wherever they are, have universal entitlements. Rights address states in the first place; some states take these rights seriously, while others ignore them. But regardless of this legal and political response, they provide individual citizens and groups of citizens with a normative ideal with which to criticize and transform specific local practices precisely because human rights embody global values.

GLOBAL CONSCIOUSNESS

The rise of global consciousness is another dimension of globalization relevant to global bioethics. Every year, the British Broadcasting Corporation undertakes a survey among citizens of eighteen countries to explore how they identify themselves. Do they regard themselves primarily as citizens of the world or as national citizens? Remarkably, in 2016 more than 51% of the world population regarded themselves as citizens of the world, up from 45% in 2001. In Nigeria, the proportion was 73%; in China, 71%; in Peru, 70%; and in India, 67%. In emerging economies, the majority of the population identified themselves as citizens of the world. In industrialized nations, the trend was in the other direction. In the United Kingdom and the United States, only a minority (40%–45%) defined itself from a global perspective. In Germany, 30% saw themselves as global citizens (down from 60% in 2002). The strongest sense of national citizenship existed in Russia; where only 25% of all respondents regarded themselves as citizens of the world. [12]

This information is interesting. Of course, it is not clear what exactly people understand the notion of "citizen of the world" to mean. It may focus primarily on economic issues; when there is free trade across borders and people believe they are participating in or benefiting from this trade system, people in developing countries especially can feel that they profit economically, that poverty is declining, and the burden of disease diminishing. People can also relate the notion to the awareness that humanity is facing similar problems of climate change and inequality that can only be addressed by common action. Global citizenship for many is first of all experienced through social media and communication. One may feel a strong sense of global citizenship through being continuously connected with family, friends, colleagues, and even total strangers. Finally, global citizenship is experienced every day through mobility. It is easy to travel almost everywhere without significant obstacles from borders. Foreign countries are not really foreign anymore. Many people are migrating. In the field of global bioethics, networks of experts, patients, and scientists are active in and across countries.

However, the survey indicates problems and changes. Western countries that have initiated the processes of globalization and have most benefited are now engaging in de-globalization. Populist movements blame globalization for many challenges in their countries. A substantial number of human beings are on the move, without home or shelter, and are blocked by new borders and fences. Instead of growing cooperation and solidarity, there are an increasing number of self-interested, autocratic regimes. Narcissistic leaders denigrate human rights discourse. This is not solely a Western phenomenon, but it is remarkable that the idea that we are all connected as inhabitants of our planet is now more shared by people in developing countries than in some parts of the developed world.

The notion of world citizenship is part of the philosophy of cosmopolitanism. Basically, this is the idea that there is a global moral community. All human beings are members of that community because they share common values and responsibilities. [13] Beginning with Stoic philosophy, it has been acknowledged that human beings are born, and therefore rooted, in a specific place; they are localized in a native community and state. They share a common origin, language, and culture with fellow national citizens. At the same time, they are inhabitants of the same planet; they are situated in a similar space. They share the same dignity and equality as members of humanity. As citizens of the world, they can transcend their localization and boundedness of culture, tradition, community, and history. Being born in a specific place and within a particular culture with its often restrictive and traditional customs, laws, and morals can be overcome through a cosmopolitan perspective. Cosmopolitanism is the aspiration to live beyond bounded horizons. Human beings are not defined, and should not be defined, by a particular location, community, culture, or religion. From this philosophical perspective, much older than the modern processes of globalization, boundaries have no moral significance; the focus should be on what human beings have in common. Cosmopolitanism expresses the moral ideal of a universal community that includes the whole of humanity. All human beings therefore have equal moral status.

Cosmopolitanism is associated with globalization, although it is contested. For critical scholars, being a citizen of the world is an easy identification, but it has no content and implication. It is merely a metaphor without substance, since there is no world community to which these citizens belong. Addressing her political party in October 2016, British Prime Minister Theresa May exclaimed, "If you believe you're a citizen of the world, you're a citizen of nowhere." [14] Indeed, there is no world state to which citizens belong. But that does not rule out that people can be members of a specific community (Scotland) and at the same time a more encompassing community (Britain and even Europe). Being a citizen is not simply being associated with a particular city, community, or nation-state; it also determines an external relationship to outsiders and what is foreign. People today are more than subjects of a state. The fact that for the first time a majority of people around the globe identify themselves as global citizens does not mean that they share a single, global identity. It shows that people recognize that they share a common fate and belong to a global community that is facing the same challenges as the rest of humanity. This is the context in which global bioethics is emerging and is engaged in the search for responses. The environmental crisis and particularly the erosion of biodiversity play an important role in the articulation of global consciousness.

Global consciousness started to develop when humanity was confronted with the image of the globe. Pictures from the *Apollo 11* spacecraft on its way to the first manned moon landing, taken shortly after liftoff on July 16, 1969, from a distance of 180,000 kilometers, showed Earth as a lonely planet in outer space. Although parts of the globe were covered by clouds, the images were very clear, with Africa in the center and parts of Europe, the Middle East and western Asia clearly visible. Later images taken from *Apollo 17* in December 1972 were even more impressive. They showed Earth in full as a perfectly rounded globe with recognizable continents. These images, called the "blue marbles," became the symbols of environmentalism. [15] Human beings had never before been able to see the planet that they inhabit. The confrontation with the image of the planet, or "spaceship Earth," made the general public aware of how vulnerable and isolated the earth really is. The planet is now fully explored; all places in the world are known; there no longer is any *terra incognita*. Our planet is a finite one. It is only a tiny part of an immense universe amidst an unknown number of other planets. If this is the home of humanity, all human beings share the same fate. They should have the same concerns about the future. If this planet is ruined, every human being will suffer and perish. The *Apollo 11* photo inspired Van Rensselaer Potter, who used it on the cover of his book on bioethics in 1971.

One of the effects of the images of Earth is that the "global" is associated with space. The globe is out there, at a distance from us within an infinite atmosphere. It is also separated from us. It refers to a world apart from human life. It can be an object of contemplation, admiration, or discontent, but it determines the place that is our habitat. This image has oriented attention toward the planet and its significance for human survival. Since the global is associated with space, borders are irrelevant, and globalization is similar to de-territorialization. Local places are sites of application and implementation rather than locations where the global is in fact constituted, produced, and invented. [16] A different image is that of the sphere. Instead of being distanced from the globe as a separate entity, we are embedded in spheres. They form our lifeworld. Here we experience life from within; what is "local" rather than "global" is constituted as our "world." One of the important spheres is the domain of biodiversity, the biosphere. Humans are part of this sphere; they live and dwell in it. From the perspective of the globe, we can observe, examine, and manipulate the planet; from the perspective of the sphere, we are part of it and actively engaged. [17] Global bioethics needs to be implemented within this human sphere.

Importance of Economics

The third dimension of globalization is the stance that global processes are primarily driven by economics. These processes are the result of the emphasis on development that was a major concern in international relations following the Second World War. After the experiences of war and in an era of decolonization, policies were aimed at human progress and material improvement. Human ecology was not a serious consideration. The concept of development was strongly associated with economic growth. It also gave priority to the principle of the market "as the vehicle of social change." [18] For a long time, however, markets were embedded within social and public controls, regulations, and constraints. This changed in the 1970s with the emergence of global production systems and international market networks. [19] Subsequent globalization promised global prosperity. Deliberate policies and international organizations promoted global trade, assuming the role of free markets. New policies were necessary. Markets do not exist; they have to be created, secured, and protected. From the perspective of global bioethics, it is important to observe that the dominating role of free trade in current globalization is not the result of natural and inevitable processes. It is the outcome of a powerful ideology that emerged in the 1970s and 1980s. This neoliberal ideology promoted policies of privatization and deregulation, downgrading the role of the state and of society. It is not a neutral and technical approach but basically a set of normative views regarding individuals and their relationships. Neoliberalism assumes particular values: the primacy of self-interest, competition, and private ownership. It is driven by a specific philosophy of individualism. It assumes that individuals are free and responsible for their own well-being. An unrestricted market will offer all of them opportunities for development and advancement. Social safety networks and protection are unnecessary and counterproductive. [20] Market ideology, in the words of Pankaj Mishra, "offers a dream of individual empowerment to all" and has therefore been extremely attractive in the modern world. [21]

The dominance of neoliberalism as the policy framework for globalization is currently severely criticized, as illustrated in previous chapters. [22] It significantly contributes to deterioration of biodiversity. It obstructs efforts to improve global health, for example, by making access to medication less affordable or by privatization of water resources. Because of these impacts, it has engendered many countermovements. This critical climate opens up practical opportunities for global bioethics. Because neoliberalism is basically a normative discourse, it can and should be the subject of ethical criticism. Mainstream bioethics with its individual perspective, however, usually does not criticize neoliberal globalization. In many cases, it even tries to quiet and remove the ethical queries raised by the social, cultural, and economic context of ethical problems. When new global problems are on the agenda because of neoliberal practices, a new approach to ethics is unavoidable. As a consequence, global bioethics should analyze and scrutinize the underlying neoliberal value framework. It presents a new ethical horizon to probe global processes. [23]

Inequality

As argued in chapter 3, the major moral challenges of globalization are now related to inequality. In fact, it is a symptom of a more fundamental problem: globalization is basically a positive process, but its neoliberal orientation favors only an elite minority of the world population. For many, Mishra's dream is not realizable;

their situation has deteriorated, and the result is a growing, active nihilism regarding the potential benefits of globalization.

It cannot be denied that global efforts have positive effects. They have reduced poverty. Diseases such as smallpox have been eradicated; some other diseases, such as Guinea worm, are on the road to extinction. In the past 25 years, child mortality has been reduced by 50%. But decades of "development" have not ended human misery. Each minute, 24 people are forced to flee from their homes because of war, violence, or persecution. [24] Each day, 16,000 children die before their fifth birthday, mostly from preventable causes. Globally, 1.8 billion people use contaminated drinking water. One in every nine people in the world is hungry; 1.8 million children die each year from diarrheal diseases related to contaminated food. Tuberculosis, a curable disease, is now the fifth most common cause of death. Today it kills more people than any other infectious disease, surpassing HIV/AIDS. Globalization therefore has two faces: prosperity for a minority and continuing misery for large populations. [25]

This is reflected in what is called the paradox of global health. Global health has significantly improved, but at the same time disparities have grown. Overall life expectancy has increased. In developed countries, death rates from infectious and cardiovascular diseases have decreased. In developing countries, child mortality is lower. The average age of death at the global level increased from 46.7 in 1990 to 59.3 in 2013. The global average life expectancy for both sexes was 71.5 years in 2013, compared to 65.4 years in 1990. [26] There is certainly reason to speak about a "health revolution." [27]

At the same time, in 2015, life expectancy at birth in Sierra Leone was 50.1 years, compared with 83.7 in Japan. Disparities exist in many other areas, for example, in access to essential medication. Healthcare workers are unequally distributed; the majority work in the Americas and Europe, areas where the burden of disease is the lowest on a global level. [28] Disparities are also growing within countries. Following the introduction of Medicare and Medicaid in the United States, healthcare became more equal; healthcare expenditures for lower-income Americans were greater than those for the wealthy. Since 2004, however, the wealthy have received 43% more healthcare, measured in terms of expenditures, than the poor. [29]

One associated phenomenon is social exclusion. Globalization produces vulnerability. Although we strive for security of existence and protection of human life, vulnerability is everywhere. [30] It is manifested at the individual level, but its roots are in structural conditions and circumstances. Over the past few decades, neoliberal policies have made life more precarious for most human beings.

They have created a context of structural violence and multiplied opportunities for injustice and exploitation. Emphasizing efficiency, private responsibility, and reduction of social safety nets, these policies have applied a "politics of disposability": a growing number of individuals and groups are becoming redundant, superfluous, and unworthy of care and protection; they are the "undesirables" and "deplorables." [31] For many people, globalization does not offer opportunities. These losers in the global game form a "new precariat." In fact, they are not even necessary for development and economic growth. In 2010, one in nine Americans was living on food stamps. [32] According to the neoliberal ideology, they are blamed. Unemployment and poverty are a matter of individual responsibility. Lives are precarious because of our own choices. This view is highly problematic from an ethical point of view. What good is the dream of individual autonomy when you go to bed hungry, drink dirty water, and do not have a home, basic healthcare and medication, or a reliable job? Unequal power structures and structural violence are simply ignored in many ethical and political discourses.

Inequality is furthermore reflected in the environmentalist's paradox: the impact of globalization leads to deterioration of the environment. [33] The benefits of globalization are obtained at the expense of increasing loss of biodiversity. The world has already lost half of its ecosystems. One factor is the intensification of agriculture, necessary to feed a growing human population; more productive food systems have indeed promoted human health but have also degraded the global ecosystem. Environmentalists such as Ramachandra Guha, argue that inequality is at the root of biodiversity loss. [34] The problems related to many environmental resources are not scarcity or misuse but rather unequal use. Biodiversity itself is not equally distributed. The transformation of natural products such as medicinal plants into beneficial drugs is driven by unequal processes. Assumptions about blame and responsibility are also skewed: it is perceived that environmental problems are mainly produced in the global South while solutions are created in the North, whereas in fact lifestyles in the North are causing global effects that primarily damage the South. Those most vulnerable to climate change are the least responsible for it.

Disappearing Islands

Five of the Solomon Islands have disappeared into the Pacific Ocean as of May 2016. Owing to rising sea levels and erosion, as an Australian study

continued

points out, the uninhabited islands vanished. Six other islands were affected; a dozen houses have been covered by the sea, and two villages have been displaced to higher ground on the islands. The capital of the province of Choiseul, Taro, will also be relocated to higher territory. [35]

Inequality is also the hallmark of climate change. While biodiversity is already under stress, climate change produced by human activities is accelerating its destruction. It is currently growing into the biggest threat to biodiversity. Soon it will become a more important threat than habitat loss. If emissions of greenhouse gases continue unchecked until 2050, the consequences for biodiversity will be disastrous. The effects of climate change are no longer in the future; they are under way and perceptible within this generation. The facts can no longer be denied. On the contrary, there is a remarkable scientific consensus about the warming of the climate and about human activity as the cause of it. Since the establishment of the Intergovernmental Panel on Climate Change in 1988, regular assessment reports have been published by the international scientific community. Each subsequent report paints a bleaker picture. These reports present observations and predictions in ever stronger terms, making it irrefutable that the mean annual temperature at the surface of the earth has been increasing in the past two centuries. [36]

It is now clear that climate change has a wide range of effects on natural, social, and human systems. It first of all forces migration. Species will move to warmer areas. New types of bioinvasion will occur. Also, diseases will emerge in regions where they have been unknown. Tropical diseases will be more widespread. Furthermore, human beings will move. Rising sea levels will cause some small island nations to disappear. Ancient and traditional cultures will be lost, and people will be displaced and forced to move to safer areas. Temperature change will impact precipitation and weather patterns, triggering floods, inundations, and heat waves in some parts of the globe, and drought, desertification, and extreme cold weather in other parts. Global vegetation will change, leading to important difficulties for agriculture and impacting food production and water availability. Crops will fail. Forest and bush fires will increase. Extreme weather events will be more common. Ecosystems will break down. Species extinctions will be more widespread than they already are. Permafrost will melt, releasing not only age-old microorganisms but, more important, huge amounts of carbon that will accelerate the warming of the planet. These effects of climate change will lead to significant economic losses in the near future. But for the survival of humanity, it is perhaps even more important that they will seriously affect human health. The result of climate change is "inexorable and unmanageable loss of biodiversity." [37] The damaging effects will be very unequal. While the major contributors to climate change are living in the developed world, the existence and health of people in developing countries will be first and most affected. Many places will become uninhabitable. For example, with rising sea levels, Bangladesh will lose 20% of its habitable land. [38]

These unequal effects are visible in the phenomenon of climate or environmental refugees. Since 2008, an average of 21.5 million people have been displaced each year by weather-related hazards such as storms, floods, wildfires, and extreme temperature. [39] It is estimated that in the next decade tens of millions of people will be driven from their homes as a result of climate change. Worries about a current migration crisis will be completely overshadowed by what can be expected in the near future. Although all countries are at risk, low- and lower-middle-income countries have the most displacement linked to climate change. In worst case scenarios, up to two billion people could be displaced by 2100—one-fifth of the world's population. [40]

Criticizing the Roots of Problems

How can we explain these paradoxes? Globalization has produced within a relatively short time an enormous improvement of global health. This is an impressive benefit. But not for everybody, and not everywhere. Some are reaping the benefits of the processes of globalization while others do not; they are marginalized, neglected, and harmed by the same processes. Nowadays, inequality is recognized as a major social, political, and ethical problem. [41] Even IMF experts conclude that at least some neoliberal policies have overplayed growth and increased income inequality, jeopardizing the economic system. [42] Inequality is blamed for political and social instability, the undermining of democracy, lack of social cohesion, lack of solidarity, erosion of a sense of community, and destruction of a sense of fair play and justice. For a long time, inequality has been ignored and deprioritized as the inevitable consequence of globalization, using the argument that it will diminish in the long run as market forces take effect. Today it is no longer possible to deny that it exists and is growing. The market is not solving this problem but rather amplifying it. Recent polls in the United States show that two-thirds of the population are dissatisfied with the distribution of wealth and income. [43] In the United Kingdom, this number is even higher: 82%. Many people feel that something is fundamentally wrong and unjust; the economic system has brought them into an undignified situation. But global policies are hard to change. [44] Joseph Stiglitz summarizes the injustice in one sentence: "We created for the banks . . . a much stronger safety net than we created for poor Americans." [45]

Neoliberal ideology has for a long time argued that the free market is a self-regulating and irresistible force. The world has become one single global market. The market is the most fair and efficient form of social organization. It allows individual liberty and freedom to flourish. Competition is its core value. There is no need for regulation or protection. In the end, everybody will benefit since these benefits will "trickle down." Inequalities will only be transitional ("a rising tide will lift all boats"). [46] This ideological discourse is currently meeting with severe criticism. Alternative approaches regarding drugs, food, and water are imagined and implemented, especially at local levels. Growing awareness that biodiversity is being irreversibly lost plays a major role. For many people, the fundamental challenge is whether another type of globalization is possible. [47] Economists themselves are now deconstructing the myths of neoliberal ideology, arguing that wealth is not the result of hard labor but is inherited and the result of returns on capital; that the trickle-down effect does not exist at all; that the market is not efficient but depends on subsidies and creation of monopolies so that competition in practice is limited. [48]

The most important conclusion of critical studies is that change is possible. The free market is not free at all, and market forces are not autonomous. The market has been shaped by governmental policies. It is not beyond our control. Politics has promoted neoliberal practices and policies, and has therefore also created the inequalities. History tells us that inequality is not a fact of life. Between 1945 and 1970, there was a marked reduction of inequality in Europe because of deliberate policies that expanded the welfare state and social provisions, financed by progressive income taxation. Anthony Atkinson argues that changing the tax structure (lowering income tax) has sharply contributed to the rise of inequality since the 1980s. After decades of decline, inequality has risen in many countries that started to change the social and economic structures, privatizing public goods such as healthcare and education, and dismantling social safety nets. This "inequality turn" in the 1980s was not a natural event but the result of deliberate neoliberal policies. [49]

For global bioethics, these critical discourses are creating new avenues for ethical contributions. The previous chapters demonstrate the significance of biodiversity for global health. They show the connections of health, environment, and society requiring a broad range of ethical principles. It is argued that we need a different normative grammar to preserve and improve human well-being for the future. It is also clear that in safeguarding health, drugs, food, and water, as well as in eliminating disease, two general moral themes need priority attention: commons and justice.

The Fundamental Challenge of Climate Change

Before addressing the themes of commons and justice it is necessary to return to the issue of climate change. Global warming is such a conglomerate of global and fundamental processes that it goes far beyond any piecemeal engineering. It is intrinsically connected to processes of globalization and inequality. Climate change clearly causes extreme harm, not only to present but also to future generations. It engages ethical considerations regarding justice, vulnerability, and social responsibility. But in fact, the problem goes beyond each of these separate moral considerations, since it is rooted in our modern way of life. The question of how annihilation of biodiversity induced by climate change can be prevented or mitigated therefore presents a fundamental challenge to global bioethics.

The Earth Summit in Rio de Janeiro in 1992 that adopted the Convention on Biological Diversity also adopted the UN Framework Convention on Climate Change. This Convention acknowledges the fact of climate change and the significance of human conduct in it. It also acknowledges that responsibilities for action are not equally divided: developed countries have a larger responsibility than the developing world. Nonetheless, action has been very weak. Targets to reduce greenhouse-gas emissions have been voluntary. The Kyoto Protocol, which was signed in 1997 and became effective in 2005, placed specific targets on participating countries with the goal of reducing emissions to at least 5% below 1990 levels. The protocol has been lauded as a success. Between 1990 and 2012, parties reduced their carbon dioxide emissions by 12.5%. Parties committed to reduce emissions by at least 18% below 1990 levels in the period from 2013 to 2020. However, the success is very limited. Developing countries have not committed to emission reductions. Worldwide emissions have increased by 50% since 1990. Emissions from China especially have steeply increased. While 191 countries have now ratified the Kyoto Protocol, it has not been ratified by the United States, the world's largest economy and one of the biggest emitters of greenhouse gases. The impact of the protocol on the climate is therefore very limited.

One of the reasons for failure from an ethical perspective is the lack of a comprehensive ethical framework for action. Negotiations among governments were not based on any ethical principle. One proposal for a moral frame, advocated by Peter Singer, is the emissions entitlements approach. [50] The idea is to stabilize carbon emissions at present levels. Emissions should be shared equally per capita. That appeals to a sense of global justice. Each person on the planet should be entitled to one metric ton of carbon emissions per year. [51] The implication is that developing countries can increase emissions whereas developed countries need to make large reductions. The ethical problem with this proposal is that it ignores historical responsibility. The assumption is that poor countries will not take the past into account. Developed countries have become rich because they have emitted and still are emitting far more gases than developing countries. They have a large responsibility for causing climate change and should thus have more duties to reduce emissions. It is therefore not fair that they claim equal per capita shares. They have already used up their shares. It is also not fair that emissions are produced for different purposes. In several countries, they help to sustain a luxury lifestyle, while in others they support a rapidly growing population.

The challenge of climate change is fundamental for two reasons. First, it is intrinsically related to a specific way of life and, second, there is at the moment an insufficient basis for change. Both reasons refer to the belief that there is no scientific solution for global warming but that addressing the problem is first of all a matter of human values. Numerous studies make clear that drastic cuts in carbon emissions are necessary. This requires enormous changes in the use of energy, particularly fossil fuels, and thus in the way of life, especially in high-income countries. The only alternative to global warming is making societies that are carbon-neutral on a global scale. This should be done as soon as possible. Even if sufficient measures are taken now, the damaging effects of climate change will carry on for some time. The planet will continue to be warmer for decades to come. It is clear that addressing climate change requires a much more ambitious transformation than developing another type of globalization, which would in itself demand a huge undertaking to redress the problem of inequality. Climate change, however, confronts us with some very basic moral problems: how we ought to live, what kind of society we want, and what kind of person we should be. This problem is related to the existence of different types of environmentalism, discussed in chapter 2. Environmentalism in the developed world is concerned with wilderness preservation and protected areas, distinguishing between human beings and nature. It does not focus on the roots of environmental degradation that are often associated with consumption patterns, social injustices, and destructive lifestyles. Environmentalism in developing countries, on the other hand, highlights the imbalances in human use of natural resources, the power differentials and exploitative mechanisms that exist across the globe. It criticizes the dominance of a consumerist culture where nature itself is regarded as one more good to be consumed. It points out that a transformation of values is required as well as a rethinking of the idea of development. [52] This is also the thesis of the "environmentalism of the poor": the unavoidable clash between economy and environment manifesting itself in global warming and biodiversity loss necessitates critical assessment of unjust, exploitative, and polluting ways of life, along with unequal patterns of production and consumption. [53] Against this background, it is not helpful to focus on specific moral problems such as population control, food scarcity, and water contamination. What is necessary is a fundamental and critical ethical examination of human existence, a focus on what James Garvey calls "the whole of life." [54] The challenge is: How to make a more humane world in which human beings can survive?

The second challenge is that, theoretically at least, change is possible but that in practice almost nothing is done. Various reasons for inaction are provided: there is uncertainty about the impact of global warming; the costs of mitigation and adaptation are prohibitive; and technological innovation will come to our rescue. There are also important strategies of denial at work. People, especially politicians, argue that they do not believe in climate change. They think that innovative and green technologies will provide solutions. They assume that severe impacts will not occur during their lifetimes. Or, they hold that minor lifestyle changes will be sufficient. At the same time, at the level of governments, there is a stalemate in political discourse and negotiations. The level of trust among developed and developing countries is low. Commitments are not honored, and reciprocity is absent. [55] In this atmosphere of distrust, climate change mitigation is considered as an argument to impede the economic development of resource-poor countries while the consumption culture of rich countries cannot be questioned. The fact that climate change is reinforcing patterns of global inequality does not promote international cooperation. As long as there is no shared understanding of what solutions will be fair, collective action will be difficult. The irony is that while climate change is certain, the political will to take effective measures is uncertain. [56]

These two challenges could paralyze any action, but from the point of view of global bioethics, they should not. There are three ethical considerations that will influence approaches toward climate change, even if global governance is weak and politics reluctant. The first is the global nature of the problem. It is perfectly clear to anybody that this problem cannot be solved at the domestic level. Of course, individual citizens can take action, change their lifestyles, and use green technologies. But they can never make a significant change at the global level unless there is collective action on a massive scale. The fact that global warming is unavoidably a global problem implies that it can only be tackled by international cooperation. This requires collective action but also stronger global institutions. Ethics discourse can no longer be based on national boundaries but must take into account the notion that human beings can only survive on this one planet. There is no better argument for global cooperation and action than the need to reduce and remediate the impact of human activity on the atmosphere that we all share. Singer argues that if rich nations in particular do not take a global ethical viewpoint, they are not only morally wrong, but they also endanger their own security and survival. [57]

The second ethical consideration is that climate change is not a governmental problem or a specific challenge for politicians and policymakers. It is everybody's problem. Even if some will suffer more, in the end everybody will suffer. It means that addressing the problem cannot be left to politics, particularly when this leads to weak and inconsistent responses. We know that change is possible. Human behaviors have developed and shifted significantly. Global bioethics discourse should look into the various components of ethical argumentation and make a strong case for addressing climate change that may convince as many citizens of the world as possible. It should focus on Aldo Leopold's idea of the community of life (discussed in chapter 2). He was one the first to advocate for a redefinition of the human relationship with the environment and biodiversity. The problem of climate change is peculiar, since it goes beyond the dichotomy of anthropocentrism and biocentrism (see chapter 1). Human and nonhuman interests are not antagonistic here. Addressing climate change is not only beneficial from the point of view of human beings but advantageous for every living being as well as biodiversity as a whole. [58]

The third consideration is that addressing climate change is not a matter of pragmatism. It is first of all an issue of inequality and injustice: poor countries are suffering the effects while they have contributed almost nothing to the problem. Not all nations are equally responsible for climate change; not all nations are suffering equally from the effects of global warming. There are significant differences in vulnerability and responsibility. The disparity in harm and suffering is clear if one looks at natural disasters. They occur more often in poor nations of the global South. The overwhelming majority of related deaths is in these countries. The harms of global warming are not only unevenly spread across the globe. There is also intergenerational harm. This point of view implies that responses to climate change are also fundamentally about global inequality. [59] Global bioethics addresses the root causes of vulnerability, often associated with structural disadvantages. Disaster prevention and emergency relief are palliative; they treat the symptoms but do not address the political and economic structures that produce vulnerability. The context of inequality furthermore implies that the moral demand for action is not the same for everyone. Because of the inherent inequality in causes and effects, there are different responsibilities. Some countries and cultures bear more responsibility for changing the climate; they have more responsibility and therefore stronger obligations to act. [60] Finally, the need for action demands trust-building. Trust is not a pragmatic device. It is not simply an instrumental value that can deliver results in negotiations. Trust is most of all a moral commitment. It indicates that fundamental values are shared. The contribution of global bioethics is especially relevant here. The remainder of this chapter specifically argues that human beings, as citizens of the world, share values. One is the basic idea that the global atmosphere is the common property of humankind and that climate change is an example of the tragedy of the commons. The other is the value of justice.

Commons

Globalization reactivated the interest in commons. The report of the World Commission on Environment and Development articulated in 1987 that the environment cannot be separated from human interests. It denounces the inability to promote the common interest. [61] The report has a separate chapter on managing the global commons (oceans, space, and Antarctica). [62] The Commission on Global Governance in 1995 also discussed the global commons, not as a tragedy but as an opportunity. They are resources for the survival of humanity; they are the global neighborhood that is the home of future generations. [63] The Convention on Biological Diversity (CBD) finally affirms that the conservation of biological diversity is "a common concern of humankind." [64] At the same time, these reports and agreements turned the environment into a commodity. This assumes that humans are independent from biodiversity and can exploit its resources. [65]

The new interest in biodiversity in the 1980s, discussed in chapter 2, illustrates the ambiguity of the discourse of commons. The notion of common heritage of humankind was applied two decades earlier to global commons as the ocean, air, and outer space. [66] It was argued that biodiversity should also be regarded as common heritage. But this terminology was rejected in the drafting of the convention. Particularly developing countries were afraid that the tragedy of the commons would lead to exploitation, with open access for international corporations. Because there is no global authority to protect common heritage, the convention allocated ownership of biodiversity to national states. They have "the sovereign right to exploit their own resources." [67] This was exactly the opposite of the proposal of the World Commission on Environment and Development, which argued that there should be a global body responsible for the management of the commons in behalf of all nations. Demoting the notion of common heritage was the effect of the convention's neoliberal approach, coupling environmental concerns to economic development. Biodiversity is regarded as an economic asset, a package of potential benefits. Ecosystems, as Charles Perring affirms, are not museums but factories. [68] The assumption, from a policy standpoint, is that states will protect "their" biodiversity because it is economically valuable. In this approach, market policies are promoted as the best way to make people concerned about the environment.

Since the adoption of the CBD in 1992, biodiversity has continued to decline. Today, it is concluded that more than 50% of all animal species have been extinguished during the past few decades. Many species are rapidly losing their habitat. In fact, a "biological annihilation" is taking place. The sixth mass extinction is already happening now. The window for effective action is probably two or three decades at most. [69] Many scholars argue that the bioeconomic approach to conserve biodiversity is not working (see chapter 2). On the contrary, it leads to continuing destruction of our common world. Commons are increasingly destroyed by powerful drivers such as pharmaceutical companies using patenting systems for medication, the food industry, and the private water lobby, all turning commons into private property. What we need now is to reclaim the commons. [70]

As discussed in chapter 6, commons discourse is often presented as an alternative vision for the current policy options of regulation by state intervention or the liberalism of free markets. Both policies assume that commons are simply resources that can be appropriated, harvested, used, and regulated. In the current policy climate, this implies that commons are privatized so that they can be efficiently exploited. But the idea that commons are collective property, that they belong to all persons on earth (for global commons), to transnational communities (for cosmopolitan commons), or to local communities (for local commons) implies that people are always involved. Commons are developed, sustained, and managed by social practices of communities who depend on them for daily subsistence, survival, and flourishing. That was true for historical commons in most countries. It is true today for commons in developing countries that are usually managed by indigenous populations. It is also true for many new forms of common property that have emerged, such as the World Wide Web; free software; and initiatives such as Creative Commons, Public Library of Science, and General Public Licensing that promote sharing of knowledge and information. It is visible in phenomena such as fair trade, economic cooperatives, and crowdfunding. New commons are also emerging within science and healthcare ("genetic commons," "science commons," and "biomedical commons"). These phenomena are characterized by an ethics of sharing. They also revive the interest in sharing economic, social, and cultural rights, such as the right to health, food, water, and housing. [71] Social sharing practices are rapidly developing in present-day society, turning private goods into shareable goods. [72]

Commons are produced by human efforts of cooperation. They arise from a view that differs from neoliberal ideology: emphasizing that human beings are not isolated, self-centred individuals; that agency is not only individual but also collective; that people often work together and engage in collective enterprises because they are concerned about the common good; and that the goals of production and consumption are to meet basic human needs rather than consumer preferences or economic growth. Economy cannot be separated from social relations and connectedness with the land and the environment. This alternative vision articulates that commons are necessary for the flourishing of present as well as future generations. Commercialization and commodification threaten this flourishing, since they make commons no longer accessible for many people. This is the point of the commons argument in the area of healthcare. Significant determinants of health, such as clean air, healthy water, and safe food should not be privatized, as that would jeopardize the public health commons that have been created over time (as systems of providing hygiene, sanitation, clean water, sewage disposal, and vaccination) and that are accessible to all members of a population, not as a privilege or a commodity but as a right. The previous chapters demonstrate that the commons are a fundamental issue in regard to biodiversity loss. It is perhaps most obvious in connection with water. Groundwater and aquifers have long been regarded as commons, as shared resources, but are now privatized.

Hundreds of millions of people lack access to safe drinking water; 40% of the world's population has no access to reliable sanitation. Approximately 10,000 people are killed every day by preventable diseases caused by unsafe water and poor sanitation. These data are used to argue that water should be a commons, because that is the only way to make it accessible to all people. Resistance to privatization of water resources has been strong and growing, with the result that many neoliberal projects have been cancelled. Commercialization and commodification of natural medicinal products is also fiercely criticized in the discourse of biopiracy. The problem in this area is the existence of an intellectual property right regime that is imposed at the global level. This has created an unfair system backed by powerful organizations and stringent implementation mechanisms. It prioritizes trade over everything else. But it has not been impossible to make this regime somewhat flexible, particularly by emphasizing the right to health and the need for justice. Similar developments have taken place in regard to food. Biodiversity has always been a reservoir for food. Agriculture is one of the major drivers of biodiversity loss. Patenting of seeds, genetic resources, and food products has privatized the production of food and put it within the domain of transnational food corporations. With their structural and discursive power they have created a global food regime that does not care much about health and environment. The food sovereignty, fair trade, and organic food movements argue that food is not merely a commodity; it is the product of commons.

The discourse of commons is important for the application of global bioethics, since it introduces a different normative horizon for everyday activities. Instead of individualizing, privatizing, and instrumentalizing human action, it articulates the connections between human beings, the environment, and the biosphere. Commons are not simply material resources or commodities. They are not opportunities in nature that present themselves to humans for exploitation. We should recognize that behind each commons is a community. Commons are social practices, linking a community with a specific area, domain, or resource that is shared as common property. Their management requires social cooperation and individual participation so that social norms are developed, boundaries set, systems monitored, and sanctions applied. The purpose of collective governance is economic production but within the context of sustainability. Care is taken not to exhaust the common resources. Commons are therefore associated with an ethics of sufficiency. They can only exist and survive as long as collective governance assures their conservation and preservation. They can be economically exploited but only if norms and practices guarantee that they are not exhausted or ruined. Traditionally, they are associated with indigenous populations that used to put emphasis on the community, social reciprocity, and respect for natural ecosystems. It is heartening that currently, especially in Latin America, new worldviews are being promoted that reactualize traditional connections between living beings and nature. [73] The concept of "living well" (Buen Vivir or Sumak Kawsay) is used to encourage a vision of life that differs from the neoliberal ideology. A basic element is respect for the integrity of nature as the source of life. This vision has been incorporated in the constitutions of Ecuador (2008) and Bolivia (2009), countries with large indigenous populations. At the initiative of Bolivia, the United Nations proclaimed April 22 as International Mother Earth Day, acknowledging that the earth and its ecosystems are our home. The idea of commons is then applied at various levels. At the global level, it expresses that the planet is our home. Mother Earth (or Pachamama) is the basis for our existence. It cannot be owned. We, as human beings, belong to it and we all share it. At the local level, it states that human beings are part of communities. Since these communities are intimately connected with nature, human beings are responsible for respecting all living creatures within their communal scope.

Reclaiming the commons has practical consequences. Arguing that the biosphere is a global commons or common heritage of humankind implies that there are limitations to what humans can do with it. Common heritage is outside the sphere of commerce. Commons refer to collective property and sharing of benefits. Global commons are vital for the continuation of human life. They demonstrate that there are interests beyond borders, since humanity is at stake. It is therefore necessary to protect them against exploitation. Commons cannot be transformed into commodities that are tradable and have exchange value. Their value is different. What makes the commons valuable is that they are necessary for the survival of the human species (e.g., water), that they safeguard the free development of human beings (e.g., knowledge and information), that they are essential for active life and human flourishing (e.g., health and education), and that they determine the context for creativity and innovation (e.g., digital commons). Commons discourse finally is connected with other global ethical principles such as solidarity and cooperation, sharing of benefits, protecting future generations, and protection of the environment. [74] It is particularly relevant to address the major global problem of inequality. Enclosure of commons necessarily creates inequality. Shared resources are appropriated by powerful actors. The essence is that access is restricted and that others are excluded from using the commons. The notion of common heritage, for example, was proposed precisely to counter such processes of dispossession. Technical progress will make it possible to harvest minerals from the deep seabed. But advanced countries will reap most of the benefits of technological progress, while the exploited resources actually belong to everybody. Emphasis should therefore be on a more equitable use of resources and distribution of benefits. This demonstrates the practical relevance of the commons discourse: it opens up a different moral vision articulating collective rather than individual interests; cooperation rather than competition; sharing in a community rather than exchanging in a market; citizens caring for the commonwealth rather than consumers and producers; preservation and conservation rather than usage and exploitation; future rather than current needs; inclusion rather than exclusion; belonging, collaboration, and networking rather than individual autonomy and self-sufficiency.

Justice

Another common thread that runs through previous chapters is the challenge of justice. There are several reasons why justice has become a dominant concern. First is that globalization is an asymmetrical phenomenon. It has not led to one world; it has instead produced segregated worlds. It is a minority that benefits, not only in developed countries but also in countries that are emerging and developing. This minority often controls the economic and political decision making. Most people, however, live in other worlds pervaded by hunger, austerity, unemployment, growing insecurity, and vulnerability. Globalization has created what Andrew Dobson calls "communities of injustice"—populations that do not profit from global development and are in fact irrelevant: the dispossessed, marginalized, and excluded. [75]

A second reason for a focus on justice is that biodiversity is not equally distributed. Biodiversity-rich countries are generally poor and not powerful within the global system. Their resources are often exploited by more developed countries, as the debate on biopiracy illustrates. Destruction of biodiversity, on the other hand, is often the result of lifestyles and policies in developed countries, while the negative impact affects people in developing countries.

A third reason is that the consumption of natural resources is unequal. Total consumption of all resources produces an ecological deficit. We consume more than the earth can sustain. But the "we" in this case are living in a limited number of countries. People in most other countries do not have a just share of the planet's resources. They suffer from disparities in health and disease, food and water.

Ecological Debt

Use of biodiversity benefits richer countries, while ecological damage mostly burdens poor populations. Human activities have very different ecological impacts in different parts of the world. Developed countries have for decades produced high carbon dioxide emissions and thus promoted climate change, with deleterious effects in developing countries. Whereas these latter countries have high financial debts, developed countries have a different type of debt in terms of environmental costs. In this context, the term *ecological debt* has been introduced. During the 1992 Earth Summit in Rio de Janeiro, nongovernmental and grassroots organizations adopted alternative treaties, including the Debt Treaty. It underlines the existence of a "planetary ecological debt" of the global North, which is responsible for global environmental deterioration. [76]

Yet another reason to make justice a focal issue is that policies to protect biodiversity can have unjust consequences. Establishment of protected areas and national parks lead to displacement, especially of indigenous peoples who took care for these areas for centuries. The emphasis on wilderness preservation in environmental policies is devastating for people in developing countries, as argued by Guha. [77]

A fifth reason is the concern about the future that is implied in the notion of biodiversity. Justice requires the conservation of biodiversity for the sake of future generations. How can it be preserved for future generations? Biodiversity emphasizes that contemporary and future living beings are connected. They all need a healthy biodiversity to survive. But whose interests will prevail? How do we weigh intragenerational and intergenerational justice?

A final reason for considering the issue of justice is the bioeconomic paradigm that has prevailed in environmentalism since the adoption of the CBD. In practical policies, biodiversity is regarded as a global resource that generates benefits. Nature is an economic asset. Green neoliberalism argues that the best means for conserving biodiversity is to embrace the market approach and private property rights, thereby enclosing the biodiversity commons. But as argued above, this approach results in appropriation, lack of access, and exclusion, thus raising questions of justice. It is the perspective of imperialism: biodiversity is exploited under the assumption that human beings have dominion and power over nature. Human beings are the masters of biodiversity; they do not belong to the community of life, contrary to Aldo Leopold's postulations. Yet the concept of biodiversity at the same time also includes the arcadian perspective: living beings are interconnected and interdependent; humans are not separated from nature but embedded in it. Instead of exploring the possibilities of its use and exploitation as individual entrepreneurs, considering the natural environment as a source of goods and services, humans come to see themselves incorporated within biodiversity, as their source of life. Biodiversity is not itself a resource but an essential environmental condition. [78] That means that biodiversity is important for everybody, not just for powerful agencies and corporations. While the bioeconomic paradigm favors an individual perspective, the notion of biodiversity itself encourages a relational and social ethics perspective, an ethics of sharing, with global justice as a paramount principle.

The principle of justice is an important normative tool in the implementation of global bioethics discourse. It transcends the usual emphasis on the autonomous individual and focuses critical attention on the underlying structures and power constellations that determine the social, economic, and environmental conditions in which people live. It not merely criticizes the inequalities produced by neoliberal policies and practices but most of all argues that the benefits of globalization should accrue to everybody.

The abstract principle of justice is translated into social movements and activism across the world, calling for attention to health justice, environmental justice, food justice, and water justice.

HEALTH JUSTICE

Chapter 4 discussed how new infectious diseases are emerging mostly from world regions with deteriorating biodiversity. They also strike countries that are the least able to defend themselves. Pandemics such as Ebola and Zika bring havoc to countries with inadequate health infrastructure and an insufficient number of health professionals. Similar disparities occur for many other diseases. Matthew Sparke and Dimitar Anguelov analyzed inequalities in connection with the H1N1 pandemic in 2009. They distinguished inequalities in blame ("Mexican flu"), in risk management (prioritizing H1N1 over more frequent diseases already affecting the poor), in access to medicines (wealthy countries stockpiling vaccines and drugs), and in explanations (industrial factory farming was not addressed in global responses). [79] A harrowing example of unequal care is tuberculosis. Over 95% of the 1.8 million people who died from this disease in 2015 lived in low- and middle-income countries. Health justice scholars and activists argue that these health disparities are the result of an unfair system of global governance that gives priority to security rather than health. The current dominance of biosecurity focuses on the dangers of contagion and invasion. When a mortal infection is raging in developing countries, the first concern of other countries is not assistance but protection; they send military, not health professionals. They regard themselves as vulnerable, not the people in affected countries. The security framework emphasizes the need for control and containment instead of prevention. It does not address the roots of threats such as public health systems destroyed by decades of neoliberal policies or the unregulated bioindustry and unsafe food production systems that give rise to new viruses.

Health justice is also important for another dimension of global governance (see chapter 5). It concerns the intellectual property rights regime. Patent activity is concentrated in a relatively small number of companies in the global North. This is especially clear in biotechnology: 90% of all patents on life forms are held by Northern companies. [80] The global intellectual property regime not only reinforces inequalities but is itself the result of an unfair global institutional order (as argued by Thomas Pogge). [81] The establishment of the World Trade Organization and the Agreement on Trade-Related Aspects of Intellectual Property Rights (TRIPS) were the outcome of coordinated pressures exerted on developing countries by Western countries and international businesses as the owners of property rights. There was no fair representation of the countries involved, no sharing of full information, no democratic bargaining—only a mixture of political and economic threats and coercion. Public involvement was also absent; all negotiations were behind closed doors. Countries were pressured to comply through bilateral trade agreements. WTO's dispute settlement system can apply punitive measures if countries violate the rules. But it is not only the process that is unfair. The globalization of intellectual property has primarily benefited Western countries. The international legal context has been created by owners of intellectual property, and in their primary interest. The intellectual property rights regime is an illustration that the global context of health and healthcare itself is unfair. The trade system favors developed countries. By guaranteeing that all fields of technology are patentable, TRIPS not only introduced the patenting of pharmaceutical products across the world but also burdened many developing countries with the need to establish legislation and bureaucratic systems. Because they lack adequate infrastructure, many countries are not able to take advantage of intellectual property protection. There is not much evidence that this protection promotes innovation in developing countries. And in the occasional cases when it does, the focus is on the needs of developed countries as the most attractive market. At the same time, developing countries face disadvantages, since TRIPS makes it very difficult to use generic medication as an alternative to pricy, imported, patented drugs.

ENVIRONMENTAL JUSTICE

Since the 1970s, the environmental justice movement has criticized the unjust distribution of environmental benefits and harms. Toxic waste used to be dumped and processed close to vulnerable populations, specifically the poor neighborhoods of people of color. Local and grassroots movements denounced the connections between poverty, race, and environmental hazards. Instead of the usual emphasis placed by environmentalists on wilderness and nature, they focused attention on the circumstances in which people are living. Activists criticized the context of oppression, disrespect, and dehumanization that exposed them disproportionally to risks. [82] The success of this movement in the United States was followed by increased outsourcing of toxic waste. Dumping in poor and developing countries generated similar environmental justice movements around the globe. The sense of injustice at the global level deepened. The dumping sites are countries that are mostly not responsible for pollutants and toxic waste and that generally do not benefit from the lifestyle and consumption patterns associated with waste production. [83]

FOOD JUSTICE

As an essential determinant of human life and health, food should be available to everybody. Nonetheless, people are hungry not because there is a lack of food but because global distribution is unequal. People die of starvation in countries that are exporting food. Intensive agriculture has replaced traditional ways of growing food. Numerous small farms in rural areas have been eliminated and their farmers driven into slums and poverty. Powerful agribusinesses control the production, transport, processing, and marketing of food. Their interest primarily is in trade, less in health and environmental sustainability. Food justice movements have emerged all over the world (fair trade, slow food, local food), thereby influencing the global food regime. They attack the power and shortsightedness of corporations and international organizations that transform food into commodities and that are not concerned with biodiversity and global health. They develop alternative approaches to the dominant agricultural model. Social movements such as La Via Campesina reclaim sovereignty over food production. Consumer movements encourage critical choices. Food is a basic and daily concern for everybody. Many people also care about health and biodiversity. They prefer food that is healthy and green. Since food is not merely a commercial commodity, issues of justice go beyond questions of distribution, scarcity, and access. They also concern respect for diversity and recognition of cultural identity, reducing inequalities in power and control over food production and consumption, and participation of people in decision-making processes, not as consumers but as citizens. [84]

WATER JUSTICE

Most people regard water as a prime example of commons. It is a good that should benefit all; everybody should have equal access, and its management should be a shared responsibility. Nonetheless, continuous efforts are undertaken to enclose these commons and privatize water resources and services. When governments and the United Nations were reluctant to support such measures, the same drivers of the intellectual property rights regime in the fields of drugs and food started to promote the idea that water is a commodity The World Bank joined with private water companies to establish nongovernmental organizations like the World Water Council. They constructed a "water crisis," arguing that there is a growing scarcity of drinking water and that the only remedy is a market approach of privatization. This neoliberal approach generated vigorous responses and resistance, focusing specifically on injustices. The idea of scarcity is contested. The main ethical problem is lack of access to water. Privatization will only increase inequalities in its availability and affordability. Water justice movements (see chapter 7) argue that water requires democratic governance (because it is a common good) and should not be controlled by powerful global corporations and agencies. They emphasize that the water crisis can only be solved by common action. Individual water-saving activities will have limited impact. Clean water is a collective challenge. [85] These movements have been successful in turning back the drive toward privatization of water resources, not only in Cochabamba, Bolivia, but in many other places around the world.

INJUSTICES

Justice is an abstract notion. Different theories of justice are not easy to apply. In practice, the emphasis is often on injustices—for example, a tuberculosis patient who can be cured but does not receive treatment because it is too expensive; a child who is hungry because food is exported rather than made available in his region; a village that runs out of water because of massive extractions from its aquifer by a bottled water company. These are concrete examples that ask for responses. They can also be cases for theoretical debate over justice and various theories. But focusing on injustices is not a theoretical exercise. Injustices have many faces. They direct attention to human responsibility; they are not misfortunes but the result of human activities. Human beings have created and are supporting systems, structures, and regimes that are unfair; these systems do not benefit everyone; they are often imposed and can be violent; they victimize vulnerable groups and populations. Judith Shklar has argued that civilizations advance when what is commonly perceived as misfortune becomes considered injustice. [86] She explains how we might address injustices. Actively unjust agents violate the rules of justice and therefore produce injustices. Of course, these violations should be examined, prosecuted, and punished. Perhaps even more important is passive injustice. [87] This occurs when people do not prevent or oppose wrongs when they are able to do so. They remain indifferent. According to Shklar, this is a failure of citizenship, a civic vice. The sense of injustice should instigate citizens to raise their voices, to protest, and to resist. This is what many are doing when confronted with increasing loss of biodiversity and threats to health, drugs, food, and water.

Justice Is More Than Sharing

To address the inequalities of biodiversity, the most frequently used concept of justice is one that emphasizes fair exchange between two parties. The CBD has defined equitable sharing as one of its goals. Justice is regarded as equal distribution of benefits and harms. In chapter 5, it is argued that this concept of exchange justice (or commutative justice) is not adequate. Biopiracy, for example, is not simply a matter of benefit sharing. Its injustice is not adequately addressed just because it is compensated. It is doubtful whether any compensation can be fair if one of the two exchanging parties is much more powerful than the other. The question also is what kind of provision can compensate biodiversity resources. For these reasons, the concept of distributive justice is more appropriate. This concept acknowledges that parties usually are in unequal positions. It defines the criteria to establish a fair distribution of goods and services. The question, however, is whether the concept of distributive justice is sufficient to address the global context and practices of injustice. The emphasis on distribution accommodates very well to neoliberal ideology and regimes. It proceeds with a similar interpretation of biodiversity as a collection of resources that can be used for human needs. These resources can be owned by private agents and agencies. The main concern is that benefits and harms are shared. However, the question of whether it is fair to privatize such common resources in the first place is not posed. Traditional knowledge is not recognized and protected. Indigenous populations are not respected, and their rights are ignored. In decisions about distribution of benefits, weaker parties such as the poor, minorities, women, and indigenous people are usually not involved. Neoliberal policies and regimes that promote the concept of distributive justice to address the problem of global injustices are in fact applying as the solution the same approach that created the problem. Under this notion of justice, the underlying causes of maldistribution do not need to be taken into consideration.

That injustice is more than maldistribution is particularly clear in environmental justice movements. They articulate other dimensions of justice: recognition, participation, and capabilities for human functioning and flourishing, not only for individuals but also for communities. They protest against dumping toxic waste in their neighborhoods because it does not respect them as citizens. Populations of color and the poor feel undignified and abused because their health and well-being are disregarded, as happened in Flint. They are subjected to power that ignores their interests and denies their rights. They are also denied any participation in decision-making processes about their own environment. Respect and dignity as preconditions for distributive justice refer to a context of dominion and oppression that first must be addressed. This is the message of the critical discourse of biopiracy. Biopiracy is not simply theft of biodiversity resources that can be compensated by fair distribution and benefit-sharing arrangements. Basically, it is a phenomenon of inequalities of power and not so much of distribution of resources. It is not a market failure that can be addressed by reallocating resources. In the first place, it is a consequence of an ideology that does not recognize differences and identities, that excludes many people from political participation, and that undermines capabilities of individuals and groups. Stories of injustice are motivating movements in environmental justice, but health, food, and water justice are also based on a plurality of notions of justice. [88]

Global Bioethics Practices

Before exploring how global bioethics discourse can be applied in practice—and has been applied in addressing issues of biodiversity—it is important to clarify that the neoliberalism that is its main target is not a well-defined notion. It operates at different levels.

- *Neoliberal ideology.* This is the liberal philosophy that the individual is a rational, self-interested, and self-responsible decision maker. This philosophy is combined with economic and political theories that the market is the most fair and efficient organizer of social life. The market is a self-regulating force that provides free, productive, and competing individuals with all opportunities for development, benefits, and flourishing. Individual entrepreneurship and choices should have priority. The only rational ethical principle for all human relationships is the principle of trade. [89] At the same time, neoliberalism presents itself as a nonmoral order based on a technical principle of efficiency and a policy of pragmatism.

- *Neoliberal regimes.* The regimes of policies and practices created over the past few decades emphasize removal of constraints on the free market, scaling back the power of states, breakdown of safety nets and social security, deregulation, contraction of the public domain, and promotion of privatization. However, there is substantial difference between ideology and implementation. In chapter 6, the example of agricultural subsidies shows that the same regime of free trade does not have the same consequences for all countries. Developing countries were forced to withdraw public support for farming and to compete with food (over) produced in Western countries supported by substantial government subsidies. Neoliberal policies are also different. They were harshly forced upon African states through structural adjustments that dismantled state support for healthcare, transport, education, water, and agriculture, while Western countries did not generally privatize many welfare provisions and continued to provide at least some basic protection to precarious and vulnerable citizens. [90]

- *Neoliberal practices.* Neoliberal ideology and regimes are driving global governance, but they also manifest themselves at local levels. States and civil society do not apply them in similar ways. Major discrepancies and variations are noticeable. Vicente Navarro points to the example of the Reagan administration. In the United States during the presidency of Ronald Reagan—one of the major promoters of neoliberalism—federal public expenditures increased as well as some taxes. The role of the state was not reduced but rather was changed. Navarro concludes that interventionism of the government has increased over the past 30 years. [91] It is evident that the role of the state is not generally reduced as demanded by neoliberal ideology. Creation of new markets and enclosure of commons require substantial state interventions and re-regulation instead of deregulation. Ultimately, in 2008, the financial system could only be saved by massive state interventions.

These different levels of neoliberalism have consequences for the practical application of global bioethics discourse. If we believe, with Lisa Forman and many others, that the power of ideas can change the world, there are very different potential targets. [92] The previous chapters have discussed examples of possible practical actions in the areas of governance, framing, awareness raising, advocacy, and grassroots activities.

Governance

Neoliberal practices are promoted and regulated by specific policies. The global nature of problems as well as the diminishing influence of national governments has led to regimes of governance, particularly at the global and regional level. Governance is a form of collective action, ideally motivated by a sense of solidarity in the face of common threats. In practice, governance is not a coherent and systematic activity producing clear-cut results. Governance is cooperation among a wide range of actors with different views and interests. Global governance is most often focused on two activities: setting standards and creating agendas. The first activity results in agreements and treaties. The adoption of the Convention on Biological Diversity is an example. It is the result of negotiations among countries recognizing that loss of biodiversity is a global threat to everybody. Like many other international agreements, it has been criticized for being ambivalent and weak. Reducing biodiversity loss will require stronger implementation and better-defined policies. The importance of these normative instruments, however, is idealistic and aspirational. They present a framework of principles and reject the idea that practice determines what should be done. Some agreements will be more effective than others, but the goal of saving biodiversity and therefore humanity should never be abandoned. An often-used example of a successful approach is the Montreal Protocol to reduce ozone pollution. [93] It was the first treaty to achieve universal ratification. It has led to a significant reduction of ozone-depleting substances such as chlorofluorocarbons. Another positive example is the international cooperation to ban toxic waste dumping that resulted in the Basel Convention, although its effects in practice are limited. [94] An issue addressed in an international treaty moves into the spotlight of global attention even if the treaty is weak.

Some governments have not committed to international agreements. Many governments will not apply treaties (although it is their primary duty after they have ratified them). Often there are no sanctions and legal or economic repercussions following lack of implementation. These failures do not diminish the moral and political importance of the challenge that the treaties address. The fact that, for example, waste has become the subject of international negotiation and regulation will set the global agenda not merely for governments but for all concerned citizens of the world. The same is true for global bodies such as the G-7 and G-20. For a long time, it was argued that tuberculosis and antimicrobial resistance are global problems, but little action was taken. For the global debate, therefore, it is important when these issues are included in international statements such as the final declaration of the G-20 meeting in Hamburg, Germany, in July 2017. [95]

Similar arguments have been used for human rights discourse. The fact that many governments neglect, deny, or violate human rights does not mean that international human rights discourse is useless or ineffective. Human rights should primarily be implemented by governments, but even if they fail to do so, it presents a framework for everybody, for all citizens of the world. They can complain of the shortcomings of their governments and address problems themselves in their own regions and communities. Previous chapters have shown how this frequently happens with movements articulating the rights to health, food, and water. They do not depend on global governance or national regulations. This means that more attention should be paid to local governance. Ethical issues emerge within the context in which people live and work. This is the message of the environmental justice movement and cases such as Flint. In this context, people experience the connection between health and environment. They can take action on the basis of human rights discourse precisely because this applies to everyone. This is also the lesson of social movements as La Via Campesina. Rural farmers struggle under the negative pressure from global policies promoting industrial agriculture and powerful agribusinesses. They discover that many farmers in various countries have similar experiences. They decide to unite and engage in collective action. They refer to the right to food endorsed by many international documents. And they are successful. They influence policies. They also convince a growing number of people that alternatives are possible to the current food regime. The same processes are visible in other areas, not everywhere and not profound enough, but showing that change is possible. Grassroots resistance to privatization of water resources is an example. Toxic waste is another. Some countries do better than others. Sweden, for example, now needs to import rubbish from other countries to keep its recycling plants going. Since 2011, only 1% of the country's waste goes to landfills. [96] Labeling genetically modified food was rejected by food companies for decades; now it is legally required, even in the United States.

This dialectic interaction between the global and the local has consequences for the application of global bioethics: the target of ethical activities should focus first of all on local levels, inspiring individual citizens to engage in collective action. Global solidarity is undermined by the neoliberal economic system. [97] At the same time, solidarity is often revived within local settings. Here people learn that as autonomous individuals they face similar global challenges. But they also learn that private acts can have public implications if they join in collective activities. [98] The conclusion is that to address global problems, the focus in practice should be on local settings.

This conclusion is supported by new approaches in governance. As argued earlier, neoliberal governance is criticized because it assumes an opposition between market and state. But in practice there is an intermediary space between the two. The language of commons and justice activates a third option in-between the neoliberal dichotomy of private and public goods, of market and state. It argues that there are common areas, domains, and resources that should be publicly accessible because they play a special role for all human beings. It furthermore argues that communities themselves are able to manage these areas, domains, and resources. Commons require the participation and involvement of citizens. There is often no need for state intervention. The example of water management demonstrates how governments often endorse the right to water but fail to provide quality water to their citizens. The emergence of civil society has created another way to address global problems. This reflects the real importance of the notion of the global citizen. It refers to horizontal relationships. It emphasizes the relationships between citizens, the value of participation, and community interests. The idea that there is no alternative between state and market is wrong. In everyday practices, there is space between the two that goes beyond the perspective of the individual. Civil society is the area where social change is happening.

Global health governance in particular is changing. The dominant system of governance is not able to cope with the loss of biodiversity that threatens planetary health and human survival. [99] It does not adequately address challenges such as bioinvasion and emerging infectious diseases or the impacts of food and water regimes on human health. It is argued that we urgently need a different type of governance focused on planetary health, one that recognizes that we can no longer separate economy and nature, and that health requires healthy ecosystems (see chapter 4). First of all, health governance should be comprehensive, integrating various disciplines and approaches (such as One Health). It should also oppose the dominant discourse, with its focus on development, growth, economy, and trade. [100] It should argue that the economy must support planetary health. Market approaches should be embedded in noneconomic values such as solidarity, community, equality, and justice. Different approaches to planetary stewardship are urgently needed. A new, radical imagination is developing that can mobilize alternative forms of governance and social agency. [101] Finally, it requires support for the World Health Organization, the only actor in the global system that represents all states. [102] Its primary goal, in distinction to many other actors, is promotion of health. It can be and should be a major driver in promoting common goods (such as essential medicines), in mobilizing global solidarity, and in stewardship. [103] These are all normative challenges to which global bioethics can make significant contributions. Of course, this will be a long and continuing struggle, with many failures.

Social Responsibility?

Scientists working for the US Department of Defense are busy trying to develop a vaccine against the Zika virus. They are partnering with the pharmaceutical company Sanofi Pasteur, which received millions of dollars in grants from the US government to do clinical trials. The army intends to grant the company an exclusive right to sell the vaccine in the United States. But the company refused to promise a fair price. While American taxpayers have provided millions of dollars to develop the vaccine, they will probably pay a high price for it once it becomes available. [104]

Framing

Global bioethics may have practical influence through the activity of framing. An influential example is the debate about access to retroviral medication. This was framed in terms of protection of property rights versus human rights. Activists also introduced another framework, public health, which contrasted pharmaceutical profits against dying HIV/AIDS patients. Frames are schemes of interpretation. [105] They present an issue within a different perspective so that new light can be thrown on it and other possible approaches become possible. Framing scarcity of food, for example, as a security concern defines it as a specific problem and highlights relevant dimensions and contexts primarily as issues of risk and safety. Food "insecurity" is a particular diagnosis of the conditions and factors that produce the problem of scarcity. It also suggests a solution. The frame unites different actors under a similar banner, since policymakers share a similar concern. Replacing the terminology of hunger and famine, and its moral connotations of aid and assistance, with the frame of insecurity assumes that food is scarce; it presents scarcity first of all as an economic, not social or moral problem. This example illustrates that frames direct attention to issues but do not necessarily reflect the significance of the issue in the real world. For example, donor funding received for diseases is often not proportionate to the burden they impose on countries. Enormous resources were spent on the pandemic of SARS even though mortality was low. It is much more difficult to address maternal mortality. Diarrhea and tuberculosis are significant killers in many countries but do not attract much attention and funding. [106]

The theoretical framework of global bioethics has two practical opportunities. First, it can be used to critically examine the frames that are employed in neoliberal governance. Water is a clear example. NGOs established by the World Bank and private water companies promoted a discourse of "crisis." They framed water as scarce commodity. The solution, then, is to privatize water resources. This frame also presents the problem as a technical issue that should be addressed by experts. The main concern should be cost-effectiveness. Critics argued that this frame is wrong; rather than scarcity, there is lack of access to and affordability of water. The frame of commodity includes a restricted and limited perspective on water. Using the alternative frame of commons opens up other possibilities and perspectives of change in water governance. Frames therefore have a use other than criticism. The second application of framing is presenting alternatives. Framing water as commons appeals to sensitivities of communities, to tradition and history, and to citizenship. It evokes a different, democratic way of managing water, motivating people to cooperate for the public interest. It is related to other frames such as "water justice" and "water rights."

Global bioethics has the normative grammar to criticize dominant frames as well as introduce other frames in order to open up the debate and to steer it into other directions. New frames such as "ecological debt" and "ecological citizen" orient attention toward inequalities among affluent countries and the global poor, as well as responsibilities toward the environment. [107] In the debate on the property rights regime and patenting, the usual interpretation of property rights as private rights was challenged with the frame of collective rights. The dominant frame neglects the role of local communities, especially indigenous peoples. Activists pointed out that injustices such as biopiracy mostly affect not individuals but groups. They articulated the frame of "group rights" or "indigenous rights" to do justice to communities. [108]

Awareness Raising

Ethical issues will not draw attention if relevant information is not disseminated. Mainstream bioethics was strongly stimulated by cases that became the focus of public debate, for example, the Karen Ann Quinlan "right to die" case in the United States in 1976. Although problematic cases were arising in other local and community hospitals, the legal battle and the involvement of the New Jersey Supreme Court catapulted the case into the public and political domain, framing it as one of the paradigms of bioethics. In global bioethics the same mechanisms exist. The Trovan case in Nigeria (1996) drew international attention to practices of unethical clinical trials. The basmati rice case did the same for the injustices of patenting (see box, p. 128, bottom), as did the Plachimada case for the privatization of water resources (see box, p. 210). These cases triggered the awareness of global bioethical challenges. They demonstrated the impact of economic values on health and biodiversity. It is therefore important that relevant and accurate information is widely available. Investigative journalists, social movements, NGOs, and action groups at local and domestic levels identify cases and problems and bring them into the public domain so that they become known in the global community. Scientists specifically contribute objective and reliable information. A recent example is the extensive study by the Rockefeller Foundation–*Lancet* Commission on planetary health, which provides detailed information about the interconnections between health and biodiversity. [109] Luckily, many more scientific studies are available on, for example, inequality (see the work of Michael Marmot, discussed in chapter 3) and the consequences of biodiversity loss for health and healthcare. Many NGOs are monitoring and criticizing global developments. [110] Specific sources of information on global bioethics are, for example, BioEdge and the Global Health Network. [111] Information and awareness raising are the first steps in providing opportunities for action.

Nonetheless, powerful regimes are reacting. Not all NGOs can be trusted; many of them are not independent. Not all information is reliable. Private companies try to dominate the debate through influential organizations and institutions. Critical scholars and physicians have been discredited. Marion Nestle has shown how the food industry is obstructing government and expert advice, sponsoring research to show that there are no risks and harms, and arguing that there are no bad foods and that personal responsibility is all that matters. [112] At the same time, important changes are taking place. Citizens are increasingly concerned with food. Consumers prefer healthy, fresh, and local food; sustainable agriculture; and food justice.

Globalization has made the sharing of information across the world much easier and widespread; it has furthered awareness raising but also the growth of social movements. The availability of information is also changing the usual mechanisms of governance. David Fidler argues that the SARS outbreak in 2002–03 was the start of a new era of global governance. China's first response was secrecy. Based on the traditional notion of national sovereignty, it managed the disease as a domestic affair, and thus as a state secret. But the information could not be suppressed owing to modern information technology. The secrecy policy had to be reversed and cooperation with WHO acknowledged. Information has no borders. Transparency, openness, and accountability are unavoidable for governance. Nonstate actors have a stronger role now that epidemiological information is no longer dependent on governmental sources alone. [113]

Advocacy

Advocacy is a well-known component of healthcare. Pediatricians operate as advocates for children, as Hanna-Attisha did in Flint. Nurses assume a significant role in advocacy for patients. Patients have established advocacy organizations. Advocacy means that there is a specific cause combined with activities. An example is environmental justice. Confronted with toxic waste dumping, people may unite and establish organizations or start movements in order to advocate a clean and just environment. They engage in activities to support this cause: arguing, promoting, championing, campaigning, lobbying, pressuring, and influencing policy processes. At the global level, NGOs such as Greenpeace and Friends of the Earth are active for the sake of biodiversity. It is argued that WHO should be the advocate of global health, given its tendency to criticize neoliberal policies. In fact, there is an advocacy role for all actors involved in planetary health: national governments, UN organizations, philanthropic organizations and foundations, academic institutions, private industry, professional organizations, and civil society and nongovernmental organizations. [114] Advocacy takes place at various levels. The Basel Convention was much influenced by Greenpeace. At local levels, neighborhoods contaminated by waste were frequently successful in challenging existing practices. Advocacy is connected to Shklar's idea of passive injustice. It is important to identify, sanction, and correct instances of active injustice. But it is equally important to overcome passive injustice. Citizens must not be indifferent to injustices. They should protest and raise their voices. This is especially the case when the injustices are not abstract and far away, and it is the lesson of the environmental, food, and water justice movements. They affect people in their everyday contexts. Just as individual scandalous cases stimulate ethical reflection and action, advocacy is connected to concrete experiences of injustice that bring people together in collective action. They no longer accept that they have no voice and are not allowed to participate in policy and decision making. Advocacy means that people have discovered another type of agency as collectives and coalitions. They can be effective in bringing about changes through concerted action.

Advocacy networks are influential because of the international human rights framework. They can set the agenda of the global discussion, shape policymaking and the behavior of states through emphasizing rights (health and food) that have been accepted by all states in declarations and conventions, and articulate new rights (water). Advocates do not focus on the theory of human rights. They assume that these rights exist and are adopted by all governments. The main issue is their insufficient or absent application by corrupt and autocratic politicians. International organizations often do not give sufficient priority to values other than trade and commerce. But advocacy can be effective. Advocating the right to health, for example, has influenced the policies of the WTO (perhaps not enough, but sufficient to make health a relevant consideration in promoting trade). [115] The expanding interpretation and monitoring of human rights such as the right to health connect health more strongly to its social determinants. The Food and Agriculture Organization is now advocating for a right-to-food approach. Even private water companies and their sponsored NGOs recognize nowadays the right to water. Destruction of biodiversity is increasingly considered as a human rights issue. Biodiversity is indispensable for the enjoyment of a safe, clean, and healthy environment. That implies, as argued recently in the Human Rights Council, that there are human rights and obligations in regard to biodiversity. Another powerful insight is that human rights are connected. As Christiana Peppard argues, the rights to health, food, and water are related to the right to life, regarded as the most fundamental human right. Without these specific rights, this fundamental right is violated. Water, food, and health are more than basic human needs; they are essential for life itself. Bioethics therefore should do more than focus on the beginning and end of life. A social ethics perspective on the whole of human life is necessary. [116]

Biodiversity and Human Rights

The Special Rapporteur on the issue of human rights obligations relating to the enjoyment of a safe, clean, healthy, and sustainable environment concluded in his January 2017 report:

65. Biodiversity is necessary for ecosystem services that support the full enjoyment of a wide range of human rights, including the rights to life, health, food, water and culture. In order to protect human rights, States have a general obligation to protect ecosystems and biodiversity.

66. Biodiversity around the world is rapidly being degraded and destroyed, with grave and far-reaching implications for human well-being. A human rights perspective: (a) Helps to clarify that the loss of biodiversity also undermines the full enjoyment of human rights; (b) Heightens the urgent need to protect biodiversity; (c) Helps to promote policy coherence and legitimacy in the conservation and sustainable use of biodiversity. [117]

Advocacy can be a dangerous activity. In 2016, 200 environmental activists were murdered—4 every week. They had been defending land, forests, and rivers. Of these murders, 60% took place in Latin America. Brazil is the most lethal country (49 killings). [118] Indigenous people are most vulnerable. Advocacy groups are often threatened and discredited. Marion Nestle provides many examples of how global food corporations obstruct regulation, discredit research and scientific evidence, and blame individual consumers, who they claim are primarily responsible for selecting healthy and sustainable food. They promise self-regulation and social responsibility. Although some improvements have been made, continued advocacy for better practices and changing food regimes will be necessary. [119]

Another danger lurking behind advocacy networks is lack of independence. Many NGOs and advocacy groups are established, funded, supported, and dominated by powerful drivers of neoliberal governance. They are not interested in regime change. A majority of US patient groups, for example, are funded by the healthcare industry. These organizations are not transparent; the overwhelming majority have no conflict-of-interest policies. Because they are not independent, their credibility as advocacy groups is low.

Industry Funding of Patient Organizations
Two recent studies show that the majority of patient advocacy organizations in the United States take money from the medical and pharmaceutical industry. According to one study, 83% of the top patient advocacy groups receive money, with almost a quarter receiving more than $1 million each year. At least 39% have current or former industry executives on their boards. The other study, based on interviews with leaders of organizations, concludes that 67.3% receive industry funding. [120]

Grassroots Activities

Neoliberalism is based on self-interest. It is often argued that trade promotes competition, not global cooperation. [121] The priority of the market is also assumed to undermine solidarity, especially at global level. Consumers are not supposed to be interested in collective goods. There is no need for collective action, since everything is available for rational consumers. They do not need social security and protection if states are inefficient and corrupt; they can take care of themselves. In practice, this neoliberal discourse proves wrong. Instead of growing more autonomous and independent, most people have become more vulnerable. They are increasingly recognizing that as "citizens of the world" they face the same challenges and that cooperation is necessary to improve their fate. Because vulnerability is everywhere, solidarity will provide a way to overcome it. Instead of an anonymous world of individuals, globalization has given rise to many community-building efforts and new social movements. There has been an enormous growth of human rights and nonprofit organizations over the past few decades. In Kenya, for example, there is now one nonprofit organization for every 110 people living in the country. [122] The number of nongovernmental organizations in the United States is estimated at two million. [123] Global solidarity is particularly strong in regard to health and biodiversity. It is not so much visible in global discourse but rather in local and domestic practices. Thousands of environmental groups have emerged over the past decades, the majority of them in Latin America and Asia. [124] La Via Campesina, one of the largest social movements in the world, started as a limited mobilization of collectives of rural farmers in a few countries in Latin America. They did not use theoretical expositions about injustice and solidarity but responded to processes that affected their lives and that seemingly were beyond their control. The framework of "food sovereignty" was introduced to criticize the dominant food system. Experiencing the impact of neoliberal policies in the reality of their life and work, they united to respond to these policies, first in their own setting. The movement gradually brought together similar movements in other countries and grew into a transnational, global network. The environmental justice movement has a similar story. It began as a local protest activity in 1982 in the town of Afton, North Carolina. It rapidly expanded throughout the United States and later disseminated over the globe. In earlier chapters, the same processes are described for social movements in health and healthcare. Activism against toxic waste, chemicals in food, and genetically modified food usually started at the local level, where people experience the negative effects on health. Using the notions of justice and the right to health was effective to broaden the movement. The frame of security brought together countries and international organizations against the threat of emerging infections. The critical discourse of biopiracy that inspired many social movements, especially in developing countries, framed the intellectual property regime as a structural injustice and lack of respect for traditional knowledge. What started at the grass roots and evolved into social movements is therefore often the result of intensive interaction between activism and ethical principles, leading to alternative ways of interpreting the results of neoliberal policies. Movements arise because policies primarily benefit a minority, such as powerful corporations and the wealthy. In other words, they are the consequences of a lack of solidarity. On the other hand, they demonstrate the power of global solidarity. They are efforts to transcend the dominant governance regimes that are focused on protecting and promoting trade. [125] But movements are more than the mere response to the lack of solidarity that created their problems. Rising up against these problems, social crusades show that global solidarity continues to exist, and they instigate people to act together by appealing to ethical principles such as vulnerability, justice, and human rights.

The growth of grassroots activities is another opportunity for practical applications of global bioethics, First, it shows the connection between the global and local. Critique of the neoliberal roots of ethical problems encourages resistance and protest at local levels, where people experience injustice and try to develop alternatives based on global ethical principles. Second, it shows that individual and social ethics are not opposed. Neoliberalism has presented and imposed an impersonal economic order. It assumes that nothing intervenes between this order and the individual life. But in fact grassroots movements show that there are many "mediating forces and buffers" such as churches, trade unions, local government, professional associations, and citizens' societies and guilds. [126] Idealistic individuals like Vandana Shiva and Maude Barlow inspire broader movements. Human life is only partially controlled by neoliberal regimes and states. The dichotomy between state and market is wrong. It has the same negative consequences for planetary health and biodiversity as the artificial separation between natural and social worlds. Third, within this intermediary space is civil society. When neoliberal regimes affect all people everywhere and when people mobilize to articulate alternatives, citizens can play a significant role in remediating biodiversity and promoting planetary health. This is currently not the case. There is a democratic deficit at all levels of governance. Citizens are most frequently not involved in decision making regarding health and environmental policies. Neoliberalism itself is criticized for its negative impacts on democracy. It coincides with declining electoral turnout, party membership, trade union membership, and the rise of populist parties. [127] Trade negotiations are usually secret. Relevant information is not shared. International organizations are generally not open to citizens and social movements. Health governance is mostly driven by technical expertise, based on Western knowledge, and focused on specific diseases. There is little room for participation of concerned local communities or citizens. Governance regimes now define the rules of the game. They are not eager to have them contested by engaged citizens. Nonetheless, there is an increasing call for participation, transparency, and accountability. Value conflicts cannot be avoided by silence and secrecy. Mutual understanding and recognition of the common good will emerge through inclusion and deliberation. [128]

Conclusion

Biodiversity is an essential condition for human flourishing. Without a healthy planet, human beings cannot be healthy. For clean air, safe water, adequate nutrition, provision of drugs, and protection from infectious diseases we all depend on biodiversity. However, biodiversity is rapidly deteriorating, threatening planetary health. The consequence is that the future will not be an extension of the present but different and dangerous instead. The fundamental connection between health and biodiversity is increasingly recognized. This book has provided numerous examples showing that today's healthcare and medicine must take a broad perspective that includes the social and environmental contexts in which individuals are situated. Focusing on individual patient care will always be necessary, but given the many impacts of biodiversity loss on planetary health, the usual approach should be extended by such broad perspectives. This extension has consequences for the ethical debate on healthcare and medical science. Mainstream bioethics has grown substantially over the past five decades. It has empowered patients, provided consultation for clinical decision making, clarified ethical dilemmas, and supported legislation and regulation. However, its main emphasis is on individual care and treatment. Given the widening global dimension of healthcare and medicine, it is insufficient to address the environmental and social context in which ethical problems nowadays arise. My argument is that processes of globalization are affecting biodiversity and thus producing significant threats to planetary health that will also affect individual patient care. Ethical values are at stake—not only economic growth and trade but also health and protection of the environment. For this reason, a new and broad concept of bioethics is needed. The emerging field of global bioethics provides ethical principles that transcend the limited vision of mainstream bioethics. It introduces a normative horizon in which other perspectives become visible, such as justice, equality, vulnerability, and solidarity. Global bioethics thus supports health professionals, policymakers, and citizens of the world in their efforts to enhance planetary health.

Notes

A complete bibliography containing all of the works cited in the notes is available online at the Johns Hopkins University Press website, www.press.jhu.edu.

CHAPTER ONE: **Global Bioethics and the Environment**

1. Van Rensselaer Potter, *Bioethics: Bridge to the Future* (Englewood Cliffs, NJ: Prentice-Hall, 1971).

2. Van Rensselaer Potter, "Global Bioethics: Origin and Development," in *Handbook for Environmental Risk Decision Making: Values, Perceptions and Ethics*, ed. C. R. Cothern (Boca Raton, FL: CRC Lewis Publications, 1995), 359.

3. Henk ten Have, "Potter's Notion of Bioethics," *Kennedy Institute of Ethics Journal* 22 (2012): 62–64.

4. Van Rensselaer Potter, "Council on the Future: A Proposal to Cope with the Gulf between Scientific Knowledge and Political Direction," *Nation* (1965): 133–36. Included as chapter 6 in Potter, *Bioethics*, 75–82.

5. Potter, *Bioethics*, 109.

6. Catalog of Spaceborne Imaging, *Earth—Apollo 11*, http://nssdc.gsfc.nasa.gov/imgcat /html/object_page/a11_h_36_5355.html.

7. Potter could not use these photos because they were taken in December 1972, after his book was published. For the impact of the pictures, see Gregory A. Petsko, "The Blue Marble," *Genome Biology* 12 (2011): 112; Denis Cosgrove, "Contested Global Visions: One-World, Whole-Earth, and the Apollo Space Photographs," *Annals of the Association of American Geographers* 84 (1994): 270–94. See also Wolfgang Sachs, *Planet Dialectics: Explorations in Environment and Development* (London: Zed Books, 2015).

8. Anna Bramwell, *Ecology in the 20th Century: A History* (New Haven, CT: Yale University Press, 1989); Roderick F. Nash, *The Rights of Nature: A History of Environmental Ethics* (Madison: University of Wisconsin Press, 1989).

9. Van Rensselaer Potter, "Evolving Ethical Concepts," *BioScience* 27 (1977): 251–53.

10. Van Rensselaer Potter, *Global Bioethics: Building on the Leopold Legacy* (East Lansing: Michigan State University Press, 1988), 3.

11. Van Rensselaer Potter, "Global Bioethics with Humility and Responsibility," *Biomedical Ethics* 5 (2000): 89–93.

12. Potter, *Bioethics*, 179.

13. Potter, *Bioethics*, 145, 180, 184.

14. Van Rensselaer Potter, "Biocybernetics and Survival," *Zygon* 5 (1970): 243.

15. Van Rensselaer Potter, "Bioethics: State-of-the-Art Books," *BioScience* 31 (1981): 172.

16. Van Rensselaer Potter, "Aldo Leopold's Land Ethics Revisited: Two Kinds of Bioethics," *Perspectives in Biology and Medicine* 30 (1987): 157–69. This article is incorporated as chapter 4 in Potter, *Global Bioethics*, 71–94.

17. Potter, *Global Bioethics*, xiii.

18. Potter, *Global Bioethics*, 7.

19. Aldo Leopold, *A Sand County Almanac* (New York: Ballantine Books, 1970), 224.

20. Potter, *Global Bioethics*, 78.

21. Van Rensselaer Potter, "Getting to the Year 3000: Can Global Bioethics Overcome Evolution's Fatal Flaw?," *Perspectives in Biology and Medicine* 34 (1990): 90.

22. Van Rensselaer Potter and Richard Grantham, "Scientists' Responsibility for Survival of the Human Species," *Scientist* 6 (May 25, 1992): 10–11.

23. Van Rensselaer Potter, "Bridging the Gap between Medical Ethics and Environmental Ethics," *Global Bioethics* 6 (1993): 161–64.

24. Van Rensselaer Potter, "An Essay Review of: *Global Responsibility. In Search of a New World Ethic* by Hans Küng," *Perspectives in Biology and Medicine* 37 (1994): 546–50. See also Van Rensselaer Potter, "Science, Religion Must Share Quest for Global Survival," *Scientist* 8 (May 16, 1994): 12.

25. Van Rensselaer Potter, "Global Bioethics: Linking Genes to Ethical Behavior," *Perspectives in Biology and Medicine* 39 (1995): 120.

26. Van Rensselaer Potter, "Moving the Culture toward More Vivid Utopias with Survival as the Goal," *Global Bioethics* 14 (2001): 26.

27. Van Rensselaer Potter, "Global Bioethics with Humility and Responsibility," 90.

28. See, for example, Erin D. Williams, "A Legacy of Bioethical Sustainability: In Memory of Dr. Van Rensselaer Potter II," *Global Bioethics* 14 (2001): 49–57; Lainie Friedman Ross, "Forty Years Later: The Scope of Bioethics Revisited," *Perspectives in Biology and Medicine* 53 (2010): 452–57.

29. James F. Drane and Patricia Pineo, eds., "Founders of Bioethics: Concepts in Tension, Dialogue, and Development," *Theoretical Medicine and Bioethics* 33 (2012): 1–105.

30. See, for example, Tom L. Beauchamp and James F. Childress, *Principles of Biomedical Ethics* (New York: Oxford University Press, 2013). None of the seven editions since 1979 mention Potter. The same holds for Tristram Engelhardt, *The Foundations of Bioethics* (New York: Oxford University Press, 1986).

31. David J. Rothman, *Strangers at the Bedside* (New York: Basic Books, 1991), mentions Potter only in a note (p. 286) as the first person to use the term. John H. Evans, *The History and Future of Bioethics* (Oxford: Oxford University Press, 2012), as well as M. L. Tina Stevens, *Bioethics in America: Origins and Cultural Politics* (Baltimore, MD: Johns Hopkins University Press, 2000), have no references to Potter. Only European textbooks and histories make an exception. See Corrado Viafora, ed., *Bioethics: A History* (San Francisco: International Scholars Publications, 1996), where Potter is mentioned several times by the same author. See also Diego Gracia, "History of Medical Ethics," in *Bioethics in a European Perspective*, ed. Henk ten Have and Bert Gordijn (Dordrecht: Kluwer Academic Publishers, 2001), 17–50.

32. Albert R. Jonsen, *The Birth of Bioethics* (New York: Oxford University Press, 1998), 27.

33. Warren T. Reich, "The Word 'Bioethics': Its Birth and the Legacies of Those Who Shaped It," *Kennedy Institute of Ethics Journal* 4 (1994): 319–35.

34. Gaines and Juengst argue that there are at least five stories about the origin of bioethics; each story has a particular purpose, that is, defining and legitimizing a distinctive form of bioethics. See Atwood D. Gaines and Eric T. Juengst, "Origin Myths in Bioethics: Constructing Sources, Motives and Reason in Bioethic(s)," *Culture Medicine and Psychiatry* 32 (2008): 303–27.

35. K. Danner Clouser, "Bioethics," in *Encyclopedia of Bioethics*, ed. Warren T. Reich (New York: Free Press, 1978), 1:120, 125.

36. Clouser, "Bioethics," 120.

37. Reich, "The Word 'Bioethics,'" 19–34.

38. Stephen Toulmin, "How Medicine Saved the Life of Ethics," *Perspectives in Biology and Medicine* 25 (1982): 736–50.

39. Potter, "Aldo Leopold's Land Ethics Revisited," 160; Van Rensselaer Potter, "Biotechnology: An Overview and Evaluation," *World & I* (November 1987): 573.

40. Sergio Nestor Osorio Garcia, "Van Rensselaer Potter: Revolutionary Vision for Bioethics," *Revista Latinoamericana de Bioetica* 8 (2005): 98.

41. Clouser, "Bioethics," 118.

42. "I propose that every moral proposition is an intuition . . ." (Van Rensselaer Potter, "What Does Bioethics Mean?," *AG Bioethics Forum* 8 [1996]: 3). See also Potter, "Global Bioethics with Humility and Responsibility," 90.

43. Potter, "Moving the Culture toward More Vivid Utopias," 27.

44. Stenmark summarizes the need for value judgments as follows: "If the survival of a species, indeed of human beings, is desirable, it must be so for reasons which go *beyond* ecology, biology and other sciences." See Mikael Stenmark, *Environmental Ethics and Policy Making* (Aldershot, UK: Ashgate, 2002), 10.

45. The first conference on environmental ethics was organized by William Blackstone at the University of Georgia in 1971. See Roderick F. Nash, *The Rights of Nature: A History of Environmental Ethics* (Madison: University of Wisconsin Press, 1989), 125. The first systematic book is John Passmore, *Man's Responsibility for Nature* (New York: Scribner's, 1974); the first specialized journal is *Environmental Ethics* (since 1979).

46. For example, there is no mention of Potter or bioethics in Robin Attfield, *The Ethics of Environmental Concern* (New York: Columbia University Press, 1983); Paul W. Taylor, *Respect for Nature. A Theory of Environmental Ethics* (Princeton, NJ: Princeton University Press, 1986); Eugene Hargrove, *Foundations of Environmental Ethics* (Englewood Cliffs, NJ: Prentice-Hall, 1989); Joseph R. DesJardins, *Environmental Ethics. An Introduction to Environmental Philosophy* (Belmont, CA: Wadsworth, 1993); Louis P. Pojman, *Global Environmental Ethics* (Mountain View, CA: Mayfield, 2000); Peter S. Wenz, *Environmental Ethics Today* (New York: Oxford University Press, 2001); Stenmark, *Environmental Ethics and Policy Making*; Robin Attfield, *Environmental Ethics. An Overview for the Twenty-First Century* (Cambridge: Polity, 2003); Dale Jamieson, *Ethics and the Environment* (Cambridge: Cambridge University Press, 2008). The exception is Nash, who refers to Potter and his book because it is dedicated to Leopold (Nash, *The Rights of Nature*, 63, 85).

47. Paul Taylor clarifies that there are in fact two types of environmental ethics: anthropocentric and nonanthropocentric (Taylor, *Respect for Nature*, 11ff).

48. See DesJardins, *Environmental Ethics*, 147–50.

49. Paul Taylor defines environmental ethics as "concerned with the moral relations that hold between humans and the natural world" (Taylor, *Respect for Nature*, 3).

50. Van Rensselaer Potter, "Real Bioethics: Biocentric or Anthropocentric?," *Ethics and the Environment* 1 (1996): 177–83.

51. Potter, *Global Bioethics*, 22.

52. Nash, *The Rights of Nature*, 124. This view is also endorsed by Minteer and Collins, who argue that environmental ethics has been "preoccupied with abstract discussions" ("Why We Need an 'Ecological Ethics,'" *Frontiers in Ecology and the Environment* 3 [2005]: 335).

53. *The Belmont Report: Ethical Principles and Guidelines for the Protection of Human Subjects of Research*, 1979, https://www.hhs.gov/ohrp/regulations-and-policy/belmont-report /read-the-belmont-report/index.html#toc (originally published in the *Federal Register*). United Nations, *Report of the World Commission on Environment and Development: Our Common Future*, 1987, http://www.un-documents.net/our-common-future.pdf; also published as a book: World Commission on Environment and Development, *Our Common Future* (Oxford: Oxford University Press, 1987). Rio Declaration on Environment and Development, 1992, http://www.unep.org/Documents.Multilingual/Default.asp?documentid =78&articleid=1163.

54. "The legacy of Adam Smith, the invisible hand, and the right to life, liberty, and the pursuit of happiness must be tempered by global ethics" (Van Rensselaer Potter and Lisa Potter, "Global Bioethics: Converting Sustainable Development to Global Survival," *Global Bioethics* 14 [2001]: 186).

55. Potter, "Moving the Culture toward More Vivid Utopias," 26.

56. Jessica Pierce, "Can Bioethics Survive in a Dying World?," *Journal of Medical Humanities* 23 (2002): 3–6; Jessica Pierce and Andrew Jameton, *The Ethics of Environmentally Responsible Health Care* (Oxford: Oxford University Press, 2004).

57. The idea of "sustainable medicine" is particularly promoted by Daniel Callahan (*False Hopes: Overcoming the Obstacles to a Sustainable, Affordable Medicine* [New Brunswick, NJ: Rutgers University Press, 1999]). The healthcare sector itself significantly contributes to environmental degradation: in 2007 it contributed 8% of the total greenhouse-gas emissions in the United States. See Jeannette Chung and David Meltzer, "Estimate of the Carbon Footprint of the US Health Care Sector," *JAMA* 302 (2009): 1971.

58. David B. Resnik, *Environmental Health Ethics* (New York: Cambridge University Press, 2012); Cristina Richie, "A Brief History of Environmental Bioethics," *Virtual Mentor: American Medical Association Journal of Ethics* 16 (2014): 749–52. See also Robin N. Fiore and Lora E. Fleming, "Occupational and Environmental Health: Towards an Environmentally Inclusive Bioethics," *Professional Ethics* 11 (2003): 65–82.

59. Jonathan D. Moreno, "In the Wake of Katrina; Has 'Bioethics' Failed?," *American Journal of Bioethics* 5 (2005): W18–W19.

60. See, for example, Susan Sherwin, "Looking Backwards, Looking Forward: Hope for Bioethics' Next Twenty-Five Years," *Bioethics* 25 (2011): 75–82; Ross, "Forty Years Later," 452–57.

61. This claim is made by James Dwyer, who argues that it is necessary to "bring environmental concerns back into ethical discussions about health and healthcare." See Dwyer,

"How to Connect Bioethics and Environmental Ethics: Health, Sustainability, and Justice," *Bioethics* 23 (2009): 497. See also Lori Gruen and William Ruddick, "Biomedical and Environmental Ethics Alliance: Common Causes and Grounds," *Journal of Bioethical Inquiry* 6 (2009): 457–66.

62. See Andrew Light and Eric Katz, eds., *Environmental Pragmatism* (London: Routledge, 1996); Robert Hood, "The Role of Cases in Moral Reasoning: What Environmental Ethics Can Learn from Biomedical Ethics," in *Moral and Political Reasoning in Environmental Practice*, ed. Andrew Light and Avner de-Shalit (Cambridge, MA: MIT Press, 2003), 239–57; Ben A. Minteer, James P. Collins, and Stephanie J. Bird, "The Emergence of Ecological Ethics," *Science and Engineering Ethics* 14 (2008): 473–81; Ben A. Minteer and James P. Collins, "From Environmental to Ecological Ethics: Towards a Practical Ethics for Ecologists and Conservationists," *Science and Engineering Ethics* 14 (2008): 483–501.

63. Daniel F. Doak, Victoria J. Bakker, Bruce Evan Goldstein, and Benjamin Hale, "What Is the Future of Conservation?," *Trends in Ecology & Evolution* 29 (2014): 77–81.

64. Paul R. Ehrlich, "Bioethics: Are Our Priorities Right?," *Bioscience* 53 (2003): 1211. In a later publication, Ehrlich uses the concept of "ecoethics" for a broader notion of ethics (Paul R. Ehrlich, "Ecoethics: Now Central to All Ethics." *Journal of Bioethical Inquiry* 6 [2009]: 417–36).

65. Reich argues that it should be recognized that Potter coined the word *bioethics*, advocated a new discipline, and identified human survival as the basic problem. See Reich, "The Word 'Bioethics,'" 321–22.

66. Henk ten Have, *Global Bioethics: An Introduction* (London: Routledge, 2016).

67. Pierce and Jameton, *The Ethics of Environmentally Responsible Health Care*.

68. This thesis of the evolution of ethics is developed by Nash, *The Rights of Nature*, 3ff.

CHAPTER TWO: **Biodiversity**

1. When it was proclaimed in 1993, December 29 was designated; in 2000, the day was moved to May 22 to commemorate the adoption of the Convention on Biological Diversity.

2. Between 1988 and 2000, there was a growing but limited number of publications; there has been a steep rise since 2000. A PubMed search in August 2017 showed 65,762 publications with the keyword *biodiversity*.

3. United Nations, Convention on Biological Diversity, 1992, https://www.cbd.int/doc /legal/cbd-en.pdf.

4. See Yrjö Haila, "Making the Biodiversity Crisis Tractable: A Process Perspective," in *Philosophy and Biodiversity*, ed. Markku Oksanen and Juhani Pietarinen (Cambridge: Cambridge University Press, 2004), 54–82.

5. Samuel P. Hays, *A History of Environmental Politics since 1945* (Pittsburgh, PA: University of Pittsburgh Press, 2000), 94ff. Thompson mentions the fact that the UK Royal Society for the Protection of Birds has over a million members, more than all British political parties combined. See Ken Thompson, *Do We Need Pandas? The Uncomfortable Truth about Biodiversity* (Foxhole, UK: Green Books, 2010), 95.

6. Stephen Fox, *The American Conservation Movement: John Muir and His Legacy* (Madison: University of Wisconsin Press, 1981); Donald Worster, *Nature's Economy. A History of Ecological Ideas* (Cambridge: Cambridge University Press, 1994).

7. David Pepper, *Modern Environmentalism. An Introduction* (London: Routledge, 1996).

8. Max Oelschlager, *The Idea of Wilderness: From Prehistory to the Age of Ecology* (New Haven, CT: Yale University Press, 1991).

9. Ramachandra Guha, "Radical American Environmentalism and Wilderness Preservation: A Third World Critique," *Environmental Ethics* 11 (1989): 71–83.

10. Guha, "Radical American Environmentalism," 74.

11. Seventeen mega-diverse countries include 66%–75% of the world's biodiversity: Brazil, Indonesia, Colombia, Mexico, Australia, Madagascar, China, Philippines, India, Peru, Papua New Guinea, Ecuador, United States, Venezuela, Malaysia, South Africa, and Democratic Republic of the Congo. See Kevin J. Gaston and John I. Spicer, *Biodiversity: An Introduction* (Malden, MA: Blackwell Publishing, 2004), 63.

12. Vandana Shiva et al., *Biodiversity: Social and Ecological Perspectives* (London: Zed Books, 1995), 7–11.

13. Guha, "Radical American Environmentalism," 80.

14. Worster argues that Leopold had a conversion around 1935, when he founded the Wilderness Society (Worster, *Nature's Economy*, 271ff.). For Leopold, see Curt Meine, *Aldo Leopold: His Life and Work* (Madison: University of Wisconsin Press, 2010); Susan L. Flader, *Thinking like a Mountain: Aldo Leopold and the Evolution of an Ecological Attitude toward Deer, Wolves and Forest* (Madison: University of Wisconsin Press, 1974); J. Baird Callicott, ed., *Companion to a Sand County Almanac: Interpretive and Critical Essays* (Madison: University of Wisconsin Press, 1987); Julianne Lutz Newton, *Aldo Leopold's Odyssey* (Washington, DC: Island Press, 2006).

15. Oelschlager, *The Idea of Wilderness*, 209.

16. Worster, *Nature's Economy*, 294, 304.

17. Oelschlager, *The Idea of Wilderness*, 1.

18. Worster, *Nature's Economy*, 9, 19.

19. For these arguments, see David Takacs, *The Idea of Biodiversity: Philosophies of Paradise* (Baltimore, MD: Johns Hopkins University Press, 1996); Dan L. Perlman and Glenn Adelson, *Biodiversity: Exploring Values and Priorities in Conservation* (Malden, MA: Blackwell Science, 1997); Nils Eldredge, *Life in the Balance: Humanity and the Biodiversity Crisis* (Princeton, NJ: Princeton University Press, 1998); Gaston and Spicer, *Biodiversity*; Patrick Blandin, *Biodiversité: L'avenir du vivant* (Paris: Albin Michel, 2010); Thompson, *Do We Need Pandas?*

20. Dale Jamieson, *Ethics and the Environment: An Introduction* (Cambridge: Cambridge University Press, 2008), 8–10.

21. Elizabeth Kolbert, *The Sixth Extinction: An Unnatural History* (New York: Picador/Henry Holt, 2014), 4–22.

22. Richard Pearson, *Driven to Extinction: The Impact of Climate Change on Biodiversity* (New York: Sterling, 2011), 207ff.

23. Edward O. Wilson, ed., *Biodiversity* (Washington, DC: National Academy Press, 1988), v.

24. See, respectively, Paul R. Ehrlich (p. 25), Norman Myers (p. 34) and Terry L. Erwin (p. 129) in Wilson, *Biodiversity*.

25. Edward O. Wilson, *The Diversity of Life* (Cambridge, MA: Belknap Press, 1992), 280; Secretariat of the Convention on Biological Diversity, "Message from Ahmed

Djoghlaf, Executive Secretary, on the Occasion of the International Day for Biological Diversity," May 22, 2007, https://www.cbd.int/doc/speech/2007/sp-2007-05-22-es-en.pdf.

26. Irus Braverman, *Wild Life: The Institution of Nature* (Stanford, CA: Stanford University Press, 2015), 231.

27. See Michael J. Jeffries, *Biodiversity and Conservation* (London: Routledge, 2006); Kim Master Evans, *Endangered Species: Protecting Biodiversity* (Farmington Hills, MI: Gale, 2015).

28. Wilson, *The Diversity of Life*, 15.

29. Humberto D. Rosa, "The Bioethics of Biodiversity," special issue, *Human Ecology* 12 (2004): 162ff.

30. Roughly every 26 million years a major mass extinction takes place. See Kolbert, *The Sixth Extinction*, 101. For the concept of Anthropocene, see Will Steffen, Jacques Grinevald, Paul Crutzen, and John McNeill, "The Anthropocene: Conceptual and Historical Perspectives," *Philosophical Transactions of the Royal Society* 369 (2011): 842–67.

31. For the data, see IUCN Red List of Threatened Species, "Summary Statistics," http://www.iucnredlist.org/about/summary-statistics; and Food and Agriculture Organization of the United Nations (FAO), "Facts and Figures," http://www.fao.org/soils-portal /soil-biodiversity/facts-and-figures/en.

32. Wilson, *Biodiversity*, 3. See also Natalie Goldstein, *Biodiversity* (New York: Facts on File, 2011), 29–52.

33. Worster, *Nature's Economy*, 417–19.

34. Mikael Stenmark, *Environmental Ethics and Policy Making* (Aldershot, UK: Ashgate, 2002), 57ff.

35. Takacs, *The Idea of Biodiversity*, 69–75. The Society for Conservation Biology founded in 1985 in the United States has played a significant role in this change of perspective.

36. Wilson, *Biodiversity*, vi.

37. Nikos Nikisianis and George P. Stamou, "Quantifying Nature: Ideological Representations and the Concept of Diversity," *History and Philosophy of the Life Sciences* 33 (2011): 366. See also John McCormick, *Reclaiming Paradise: The Global Environmental Movement* (Bloomington: Indiana University Press, 1989), 47–68.

38. Charles Perring, *Our Uncommon Heritage: Biodiversity Change, Ecosystem Services, and Human Wellbeing* (Cambridge: Cambridge University Press, 2014), 25.

39. Wilson, *The Diversity of Life*, 311, 320.

40. Robin Kundis Craig, "Towards a Notion of Environmental Bioethics," *Western New England Law Review* 23 (2001): 177. Ecosystem services provide, for example, purification of water and air, regulation of climate, and decomposition of waste.

41. Sometimes, however, Potter approaches the environment simply as an object of interaction, as mere physical or natural environment, for example, in studying the genetic basis for differential responses to environmental exposure. See Van Rensselaer Potter, *Bioethics: Bridge to the Future* (Englewood Cliffs, NJ: Prentice-Hall, 1971), 19–22, 105ff., 129 ff.

42. Susan Flader argues that Leopold was fascinated with the concept of diversity (*Thinking like a Mountain*, 51). See also Meine, *Aldo Leopold*, 405.

43. Aldo Leopold, *A Sand County Almanac* (New York: Ballantine Books, 1970), 240.

282 Notes to Pages 28–32

44. Leopold, *A Sand County Almanac*, 251.

45. Robin Attfield, *Environmental Ethics: An Overview for the Twenty-First Century* (Cambridge: Polity, 2003), 1–2.

46. Interconnectedness was one of Potter's characteristics of bioethics; see chapter 1. See also Henk ten Have, *Global Bioethics: An Introduction* (London: Routledge, 2016).

47. Potter makes a distinction between cultural pluralism and cultural laissez-faire. Cultures that do not promote human survival should be avoided: "diversity should not be permitted to include life-styles that lead to irreversible damage to the natural world on which we all depend." See Van Rensselaer Potter, "The Ethics of Nature and Nurture," *Zygon* 8 (1973): 46.

48. Worster's *Nature's Economy* is based on the interpretation of ecology as economy of nature. See also Nikisianis and Stamou, "Quantifying Nature," 365–88; Bramwell, *Ecology in the 20th Century*.

49. Worster, *Nature's Economy*, 313 and 342ff.

50. For example, the reintroduction of the beaver in Denmark or black-footed ferrets in Colorado. See Christian Gamborg and Peter Sandøe, "Beavers and Biodiversity: The Ethics of Ecological Restoration," in *Philosophy and Biodiversity*, ed. Markku Oksanen and Juhani Pietarinen (Cambridge: Cambridge University Press, 2004), 217–36; Braverman, *Wild Life*, 119–43.

51. James K. Boyce, "Rethinking Extinction: Toward a Less Gloomy Environmentalism," *Harper's Magazine*, November 2015, 67–75.

52. John Dryzek uses the expression "Promethean environmentalism" to refer to confidence in the ability of humans and their technologies to overcome environmental problems. See Dryzek, *The Politics of the Earth: Environmental Discourses* (Oxford: Oxford University Press, 2013), 52–72.

53. McKibben mentions a new category of "planet managers." See Bill McKibben, *The End of Nature* (New York: Random House, 2006), 136.

54. McKibben, *The End of Nature*, xviii.

55. Braverman, *Wild Life*, 98.

56. Altieri concludes that given this history, biotechnology "holds more promise for environmental harm." Miguel A. Altieri, "Industrial Agriculture Is One of the Main Threats to Biodiversity," in *Current Controversies: Biodiversity*, ed. Debra A. Miller (Farmington Hills, MI: Greenhaven Press, 2008), 112.

57. United Nations, World Commission on Environment and Development, *Our Common Future*, 1987, 38, http://www.un-documents.net/our-common-future.pdf.

58. Dryzek, *The Politics of the Earth*, 38, 42.

59. Perring, *Our Uncommon Heritage*, 301.

60. Robert Costanza, Ralph d'Arge, Rudolf de Groot, et al. "The Value of the World's Ecosystem Services and Natural Capital," *Nature* 387 (1997): 253–60. The UN Millennium Ecosystem Assessment, launched in 2001, distinguishes three types of services (1) provisioning services (food and water); (b) regulating services (climate and disease); (c) cultural services (spiritual and educational). Millennium Ecosystem Assessment, *Ecosystems and Human Well-Being: A Framework for Assessment* (New York: United Nations, 2003), 49–70.

61. Wilson, *Biodiversity*, 326.

62. Perring argues that intrinsic value cannot guide human decisions (*Our Uncommon Heritage*, 109). For economic rationalism, see Dryzek, *The Politics of the Earth*, 122–44.

63. Noel Castree, "Green Development Offers a Way to Save Biodiversity," in *Current Controversies: Biodiversity*, ed. Debra A. Miller (Farmington Hills, MI: Greenhaven Press, 2008), 139.

64. Craig, "Towards a Notion of Environmental Bioethics," 183.

65. Wilson, *Biodiversity*, 16.

66. Takacs, *The Idea of Biodiversity*, 9.

67. Dirk Lanzerath, "Biodiversity as an Ethical Concept: An Introduction," in *Concepts and Values in Biodiversity*, ed. Dirk Lanzerath and Minou Friele (London: Routledge, 2014), 1–15. See also Haila, "Making the Biodiversity Crisis Tractable," 54–82.

68. For a discussion of the different types of value, see Gaston and Spicer, *Biodiversity*, 91–105. See also Perlman and Adelson, *Biodiversity*, 37–51; Julia Koricheva and Helena Siipi, "The Phenomenon of Biodiversity," in *Philosophy and Biodiversity*, ed. Markku Oksanen and Juhani Pietarinen (Cambridge: Cambridge University Press, 2004), 38–42.

69. Perlman and Adelson, *Biodiversity*, 1–17.

70. Wilson, *The Diversity of Life*, 351.

71. Paul M. Wood, *Biodiversity and Democracy: Rethinking Society and Nature* (Vancouver: UBC Press, 2000), xv, 41, 72. See also Catherine Larrère, "Biodiversity: Common Good or Common World?," *International Social Science Journal* 64 (2013): 125–33.

72. See David Tilman, "Causes, Consequences and Ethics of Biodiversity," *Nature* 405 (2000): 208–11. For a critical review, see Thompson, *Do We Need Pandas?*

73. Shiva et al., *Biodiversity*; José Roque Junges and Lucilda Selli, "Bioethics and Environment: A Hermeneutic Approach," *Journal International de Bioéthique* 19 (2008): 105–19.

74. Kevin J. O'Brien, *An Ethics of Biodiversity: Christianity, Ecology, and the Variety of Life* (Washington, DC: Georgetown University Press, 2010), 20ff.

75. See Lawrence S. Hamilton, ed., *Ethics, Religion and Biodiversity: Relations between Conservation and Cultural Values* (Cambridge: White Horse Press, 1993); O'Brien, *An Ethics of Biodiversity*.

76. See McKibben, *The End of Nature*.

77. Holmes Rolston, "The Future of Environmental Ethics," *Teaching Ethics* 8 (2007): 22.

78. Tilman, "Causes, Consequences and Ethics of Biodiversity," 210–11.

79. For elaboration, see ten Have, *Global Bioethics*.

80. Ramachandra Guha, *Environmentalism: A Global History* (New York: Longman, 2000), 98–124. See also Marco Armiero and Lise Sedrez, eds., *A History of Environmentalism: Local Struggles, Global Histories* (London: Bloomsbury Academic, 2014).

81. Wilson, *Biodiversity*, 3.

82. Perlman and Adelson, *Biodiversity*, 139ff., 158.

83. This is true in two ways. First, the value of biodiversity does not depend on individual subjects attributing value to some species rather than others. Second, the value of biodiversity does not depend on individual specimens of species that are valued (instrumentally or intrinsically) but of the species as a whole category.

84. Rosa, "The Bioethics of Biodiversity," 169.

85. See Shiva et al., *Biodiversity*.

86. These interconnections are analyzed by Elizabeth Bird: "Environmental problems represent situations in which some segments of society engage in practices that adversely affect other members of society and have the potential to injure the future quality and survivability of the planet." See Bird, "The Social Construction of Nature: Theoretical Approaches to the History of Environmental Problems," *Environmental Review* 11 (1987): 261.

87. Larry Lohmann, "Who Defends Biological Diversity? Conservation Strategies and the Case of Thailand," in *Biodiversity: Social and Ecological Perspectives*, ed. Vandana Shiva et al. (London: Zed Books; Penang: World Rainforest Movement, 1991), 77ff.

88. Andrew Brennan, ed., *The Ethics of the Environment* (Aldershot, UK: Dartmouth, 1995), xviii–xxii. For Leopold, this is inherent in the idea of being "a member of a community of interdependent parts," the holistic approach and interdependency promoted by the ecological perspective (*A Sand County Almanac*, 239).

89. See James Dwyer, "How to Connect Bioethics and Environmental Ethics: Health, Sustainability, and Justice," *Bioethics* 23 (2009): 497–502.

90. Earth Overshoot Day, July 31, 2017, http://www.overshootday.org/newsroom/press -release-english-2017-calculator.

91. See Henk ten Have, *Vulnerability: Challenging Bioethics* (London: Routledge, 2016).

92. The basic problem, according to Elizabeth Bird, is not so much our relations with nature as it is the relations among people. The question is "how our social ethics in relation to nature have translated into environmental problems" (Bird, "The Social Construction of Nature," 262).

93. See Ivan Illich, *Tools for Conviviality* (New York: Harper and Row, 1973).

94. See Leonardo Boff, *Ecology and Liberation: A New Paradigm* (Maryknoll, NY: Orbis Books, 1995).

95. Also "creation care" in the perspective of ecotheology (Dryzek, *The Politics of the Earth*, 194). Moreno emphasizes the "need to care for a fragile and increasingly ailing planet." (Jonathan D. Moreno, "In the Wake of Katrina: Has 'Bioethics' Failed?," *American Journal of Bioethics* 5 [2005]: W19).

96. For examples from China, Micronesia, Solomon Islands, and Hawaii, see Hamilton, *Ethics, Religion and Biodiversity*.

97. Dryzek, *The Politics of the Earth*, 191.

98. Dryzek, *The Politics of the Earth*, 159.

99. See Larrère, "Biodiversity: Common Good or Common World?," 125–33.

100. See Jedediah Purdy, *After Nature: A Politics for the Anthropocene* (Cambridge, MA: Harvard University Press, 2015).

101. This argument is made by Tony L. Goldberg and Jonathan A. Patz, "The Need for a Global Health Ethic," *Lancet* 386 (2015): e37–e39.

102. Resnik and Roman argue that bioethicists should not address social and economic issues; they "will waste their time . . . and anger politicians and the public." See David B. Resnik and Gerard Roman, "Health, Justice, and the Environment," *Bioethics* 21 (2007): 239.

103. Purdy, *After Nature*, 267.

104. Peter Stoett has called this separation "bioapartheid." See Stoett, *Global Ecopolitic: Crisis, Governance, and Justice* (Toronto: University of Toronto Press, 2012), 190.

105. "We have created an unjust global economic system that favours a small, wealthy elite over the many who have so little." See Richard Horton, Robert Beaglehole, Ruth Bonita, John Raeburn, Martin McKee, and Stig Wall, "From Public Health to Planetary Health: A Manifesto," *Lancet* 383 (2014): 847.

106. United Nations, *The Millennium Development Goals Report 2015*, http://www .un.org/millenniumgoals/2015_MDG_Report/pdf/MDG%202015%20rev%20 (July%201).pdf.

CHAPTER THREE: Health

1. Hippocrates, *Airs Waters Places*, in *Hippocrates*, ed. W. H. S. Jones (London: William Heineman; Cambridge, MA: Harvard University Press, 1972), 1:71–72.

2. The term *meteorological medicine* was introduced by Karl Deichgräber, "Die Stellung der griechischen Ärztes zur Natur," *Die Antike* 15 (1939): 116. See also Ludwig von Brunn, "Hippokrates und die meteorologische Medizin," *Gesnerus* 3 (1946): 151–73, and *Gesnerus* 4 (1947): 1–18, 65–84.

3. These four constituents, or humors, of humans are blood, phlegm, yellow bile, and black bile, corresponding to air, water, fire, and earth, respectively. See E. D. Phillips, *Greek Medicine* (London: Thames and Hudson, 1973), 48–58. See also Owsei Temkin, *Galenism: Rise and Decline of a Medical Philosophy* (Ithaca, NY: Cornell University Press, 1973).

4. Ludwik Fleck introduced the idea of prescientific thought-styles. See Fleck, *Entstehung und Entwicklung einer wissenschaftlichen Tatsache: Einführung in die Lehre vom Denkstil und Denkkollektiv* (Frankfurt am Main: Suhrkamp Verlag, 1935).

5. Henk A.M.J. ten Have, "Knowledge and Practice in European Medicine: The Case of Infectious Diseases," in *The Growth of Medical Knowledge*, ed. Henk A.M.J. ten Have, Gerrit K. Kimsma, and Stuart F. Spicker (Dordrecht: Kluwer Academic Publishers, 1990), 18–25. See also Carlo M. Cipolla, *Miasmas and Disease: Public Health and the Environment in the Pre-industrial Age* (New Haven, CT: Yale University Press, 1992).

6. James C. Riley uses the term *environmentalism* for this eighteenth-century movement to make the environment healthier. See Riley, *The Eighteenth-Century Campaign to Avoid Disease* (Houndmills, UK: Macmillan, 1987), xv.

7. See R. J. Morris, *Cholera 1832: The Social Response to an Epidemic* (London: Croom Helm, 1976); Margaret Pelling, *Cholera, Fever and English Medicine 1825–1865* (Oxford: Oxford University Press, 1978); Pamela K. Gilbert, *Cholera and Nation: Doctoring the Social Body in Victorian England* (Albany: State University of New York Press, 2008).

8. ten Have, "Knowledge and Practice in European Medicine," 28–29.

9. Mark D. Hardt, *History of Infectious Disease Pandemics in Urban Societies* (Lanham, MD: Lexington Books, 2016).

10. ten Have, "Knowledge and Practice in European Medicine," 31. This movement was particularly strong in England. See G. Melwyn Howe, *Man, Environment and Disease in Britain: A Medical Geography of Britain through the Ages* (Harmondsworth, UK: Penguin Books, 1972), 193ff. The leader of the movement in Britain was Edwin Chadwick (see S. E. Finer, *The Life and Times of Sir Edwin Chadwick* [New York: Barnes & Noble, 1970]). See also René Dubos, *Man, Medicine, and Environment* (Harmondsworth, UK: Penguin Books, 1968).

11. Howe, *Man, Environment and Disease in Britain*, 167–92. See also René Dubos, *Mirage of Health: Utopias, Progress, and Biological Change* (New York: Harper and Row, 1959).

12. Virchow cited in George Rosen, "What Is Social Medicine? A Genetic Analysis of the Concept," in *From Medical Police to Social Medicine: Essays on the History of Health Care*, ed. George Rosen (New York: Science History Publications, 1974), 62.

13. See Francesca Grifo and Joshua Rosenthal, *Biodiversity and Human Health* (Washington, DC: Island Press, 1997); Jessica Pierce and Andrew Jameton, *The Ethics of Environmentally Responsible Health Care* (Oxford: Oxford University Press, 2004), 8–25; Aaron S. Bernstein and David S. Ludwig, "The Importance of Biodiversity to Medicine," *JAMA* 300 (2008): 2297–99; Osvaldo E. Sala, Laura A. Meyerson, and Camille Parmesan, eds., *Biodiversity Change and Human Health: From Ecosystem Services to Spread of Disease* (Washington, DC: Island Press, 2009).

14. See, for this example, Nils Eldredge, *Life in the Balance: Humanity and the Biodiversity Crisis* (Princeton, NJ: Princeton University Press, 1998), 169–70.

15. David B. Resnik, "Human Health and the Environment: In Harmony or in Conflict?," *Health Care Analysis* 17 (2009): 272–75; David B. Resnik, *Environmental Health Ethics* (Cambridge: Cambridge University Press, 2012), 90.

16. Hagai Levine, Niels Jørgensen, Anderson Martino Andrade, Jaime Mendiola, Dan Weksler-Derri, Irina Mindlis, Rachel Pinotti, and Shanna H. Swan, "Temporal Trends in Sperm Count: A Systematic Review and Meta-regression Analysis," *Human Reproduction Update* 23 (2017): 1–14; https://doi.org/10.1093/humupd/dmx022.

17. See, in particular, Eric Chivian and Aaron Bernstein, eds., *Sustaining Life: How Human Health Depends on Biodiversity* (Oxford: Oxford University Press, 2008).

18. Contact with nature not only reduces stress but also aids the healing process and enhances recovery from illness. See Stephen R. Kellert, "Biodiversity, Quality of Life, and Evolutionary Psychology," in *Biodiversity Change and Human Health: From Ecosystem Services to Spread of Disease*, ed. Osvaldo E. Sala, Laura A. Meyerson and Camille Parmesan (Washington, DC: Island Press, 2009), 104–7.

19. American Medical Association, Human Health and the Protection of Biodiversity, H-135.955, resolution presented at the AMA annual meetings, 1995, Washington, DC.

20. Annette Prüss-Üstün and Carlos Corvalán, *Preventing Disease through Healthy Environments* (Geneva: World Health Organization, 2006), 10.

21. WHO and Secretariat of the Convention on Biological Diversity, *Connecting Global Priorities: Biodiversity and Human Health, a State of Knowledge Review* (Geneva: World Health Organization, 2015), 17.

22. See Robert Beaglehole and Ruth Bonita, "What Is Global Health?," *Global Health Action* 3 (2010): 5142; Jeffrey P. Koplan et al., "Toward a Common Definition of Global Health," *Lancet* 373 (2009): 1993–95; Sarah B. Macfarlane, Marian Jacobs, and Ephata E. Kaaya, "In the Name of Global Health: Trends in Academic Institutions," *Journal of Public Health Policy* 29 (2008): 383–401.

23. The motivation, however, is self-interested. The initiative is necessary because "the health and stability of countries around the world have a direct impact on the security and prosperity of the United States" (US Global Health Initiative, *Implementation of the Global*

Health Initiative: Consultation Document, 2010, https://reliefweb.int/sites/reliefweb.int/files
/resources/2760E0ED9BDB5FB9492576C4001C08CD-govt.usa-feb2010.pdf).

24. See Theodore M. Brown and Elizabeth Fee, "The World Health Organization and
the Transition from International to Global Public Health," *American Journal of Public
Health* 96 (2006): 62–72. Contemporary global health is actually rooted in the tradition
of colonial medicine. See Randall M. Packard, *A History of Global Health: Interventions into
the Lives of Other Peoples* (Baltimore, MD: Johns Hopkins University Press, 2016).

25. One of the early definitions of global health is given by the Institute of Medicine
in the United States: "health problems, issues, and concerns that transcend national
boundaries, and may best be addressed by cooperative actions, represent what is encom-
passed . . . by the term 'global health.'" IOM, *America's Vital Interest in Global Health* (Washington, DC: National Academy Press, 1997), 2. See also Koplan et al., "Toward a Common
Definition of Global Health," 1993.

26. See Henk ten Have, *Vulnerability: Challenging Bioethics* (London: Routledge, 2016).

27. Michael Jeffries, *Biodiversity and Conservation* (London: Routledge, 2006), 144. See
also Chivian and Bernstein, *Sustaining Life*, 55.

28. CSDH, *Closing the Gap in a Generation: Health Equity through Action on the Social
Determinants of Health; Final Report of the Commission on Social Determinants of Health*
(Geneva: World Health Organization, 2008).

29. CSDH, *Closing the Gap in a Generation*, 2, 3, 49.

30. CSDH, *Closing the Gap in a Generation*, 60–71. See also Michael Marmot and Richard G. Wilkinson, *Social Determinants of Health*, 2nd ed. (Oxford: Oxford University
Press, 2006). This book includes discussions of pollution, waste, vulnerability, green urban
spaces, and access to water as determinants of health.

31. WHO, Declaration of Alma-Ata, 1978, http://www.who.int/publications/almaata
_declaration_en.pdf.

32. Marc Lalonde, *A New Perspective on the Health of Canadians* (Ottawa, 1974), 11,
31–34.

33. Thomas McKeown, *The Role of Medicine* (London: Nuffield Provincial Hospitals
Trust, 1976).

34. Kelley Lee, *The World Health Organization (WHO)* (London: Routledge, 2009), 71–
98, 126–28.

35. Packard has called this approach the "medicalization of global health" (*A History
of Global Health*, 305ff.).

36. Michael Marmot, *The Health Gap: The Challenge of an Unequal World* (London:
Bloomsbury, 2015).

37. Louise Potvin, "Promoting Health and Equity: A Theme That Is More Relevant
Than Ever," *Global Health Promotion* 22 (2015): 3–5. See Vicente Navarro, "What We Mean
by Social Determinants of Health," *Global Health Promotion* 16 (2009): 5–16.

38. Social policy no longer works as social vaccine, protecting society against adver-
sity, according to Jalil Safaei, "Social Policy as Social Vaccine," *Social Medicine* 8 (2014):
11–22. For the negative effects of austerity politics, see Armin Schäfer and Wolfgang
Streeck, eds., *Politics in the Age of Austerity* (Cambridge: Polity Press, 2013); Salmaan Ke-
shavjee, *Blind Spot: How Neoliberalism Infiltrated Global Health* (Oakland: University of

California Press, 2014); David Stuckler and Sanjay Basu, *The Body Economic: Why Austerity Kills* (New York: Basic Books, 2013).

39. WHO, Constitution of the World Health Organization, 1946, http://www.who.int /governance/eb/who_constitution_en.pdf.

40. United Nations General Assembly, International Covenant on Economic, Social and Cultural Rights, article 12(b). 1966, http://www.ohchr.org/Documents/Professional Interest/cescr.pdf.

41. Lawrence Gostin, *Global Health Law* (Cambridge, MA: Harvard University Press, 2014), 20–31.

42. Heymann et al. found that globally 69 countries (36%) guarantee the right to health (having it included in their constitutions). See Jody Heymann, Adèle Cassola, Amy Raub, and Lipi Mishra, "Constitutional Rights to Health, Public Health and Medical Care: The Status of Health Protections in 191 Countries," *Global Public Health* 8 (2013): 639–53.

43. Henk ten Have, *Global Bioethics* (London: Routledge, 2016) 105, 152.

44. The connection between human rights and social determinants of health is discussed by Audrey Chapman, "The Social Determinants of Health, Health Equity, and Human Rights," *Health and Human Rights* 12 (2010): 17–30; see also Paula Braveman, "Social Conditions, Health Equity, and Human Rights," *Health and Human Rights* 12 (2010): 31–48.

45. Alison Katz, "Prospects for a Genuine Revival of Primary Health Care—Through the Visible Hand of Social Justice Rather Than the Invisible Hand of the Market: Part I," *International Journal of Health Services* 39 (2009): 569. See Anne-Emanuelle Birn, "Making It Politic(al): Closing the Gap in a Generation: Health Equity through Action on the Social Determinants of Health," *Social Medicine* 4 (2009): 166–82.

46. See Marmot, *The Health Gap.*

47. Daniel Goldberg argues that the moral concern with health inequities is in fact a long tradition in healthcare, dating back at least to the times of Virchow. See Goldberg, "Global Health Care Is Not Global Health: Populations, Inequities, and Law as a Social Determinant of Health," in *The Globalization of Health Care: Legal and Ethical Issues*, ed. I. Glenn Cohen (Oxford: Oxford University Press, 2013), 403–20.

48. Planetary health is defined as "the health of human civilisation and the state of the natural systems on which it depends" (Sarah Whitmee, Andy Haines, Chris Beyrer, et al., "Safeguarding Human Health in the Anthropocene Epoch: Report of the Rockefeller Foundation–*Lancet* Commission on Planetary Health," *Lancet* 386 (2015): 1978. See also Richard Horton, Robert Beaglehole, Ruth Bonita, John Raeburn, Martin McKee, and Stig Wall, "From Public to Planetary Health: A Manifesto," *Lancet* 383 (2014): 847.

49. Aldo Leopold, *A Sand County Almanac* (New York: Ballantine Books, 1970), 272.

50. Jeffries, *Biodiversity and Conservation*, 79; Ulrich Heink and Kurt Jax, "Framing Biodiversity: The Case of 'Invasive Alien Species,'" in *Concepts and Values in Biodiversity*, ed. Dirk Lanzerath and Minou Friele (London: Routledge, 2014), 81. See also Gregg Mitman, "In Search of Health: Landscape and Disease in American Environmental History," *Environmental History* 10 (2005): 184–210; Fikret Berkes, Nancy C. Doubleday, and Graeme S. Cumming, "Aldo Leopold's Land Health from a Resilience Point of View: Self-Renewal Capacity of Social-Ecological Systems," *EcoHealth* 9 (2012): 278–87.

51. Alessandro R. Demaio and Johan Rockström, "Human and Planetary Health: Towards a Common Language," *Lancet* 386 (2015): e36–e37.

52. UN, *Sustainable Development Goals*, 2015, http://www.un.org/sustainabledevelop ment/sustainable-development-goals.

53. ten Have, *Global Bioethics*, 138–56.

54. Satesh Bidaisee and Calum N. L. Macpherson, "Zoonoses and One Health: A Review of the Literature," *Journal of Parasitology Research* (2014): https://doi.org/10.1155/2014 /874345; FAO, OIE,WHO, UN System Influenza Coordination, UNICEF, World Bank, *Contributing to One World, One Health: A Systematic Framework for Reducing Risks of Infectious Diseases at the Animal-Human-Ecosystems Interface*, consultation document, October 2008, http://www.fao.org/docrep/pdf/011/aj137e/aj137e00.pdf.

55. Laura H. Kahn, *One Health and the Politics of Antimicrobial Resistance* (Baltimore, MD: Johns Hopkins University Press, 2016).

56. Anne-Emmanuelle Birn, "Addressing the Societal Determinants of Health: The Key Global Health Ethics Imperative of Our Times," in *Global Health and Global Health Ethics*, ed. Solomon Benatar and Gillian Brock (Cambridge: Cambridge University Press, 2011), 38.

57. Whitmee, Haines, Beyrer, et al., "Safeguarding Human Health in the Anthropocene Epoch," 1974, 2011–12, 2018.

58. Gemma Carey and Brad Crammond, "Action on the Social Determinants of Health: Views from inside the Policy Process," *Social Science & Medicine* 128 (2015): 139.

59. ten Have, *Global Bioethics*, 93–112.

60. This is clearly not acceptable for at least some bioethicists. David Resnik and Gerard Roman, for example, argue that an expanded scope of bioethics is necessary but at the same time advocate restricted action because of pragmatic reasons; addressing poverty and inequality is "unwise"; bioethicists "will waste their time . . . and anger politicians and the public." Resnik and Roman, "Health, Justice, and the Environment," *Bioethics* 21 (2007): 239.

61. Alastair Greig, David Hulme, and Mark Turner, *Challenging Global Inequality: Development Theory and Practice in the 21st Century* (Houndmills, UK: Palgrave Macmillan, 2007), 134. In the 1990s, more children died of diarrhea and sanitation-related disease than all the people who died in armed conflict since 1945 (ibid., 141).

62. See Robert Hunter Wade, "Winners and Losers," *Economist* 359 (2001): 93–97; Robert Hunter Wade, "Is Globalization Reducing Poverty and Inequality?," *World Development* 32 (2004): 567–89. See also Joseph E. Stiglitz, *The Price of Inequality: How Today's Divided Society Endangers Our Future* (New York: W.W. Norton, 2012).

63. Oxfam, *Working for the Few: Political Capture and Economic Inequality* (Oxford: Oxfam International, 2014), 2.

64. Feeding America, "Hunger and Poverty Facts and Statistics," 2014, http://www .feedingamerica.org/hunger-in-america/impact-of-hunger/hunger-and-poverty/hunger -and-poverty-fact-sheet.html?referrer=https://www.google.com.

65. Stiglitz, *The Price of Inequality*, 13.

66. Michael Marmot, *Fair Society, Healthy Lives* (2010), 10, Marmot Review, http:// www.instituteofhealthequity.org/resources-reports/fair-society-healthy-lives-the

-marmot-review/fair-society-healthy-lives-full-report-pdf.pdf. For the connection between health, wealth, and inequality, see Angus Deaton, *The Great Escape: Health, Wealth, and the Origins of Inequality* (Princeton, NJ: Princeton University Press, 2013).

67. Sharon Friel, Michael Marmot, Anthony J. McMichael, Tord Kjellstrom, and Denny Vägerö, "Global Health Equity and Climate Stabilisation: A Common Agenda," *Lancet* 372 (2008): 1677–83.

68. Greig, Hulme, and Turner, *Challenging Global Inequality*, 189.

69. Stiglitz summarizes the policy choices involved: "We created for the banks . . . a much stronger safety net than we created for poor Americans" (*The Price of Inequality*, 74). The economic myths of neoliberalism are deconstructed by, among others, Thomas Piketty, *Capital in the Twenty-First Century* (Cambridge, MA: Belknap Press, 2014); Anthony B. Atkinson, *Inequality: What Can Be Done?* (Cambridge, MA: Harvard University Press, 2015); and John Rapley, *Globalization and Inequality: Neoliberalism's Downward Spiral* (Boulder, CO: Lynne Rienner Publishers, 2004).

70. ten Have, *Global Bioethics*, 61ff.

71. See, for example, Marina Karanikolos, Philipa Mladovsky, Jonathan Cylus, Sarah Thompson, Sanjay Basu, David Stuckler, Johan P. Mackenbach, and Martin McKee, "Financial Crisis, Austerity, and Health in Europe," *Lancet* 381 (2013): 1323–31.

72. CSDH, *Closing the Gap in a Generation*, 26. See also Marmot, *Fair Society, Healthy Lives*, 10.

73. Resnik and Roman have argued that justice is the most important principle if health and environment are connected. David Resnik and Gerard Roman, "Health, Justice, and the Environment," *Bioethics* 21 (2007): 230–41.

74. Robert Veatch, *Hippocratic, Religious, and Secular Ethics: The Points of Conflict* (Washington, DC: Georgetown University Press, 2012), 159ff.

75. See Katherine Boo, *Behind the Beautiful Forevers: Life, Death, and Hope in a Mumbai Undercity* (New York: Random House, 2012).

76. M. Whitehead, "The Concepts and Principles of Equity and Health," *International Journal of Health Services* 22 (1992): 429–45.

77. Marmot, *Fair Society, Healthy Lives*, 3.

78. Vandana Shiva has called it "environmental apartheid"; a more common term is "environmental injustice." Shiva, "The World on the Edge," in *On the Edge: Living with Global Capitalism*, ed. Will Hutton and Anthony Giddens (London: Jonathan Cape, 2000), 112–13.

79. Greig, Hulme, and Turner, *Challenging Global Inequality*, 189.

80. Dale Jamieson argues that "justice is both conceptually and historically at the heart of environmentalism." Jamieson, "Justice: The Heart of Environmentalism," in *Environmental Justice and Environmentalism: The Social Justice Challenge to the Environmental Movement*, ed. Ronald Sandler and Phaedra C. Pezzullo (Cambridge, MA: MIT Press, 2007), 86.

81. See ten Have, *Vulnerability*. See also Tim Brown, "'Vulnerability Is Universal': Considering the Place of 'Security' and 'Vulnerability' within Contemporary Global Health Discourse," *Social Science & Medicine* 72 (2011): 319–26.

82. G. J. Teunissen, M. A. Visser, and T. A. Abma, "Struggling between Strength and Vulnerability: A Patients' Counter Story," *Health Care Analysis* 23 (2015): 288–305.

83. Sandra Diaz, Joseph Fargione, F. Stuart Chapin III, David Tilman, "Biodiversity Loss Threatens Human Well-Being," *PLOS Biology* 4 (2006): 1302.

84. Madison Powers and Ruth Faden, *Social Justice: The Moral Foundations of Public Health and Health Policy* (Oxford: Oxford University Press, 2006), 71–79.

85. Craig R. Janes and Kitty K. Corbett, "Anthropology and Global Health," *Annual Review of Anthropology* 38 (2009): 169.

86. Ignacio Ramonet, *Wars of the 21st Century: New Threats, New Fears* (Melbourne: Ocean Press, 2004), 127. See also Whitehead, "The Concepts and Principles of Equity and Health," 429–45; Chapman, "The Social Determinants of Health, Health Equity, and Human Rights," 17–30.

87. Michael Soulé, "What Is Conservation Biology?," *BioScience* 35 (1985): 730.

88. Peter Wood, *Diversity: The Invention of a Concept* (San Francisco: Encounter Books, 2003).

89. Lanzerath and Friele, *Concepts and Values in Biodiversity*, 1–15.

90. United Nations, Convention on Biological Diversity, 1992, preamble, https://www.cbd.int/doc/legal/cbd-en.pdf.

91. Wood, *Diversity*, 13.

92. Walter Benn Michaels, *The Trouble with Diversity and How We Learned to Love Identity and Ignore Inequality* (New York: Henry Holt, 2006), 199. The emphasis on cultural diversity focuses on the image and our feelings regarding a phenomenon rather than on real, material difference. "What the commitment to diversity seeks is not a society in which there are no poor people but one in which there's nothing wrong with being poor, a society in which poor people—like blacks and Jews and Asians—are respected." The problem is prejudice not poverty in this view. See ibid., 109.

93. UNESCO, Universal Declaration on Bioethics and Human Rights (Paris: UNESCO, 2005), article 12.

94. United Nations, Declaration of the United Nations Conference on the Human Environment Stockholm, 1972, principle 1, http://www.unep.org/documents.multilingual/default.asp?documentid=97&articleid=1503.

95. United Nations, Convention on Biological Diversity, 1992, preamble, https://www.cbd.int/convention/articles/default.shtml?a=cbd-00; United Nations, Rio Declaration on Environment and Development, 1992, principle 3, http://www.unesco.org/education/pdf/RIO_E.PDF.

96. United Nations Educational, Scientific and Cultural Organization, Declaration on the Responsibilities of the Present Generations towards Future Generations (Paris: UNESCO, 1997), http://portal.unesco.org/en/ev.php-URL_ID=13178&URL_DO=DO_TOPIC&URL_SECTION=201.html; Universal Declaration on Bioethics and Human Rights (Paris: UNESCO, 2005), article 16, http://portal.unesco.org/en/ev.php-URL_ID=31058&URL_DO=DO_TOPIC&URL_SECTION=201.html.

97. "Sustainable development is development that meets the needs of the present without compromising the ability of future generations to meet their own needs." UN World Commission on Environment and Development, *Our Common Future*, 1987, 41, http://www.un-documents.net/our-common-future.pdf.

98. See Emmanuel Agius, "Environmental Ethics: Towards an Intergenerational Perspective," in *Environmental Ethics and International Policy*, ed. Henk A.M.J. ten Have (Paris: UNESCO Publishing, 2006), 89–115.

99. See chapter 1 for the "blue marble" experience.

100. ten Have, *Global Bioethics*, 216–18.

101. Alain Supiot, ed., *La solidarité: Enquête sur un principle juridique* (Paris: Odile Jacob, 2015), 7–32; Bruce Jennings and Angus Dawson, "Solidarity in the Moral Imagination of Bioethics," *Hastings Center Report* 45 (2015): 31–38.

102. Organisation of African Unity, African Charter on Human and Peoples' Rights, Nairobi, 1981, articles 10, 23, and 29, http://www.humanrights.se/wp-content/uploads /2012/01/African-Charter-on-Human-and-Peoples-Rights.pdf.

103. For the present debate about new approaches toward the environment, see Daniel Doak, Victoria J. Bakker, Brice Evan Goldstein, and Benjamin Hale, "What Is the Future of Conservation?," *Trends in Ecology & Evolution* 29 (2014): 77–81; Peter Kareiva and Michelle Marvier, "What Is Conservation Science?," *BioScience* 62 (2012): 962–69.

104. "Health was an arena in which the assumed separation between the human and the nonhuman world broke down," as appropriately formulated by Linda Nash, *Inescapable Ecologies: A History of Environment, Disease, and Knowledge* (Berkeley: University of California Press, 2006), 50.

105. The richer biodiversity, the more languages. See Luisa Maffi and Ellen Woodley, *Biocultural Diversity Conservation: A Global Sourcebook* (London: Earthscan, 2010), 6–8; David Harmon, "Losing Species, Losing Languages: Connections between Biological and Linguistic Diversity," *Southwest Journal of Linguistics* 15 (1996): 89–108.

106. "Biocultural diversity comprises the diversity of life in all of its manifestations— biological, cultural, and linguistic—which are interrelated (and likely co-evolved) within a complex socio-ecological adaptive system" (Maffi and Woodley, *Biocultural Diversity Conservation*, 5).

107. Ricardo Rozzi, "Biocultural Ethics: Recovering the Vital Links between the Inhabitants, Their Habits, and Habitats," *Environmental Ethics* 34 (2012): 29.

108. Rozzi, "Biocultural Ethics," 32.

109. Rozzi, "Biocultural Ethics," 34.

110. It is estimated that 50% of all languages will disappear in the next hundred years. That means that every two weeks one language dies. See David Crystal, *Language Death* (Cambridge: Cambridge University Press, 2002), 19.

111. Maffi and Woodley, *Biocultural Diversity Conservation*, 19.

112. Philippe Descola argues that the dualism between nature and culture is typical for Western cultures and only emerged in the last third of the nineteenth century (Descola, *Beyond Nature and Culture* (Chicago: University of Chicago Press, 2013); Descola, *The Ecology of Others* (Chicago: Prickly Paradigm, 2013). Jeremy Davies argues that choosing "the Anthropocene" as the name of a new epoch demonstrates that human beings are not only biological but also geological agents; life and nonliving matter cannot be distinguished. Davies, *The Birth of the Anthropocene* (Oakland: University of California Press, 2016), 41–68.

113. See Harriet Ketley, "Cultural Diversity versus Biodiversity," *Adelaide Law Review* 16 (1994): 99–160.

114. Rozzi, "Biocultural Ethics," 39–41.

115. Rozzi uses as an example the development of the Ecuadorian shrimp-farming industry, which has destroyed the mangroves along the coast that provided the ecological

basis for the economy of local shellfish-collecting communities. As the mangroves disappeared, a whole community disappeared with them, because the vital links between the habitat and the habits of this community were eliminated. Rozzi, "Biocultural Ethics," 44–46.

116. Kareiva and Marvier, "What Is Conservation Science?," 962–69.

117. Paul R. Ehrlich and Anne H. Ehrlich, "The Value of Biodiversity," *Ambio* 21 (1992): 219–26.

118. For an overview of various values of biodiversity, see Kevin J. Gaston and John I. Spicer, *Biodiversity: An Introduction*, 2nd ed. (Malden, MA: Blackwell Publishing, 2004), 91–107. See also Holmes Rolston III, *Environmental Ethics: Duties to and Values in the Natural World* (Philadelphia, PA: Temple University Press, 1988).

119. Bryan G. Norton, *Why Preserve Natural Variety?* (Princeton, NJ: Princeton University Press, 1987); Sahotra Sarkar, *Biodiversity and Environmental Philosophy: An Introduction* (New York: Cambridge University Press, 2005).

120. Sarkar, *Biodiversity and Environmental Philosophy*, 80.

121. Sarkar provides examples of direct transformation, such as traveling through a rain forest resulting in efforts to protect tropical rain forests, as well as indirect transformation, such as Darwin's development of the theory of evolution after his experiences in Galápagos (*Biodiversity and Environmental Philosophy*, 82–87).

122. This view implies that economic approaches to biodiversity have limitations.

123. Lanzerath and Friele, *Concepts and Values in Biodiversity*, 6.

124. Marie-Hélène Parizeau, "Can Biodiversity Be a Universal Value?," *International Social Sciences Journal* 64 (2013): 139.

125. Marion Hourdequin, *Environmental Ethics: From Theory to Practice* (London: Bloomsbury Academic, 2015), 79–84. For a relational ethics perspective on the environment, see also Timothy William Lambert, Colin L. Soskoline, Vangie Bergum, James Howell, and John B. Dossetor, "Ethical Perspectives for Public and Environmental Health: Fostering Autonomy and the Right to Know," *Environmental Health Perspectives* 111 (2003): 133–37; Marion Hourdequin and David B. Wong, "A Relational Approach to Environmental Ethics," *Journal of Chinese Philosophy* 32 (2005): 19–33.

126. See Ursula K. Heise, *Sense of Place, and Sense of Planet: The Environmental Imagination of the Global* (Oxford: Oxford University Press, 2008); Marco Armiero and Lise Sedrez, eds., *A History of Environmentalism: Local Struggles, Global Histories* (London: Bloomsbury, 2014).

127. John Rodman, "Four Forms of Ecological Consciousness Reconsidered," in *Ethics and the Environment*, ed. Donald Scherer and Thomas Attig (Englewood Cliffs, NJ: Prentice-Hall, 1983), 82–92.

128. Howard Frumkin, "Healthy Places: Exploring the Evidence," *American Journal of Public Health* 93 (2003): 1451–56.

CHAPTER FOUR: **Disease**

1. Global Priorities Project, *Global Catastrophic Risks* (Stockholm: Global Challenges Foundation, 2016).

2. See chapter 2. Donald Worster, *Nature's Economy: A History of Ecological Ideas*, 2nd ed. (Cambridge: Cambridge University Press, 1994), 9, 19.

3. Annette Prüss-Üstün and Carlos Corvalán, *Preventing Disease through Healthy Environments: Towards an Estimate of the Environmental Burden of Disease* (Geneva: World Health Organization, 2006), 10.

4. United Nations, Convention on Biological Diversity, 1992, https://www.cbd.int/doc /legal/cbd-en.pdf. For a general overview of bioinvasion, see Daniel Simberloff, *Invasive Species: What Everyone Needs to Know* (Oxford: Oxford University Press, 2013).

5. L. Philip Lounibos, "Invasions by Insect Vectors of Human Disease," *Annual Review of Entomology* 47 (2002): 233–66.

6. Marianne Kettunen et al., *Technical Support to EU Strategy on Invasive Species* (Brussels: Institute for European Environmental Policy, 2008), 26ff.

7. European Union, "Regulation No. 1143/2014 of the European Parliament and of the Council of 22 October 2014 on the Prevention and Management of the Introduction and Spread of Invasive Alien Species," *Official Journal of the European Union*, November 4, 2014, L317/35.

8. Jeffrey R. Powell and Walter J. Tabachnick, "History of Domestication and Spread of *Aedes aegypti*: A Review," *Memorias do Instituto Oswaldo Cruz* 108, suppl. 1 (2013): 11–17.

9. Claus Emmeche, "Bioinvasion, Globalization, and the Contingency of Cultural and Biological Diversity: Some Ecosemiotic Observations," *Sign Systems Studies* 29 (2001): 237–61; Mark A. Davis et al., "Don't Judge Species on Their Origins," *Nature* 474 (2011): 153–54; Nigel Clark, "The Demon-Seed: Bioinvasion as the Unsettling of Environmental Cosmopolitanism," *Theory, Culture and Society* 19 (2002): 101–25; Timothy Lee Scott, *Invasive Plant Medicine: The Ecological Benefits and Healing Abilities of Invasives* (Rochester, NY: Healing Arts Press, 2010); Tao Orion, *Beyond the War on Invasive Species: A Permaculture Approach to Ecosystem Restoration* (White River Junction, VT: Chelsea Green Publishing, 2015); Fred Pearce, *The New Wild: Why Invasive Species Will Be Nature's Salvation* (Boston: Beacon Press, 2015).

10. Kettunen et al., *Technical Support to EU Strategy on Invasive Species*, 39.

11. Chris Bright, *Life out of Bounds: Bioinvasion in a Borderless World* (New York: W. W. Norton, 1998), 143–44. Simberloff gives many examples of species introduced as new food items. In his view, the main incentive for introducing non-native species is that "people are never happy with the species they have" (*Invasive Species*, 129).

12. Mark Williamson, *Biological Invasions* (London: Chapman and Hall, 1996). See also Martin A. Schlaepfer, Dov F. Sax, and Julian D. Olden, "The Potential Conservation Value of Non-native Species," *Conservation Biology* 25 (2011): 428–37.

13. As argued by Mark A. Davis, *Invasion Biology* (Oxford: Oxford University Press, 2009); Mark A. Davis et al., "Don't Judge Species on Their Origins," 153–54; Emma Marris, "The End of Invasion," *Nature* 459 (2009): 327–28.

14. Ned Hettinger, "Conceptualizing and Evaluating Non-native Species," *Nature Education Knowledge* 3 (2012): 7–12.

15. Mike Davis, *The Monster at Our Door: The Global Threat of Avian Flu* (New York: Henry Holt, 2006), 122; Juliet Jane Fall, "Biosecurity and Ecology: Beyond the Nativist Debate," in *Biosecurity: The Socio-politics of Invasive Species and Infectious Diseases*, ed. Kezia Barker, Andrew Dobson, and Sarah Taylor (Abingdon, UK: Earthscan/Routledge, 2014), 167–81.

16. Martina Carrete and José L. Tella, "Wild-Bird Trade and Exotic Invasions: A New Link of Conservation Concern?," *Frontiers in Ecology and the Environment* 6 (2008): 207–11; ENDCAP, *Wild Pets in the European Union*, 2012, http://endcap.eu/wp-content/uploads /2013/02/Report-Wild-Pets-in-the-European-Union.pdf.

17. Colin Edwards, *Managing and Controlling Invasive Rhododendron: Practice Guide* (Edinburgh: Forestry Commission, 2006).

18. Andrew Cockburn, "Weed Whackers: Monsanto, Glyphosate, and the War on Invasive Species," *Harper's Magazine*, September 2015, 58–59.

19. Charles M. Benbrook, "Trends in Glyphosate Herbicide Use in the United States and Globally," *Environmental Sciences Europe* 28 (2016): 3, https://doi.org/10.1186/s12302 -016-0070-0.

20. Scott, *Invasive Plant Medicine*, 67ff. See also Orion: *Beyond the War on Invasive Species*, 17ff.

21. Federal Interagency Committee for Management of Noxious and Exotic Weeds, ed., *Pulling Together: National Strategy for Invasive Plant Management*, 2nd ed. (Washington, DC: US Government Printing Office, 1998).

22. Cockburn, "Weed Whackers," 61.

23. A famous example is the introduction of the Nile perch in Lake Victoria. This created an ecological disaster that destroyed the native species. In the 1990s, many fishing families lost their jobs. Bilharzia and malaria became more frequent. Nile perch fishery was primarily for export, not meeting the needs of the local population. See Bright, *Life out of Bounds*, 86–92. Another example of unintended introduction is the outbreak of cholera in Peru in 1991 as a result of ballast discharge from ships from South Asia (ibid., 185).

24. Fred Pearce points out that the tropical forests in West Africa are not wilderness but the result of intensive transformations by human beings over thousands of years (*The New Wild*, 121ff.).

25. Pearce, *The New Wild*, 186ff.

26. Clark, "The Demon-Seed," 105.

27. Fall, "Biosecurity and Ecology," 173.

28. A classic reference here is the introduction of wild rabbits into Australia. In their native Spain, rabbits have never been a pest but rather a delicacy. In the 1950s, myxoma virus was used as biological control to reduce the Australian rabbit population. The rabbits developed genetic resistance, while the virus became less virulent (Simberloff, *Invasive Species*, 115–16). Simberloff argues that living organisms are fundamentally unpredictable because they move, adapt, and interact with other species (ibid., 145).

29. Mark S. Smolinski, Margaret A. Hamburg, Joshua Lederberg, and Institute of Medicine, eds., *Microbial Threats to Health: Emergence, Detection, and Response* (Washington, DC: National Academies Press, 2003).

30. WHO, "Leishmaniasis," fact sheet, March 2018, http://www.who.int/news-room /fact-sheets/detail/leishmaniasis; Rebecca Du, Peter J. Hotez, Waleed S. Al-Salem, Alvaro Aosta-Serrano, "Old World Cutaneous Leishmaniasis and Refugee Crisis in the Middle East and North Africa," *PLOS Neglected Tropical Diseases* 10 (2016): e0004545, https://doi .org/10.1371/journal.pntd.0004545.

31. DNA of the *Borrelia* bacterium was found in the mummy of a 5,300-year-old Neolithic "iceman." See Stephen S. Hall, "Unfrozen," *National Geographic* 220 (2011): 118–29, 131–33.

32. Barbour and Fish conclude that the emergence of Lyme disease is in part attributable to the "greening" of the United States. In the past, forests have been cleared to give way to agriculture. Now that agriculture is no longer profitable, forests are regaining the lands; citizens like to live close to nature and wildlife such as deer. Alan G. Barbour and Durland Fish, "The Biological and Social Phenomenon of Lyme Disease," *Science* 260 (1993): 1614. See also Brian F. Allan, Felicia Keesing, and Richard S. Ostfeld, "Effect of Forest Fragmentation on Lyme Disease Risk," *Conservation Biology* 17 (2003): 267–72.

33. Kate E. Jones et al., "Global Trends in Emerging Infectious Diseases," *Nature* 451 (2008): 990; Peter Daszak et al., "Conservation Medicine and a New Agenda for Emerging Diseases," *Annals of the New York Academy of Sciences* 1026 (2004): 1–11.

34. Boris A. Revich and Marina A Podolnaya, "Thawing of Permafrost May Disturb Historic Cattle Burial Grounds in East Siberia," *Global Health Action* 4 (2011): https://doi.org/10.3402/gha.v4i0.8482; Nick Visser, "Siberian Heatwave Sparks Anthrax Outbreak, Killing a Child and Thousands of Reindeer," *Huffington Post*, August 2, 2016.

35. Peter Daszak, Andrew A. Cunningham, and Alex D. Hyatt, "Emerging Infectious Diseases of Wildlife: Threats to Biodiversity and Human Health," *Science* 287 (2000): 443.

36. Montira J. Pongsiri et al., "Biodiversity Loss Affects Global Disease Ecology," *BioScience* 59 (2009): 945–54.

37. See Philip Alcabes, *Dread: How Fear and Fantasy Have Fueled Epidemics from the Black Death to Avian Flu* (New York: PublicAffairs, 2009).

38. Peter Washer, *Emerging Infectious Diseases and Society* (Houndmills, UK: Palgrave Macmillan, 2014), 114; see also Priscilla Wald, *Contagious: Cultures, Carriers, and the Outbreak Narrative* (Durham, NC: Duke University Press, 2008).

39. Bruce Braun, "Power over Life: Biosecurity as Biopolitics," in *Biosecurity: The Sociopolitics of Invasive Species and Infectious Diseases*, ed. Andrew Dobson, Kezia Barker, and Sarah L. Taylor (London: Routledge, 2013), 51.

40. Theresa Macphail, *The Viral Network: A Pathography of the H1N1 Influenza Pandemic* (Ithaca, NY: Cornell University Press, 2014), 127.

41. Since 1960, the number of aviation passengers has grown 9% per year. See A. J. Tatem, D. J. Rogers, and S. J. Hay, "Global Transport Networks and Infectious Disease Spread," *Advances in Parasitology* 62 (2006): 293. See also A. J. Tatem, S. L. Hay, and D. J. Rogers, "Global Traffic and Disease Vector Dispersal," *Proceedings of the National Academy of Sciences* 103 (2006): 6242–47.

42. That "the global is viral" is elaborated by Macphail, *The Viral Network*, 209ff.

43. WHO, *Emerging Diseases*. http://www.who.int/topics/emerging_diseases/en.

44. Henk ten Have, *Vulnerability: Challenging Bioethics* (London: Routledge, 2016), 29–32, 149ff. See also Carlo Caduff, "On the Verge of Death: Visions of Biological Vulnerability," *Annual Review of Anthropology* 43 (2014): 105–21.

45. See Ulrich Beck, *World Risk Society* (Cambridge: Polity Press, 2009).

46. Macphail, *The Viral Network*, 135.

47. Harvey V. Fineberg, "Pandemic Preparedness and Response: Lessons from the H1N1 Influenza of 2009," *New England Journal of Medicine* 370 (2014): 1341. The same point

is made by Alcabes, who argues that we cannot prepare for what is unlikely and unfore-seeable: "It is a waste of time and energy to prepare for what we can't foresee, and it is a lie to pretend, as the biopreparedness promoters have, that we can see what we cannot." Health officials have become "soothsayers, imagining nonevents as real threats" (Al-cabes, *Dread*, 186–87). Nonetheless, scientists are developing methods to predict where new pandemics may begin; see Kevin J. Olival, Parviez R. Hosseini, Carlos Zambrana-Torrelio, Noam Ross, Tiffany L. Bogich, and Peter Daszak, "Host and Viral Traits Predict Zoonotic Spillover from Mammals," *Nature* 546 (2017): 646–50.

48. See Michael J. Selgelid, Margaret P. Battin, and Charles B. Smith, eds., *Ethics and Infectious Disease* (Malden, MA: Blackwell Publishing, 2006).

49. This emphasis reflects the distinction, made by René Dubos decades ago, between two traditions in Western medicine. Dubos argues that Asclepius as the god of medicine had two different offspring: Hygeia and Panacea. Hygeia represented the emphasis on pre-vention, public health, and a healthy way of life. Panacea was the focus on therapies and drugs, treatment of disease, and medical interventions. See René Dubos, *Man, Medicine, and Environment* (Harmondsworth, UK: Penguin Books, 1968), 76–81.

50. See, for example, Rita Colwell, "Global Climate and Infectious Disease: The Chol-era Paradigm," *Science* 274 (1996): 2025–31; Francesca Grifo and Joshua Rosenthal, *Biodi-versity and Human Health* (Washington, DC: Island Press, 1997); Daszak, Cunningham, and Hyatt, "Emerging Infectious Diseases of Wildlife," 443–49; Eric Chivian and Aaron Bernstein, eds., *Sustaining Life: How Human Health Depends on Biodiversity* (Oxford: Oxford University Press, 2008); Osvaldo E. Sala, Laura A. Meyerson, and Camille Parmesan, eds., *Biodiversity Change and Human Health: From Ecosystem Services to Spread of Disease* (Wash-ington, DC: Island Press, 2009); Sonia Shah, *Pandemic: Tracking Contagions, from Cholera to Ebola and Beyond* (New York: Farrar, Straus and Giroux, 2016); Davis, *The Monster at Our Door*.

51. Justin Brashares et al., "Bushmeat Hunting, Wildlife Declines, and Fish Supply in West Africa," *Science* 306 (2004): 1180–83; David Quammen, *Spillover: Animal Infections and the Next Human Pandemic* (New York: W. W. Norton, 2012).

52. Chivian and Bernstein, *Sustaining Life*, 105.

53. China raises half of the pigs in the world; in 2010, 660 million pigs were raised in China. See Shah, *Pandemic*, 93. See also S. McOrist, K. Khampee, and A. Guo, "Modern Pig Farming in the People's Republic of China: Growth and Veterinary Challenges," *Re-vue Scientifique et Technique (International Office of Epizootics)* 30 (2011): 961–68.

54. Food and Agriculture Organization of the United Nations (FAO) et al., *Contribut-ing to One World, One Health*, October 2008, 17, 20, www.fao.org/docrep/fao/011/aj137e/aj137e00.pdf.

55. Food and Agriculture Organization of the United Nations, *Livestock's Long Shadow: Environmental Issues and Options* (Rome: FAO, 2006).

56. Ellen K. Silbergeld, *Chickenizing Farms and Food: How Industrial Meat Production Endangers Workers, Animals, and Consumers* (Baltimore, MD: Johns Hopkins University Press, 2016); David Kirby, *Animal Factory: The Looming Threat of Pig, Dairy, and Poultry Farms to Humans and the Environment* (New York: St. Martin's Griffin, 2010); Danielle Nie-renberg, *Happier Meals: Rethinking the Global Meat Industry*, Worldwatch Paper no. 171 (Washington, DC: Worldwatch Institute, 2005).

57. Reassortment is the mixing up of genetic information so that more dangerous pathogens may be produced. Viruses are continuously changing and mutating; they also combine genetic sequences with other viruses so that new subtypes are formed (genetic shift).

58. Robert G. Wallace, "Breeding Influenza: The Political Virology of Offshore Farming," *Antipode* 41 (2009): 921. See also S. P. Cobb, "The Spread of Pathogens through Trade in Poultry Meat: Overview and Recent Developments," *Revue Scientifique et Technique (International Office of Epizootics)* 30 (2011): 149–64.

59. "Global agribusiness appears to be a key player in the emergence and spread of new influenzas" (Robert G. Wallace, Luke Bergmann, Lenny Hogerwerf, and Marius Gilbert, "Are Influenzas in Southern China Byproducts of the Region's Globalizing Historical Present?," in *Influenza and Public Health: Learning from Past Pandemics*, ed. Tamara Giles-Vernick and Susan Craddock [Abingdon, UK: Routledge, 2010], 106). Ali Khan refers to his experiences during a poultry epidemic (H7N2 virus) in Virginia in 2002. The industrial farming companies did not take responsibility, shifting it instead onto the shoulders of small farmers who have mortgaged themselves to the companies to build their factories while the animals remain the property of the companies. The companies only want to destroy the dead animals if they receive federal compensation. See Ali S. Kahn and William Patrick, *The Next Pandemic: On the Front Lines against Humankind's Gravest Dangers* (New York: PublicAffairs, 2016), 151–63. See also Ann Marie Kimball, *Risky Trade: Infectious Disease in the Era of Global Trade* (Aldershot, UK: Ashgate, 2006), 55–59.

60. Davis, *The Monster at Our Door*, 97ff.; Isabelle Delforge, "Thailand, the World's Kitchen," *Le Monde Diplomatique*, July 2004.

61. Fineberg, "Pandemic Preparedness and Response," 1335–42.

62. The global response is "limited, uncoordinated, and dysfunctional," according to Laurie Garrett. She also points out that the major problem is not the disease but the "world's disastrous response to it." See Garrett, "Ebola's Lessons: How the WHO Mishandled the Crisis," *Foreign Affairs* 94 (2015): 85, 84.

63. Fineberg, "Pandemic Preparedness and Response," 1339.

64. Bernard Taverne, "Preparing for Ebola Outbreaks: Not without the Social Sciences," *Global Health Promotion* 22 (2015): 5.

65. Davis, *The Monster at Our Door*, 24–30; Kahn and Patrick, *The Next Pandemic*, 10, 21; Macphail, *The Viral Network*, 127ff.; Alcabes, *Dread*, 109–12.

66. See Dyna Arhin-Tenkoran and Pedro Conceição, "Beyond Communicable Disease Control: Health in the Age of Globalization," in *Providing Global Public Goods: Managing Globalization*, ed. Inge Kaul (New York: Oxford University Press, 2003), 484–515; Garrett, "Ebola's Lessons," 80–107; Benedict Shing Bun Chan et al., "Why Should We Care about Ebola in West Africa and Middle East Respiratory Syndrome in South Korea? Global Health Ethics and the Moral Insignificance of Proximity," *Bioethical Inquiry* 12 (2015): 541–43.

67. Nancy Leys Stepan gives the example of yellow fever, which has become the target of eradication campaigns, although in many countries other diseases were important for population health. See Stepan, *Eradication: Ridding the World of Diseases Forever?* (Ithaca, NY: Cornell University Press, 2011), 36.

68. Paul Richards, *Ebola: How a People's Science Helped End an Epidemic* (London: Zed Books, 2016).

69. Randall M. Packard, *A History of Global Health: Interventions into the Lives of Other Peoples* (Baltimore, MD: Johns Hopkins University Press, 2016), 14–15.

70. Ken Thompson, *Do We Need Pandas? The Uncomfortable Truth about Biodiversity* (Foxhole, UK: Green Books, 2010); Dennis Normile, "Driven to Extinction," *Science* 319 (2008): 1606–9.

71. See, for this argument, Stepan, *Eradication*, 130.

72. D. J. Rogers, A. J. Wilson, S. I. Hay, and A. J. Graham, "The Global Distribution of Yellow Fever and Dengue," *Advances in Parasitology* 62 (2006): 181–220.

73. WHO, *Yellow Fever*, June 2016, http://www.who.int/emergencies/yellow-fever/en. Alan D. T. Barrett, "Yellow Fever in Angola and Beyond: The Problem of Vaccine Supply and Demand," *New England Journal of Medicine* 375 (2016): 301–3. According to Barrett, the main problem is the lack of availability of yellow fever vaccine; worldwide there are only six manufacturers (ibid., 302).

74. WHO, *Dengue and Severe Dengue*, April 2016, http://www.who.int/mediacentre /factsheets/fs117/en.

75. Stepan, *Eradication*, 257.

76. Stepan, *Eradication*, 225ff.

77. David P. Fidler, *SARS, Governance and the Globalization of Disease* (Houndmills, UK: Palgrave Macmillan, 2004).

78. Fidler, *SARS, Governance and the Globalization of Disease*, 121ff.

79. This is, for example, emphasized by Kahn and Patrick: "To prevent the next pandemic, we need a comprehensive approach that gets beyond a focus on the microbe" (*The Next Pandemic*, 251).

80. See Linda Nash, *Inescapable Ecologies: A History of Environment, Disease, and Knowledge* (Berkeley: University of California Press, 2006).

81. Packard, *A History of Global Health*.

82. Kahn and Patrick, *The Next Pandemic*, 4, 6, 250.

83. Wallace, Bergmann, Hogerwerf, and Gilbert, "Are Influenzas in Southern China Byproducts of the Region's Globalizing Historical Present?," 109.

84. As articulated, for example, by FAO et al., *Contributing to One World, One Health*. See also Satesh Bidaisee and Calum N. L. Macpherson, "Zoonoses and One Health: A Review of the Literature," *Journal of Parasitology Research* (January 30, 2014): http://dx.doi .org/10.1155/2014/874345.

85. Macphail, *The Viral Network*, 17.

86. Rebecca E. Hirsch, *The Human Microbiome: The Germs That Keep You Healthy* (Minneapolis, MN: Twenty-First Century Books, 2017). See also Silbergeld, *Chickenizing Farms and Food*, 128–44.

87. In 2000, the UN Security Council declared HIV/AIDS to be a threat to international security. See Gregory D. Koblentz, "Biosecurity Reconsidered: Calibrating Biological Threats and Responses," *International Security* 34 (2010): 96.

88. Laura A. Meyerson and Jamie K. Reaser, "Biosecurity: Moving toward a Comprehensive Approach," *BioScience* 52 (2002): 593–600; Laura A. Meyerson and Jamie K. Reaser, "Bioinvasions, Bioterrorism, and Biosecurity," *Frontiers in Ecology and the Environment* 1 (2003): 307–14; Jeffrey R. Ryan, *Biosecurity and Bioterrorism: Containing and Preventing Biological Threats*, 2nd ed. (Amsterdam: Elsevier, 2016); Ronald J. Glanville,

Simon M. Firestone, and Simon J. More, "Biosecurity," in *Encyclopedia of Global Bioethics*, ed. Henk ten Have (Dordrecht: Springer Science-Business Media, 2016), 388–89. For the connection with neoliberalism, see Damian May, Jacqui Dibden, Vaughan Higgins, and Clive Potter, "Governing Biosecurity in a Neoliberal World: Comparative Perspectives from Australia and the United Kingdom," *Environment and Planning A* 44 (2012): 150–68.

89. Ryan, *Biosecurity and Bioterrorism*, 170–73. See also Kathryn Baker, "The Meaning and Practice of Biosecurity," *International Journal of Risk Assessment and Management* 12 (2009): 121–46.

90. Ryan, *Biosecurity and Bioterrorism*, 298ff.

91. Brian Rappert and Filippa Lentzos, "Biosecurity and Bioterror: Reflections on a Decade," in *Biosecurity: The Socio-politics of Invasive Species and Infectious Diseases*, ed. Andrew Dobson, Kezia Barker, and Sarah L. Taylor (London: Routledge, 2013), 155.

92. Joseph Masco mentions that until 2001 there was almost no public discourse about biosecurity in the United States. See Masco, *The Theater of Operations: National Security Affect from the Cold War to the War on Terror* (Durham, NC: Duke University Press, 2014), 224. See also Dobson, Barker, and Taylor, *Biosecurity*, 5.

93. Kimball, *Risky Trade*, 94; Washer, *Emerging Infectious Diseases and Society*, 145, 147. C. Raina MacIntyre mentions that in 2014 there have been four major safety breaches with dangerous pathogens in laboratories (MacIntyre, "Biopreparedness in the Age of Genetically Engineered Pathogens and Open Access Science: An Urgent Need for a Paradigm Shift," *Military Medicine* 180 [2015]: 945). See also Rappert and Lentzos, "Biosecurity and Bioterror," 151–64.

94. Koblentz, "Biosecurity Reconsidered," 115.

95. Katherine Harmon, "Nearly 400 Accidents with Dangerous Pathogens and Biotoxins Reported in U.S. Labs over 7 Years," *Scientific American*, October 3, 2011.

96. Steve Hinchliffe, "The Insecurity of Biosecurity: Remaking Emerging Infectious Diseases," in *Biosecurity: The Socio-politics of Invasive Species and Infectious Diseases*, ed. Andrew Dobson, Kezia Barker, and Sarah L. Taylor (London: Routledge, 2013), 200, 209.

97. Koblentz, "Biosecurity Reconsidered," 116.

98. Washer, *Emerging Infectious Diseases and Society*, 143. In fact, three examples of biological attack are frequently mentioned: the salmonella food poisoning in Oregon by the Rajneesh cult in 1984; the efforts of the Aum Shinrikyo cult in Japan in 1984 to weaponize anthrax and Ebola; and the anthrax letters in 2001 in the United States. See Ryan, *Biosecurity and Bioterrorism*, 116, 165–67, 170–73.

99. Stephen J. Collier, Andrew Lakoff, and Paul Rabinow, "Biosecurity: Towards an Anthropology of the Contemporary," *Anthropology Today* 20 (2004): 4–5.

100. William Aldis, "Health Security as a Public Health Concept: A Critical Analysis," *Health Policy and Planning* 23 (2008): 373–74.

101. Nicholas B. King calls for an ethics *of* biodefense, arguing that it raises many ethical questions that need further scrutiny. See King, "The Ethics of Biodefense," *Bioethics* 19 (2005): 432–46.

102. See Braun, "Power over Life," 45–57. See also Fall, "Biosecurity and Ecology," 167–81.

103. Aldis, "Health Security as a Public Health Concept," 373. See also Brendon M. H. Larson, "The War of the Roses: Demilitarizing Invasion Biology," *Frontiers in Ecology and*

the Environment 3 (2005): 495–500; Harley Feldbaum, Preeti Patel, Egbert Sondorp, and Kelley Lee, "Global Health and National Security: The Need for Critical Engagement," *Medicine, Conflict and Survival* 22 (2006): 192–98.

104. Meyerson and Reaser, "Biosecurity," 593–600; Meyerson and Reaser, "Bioinvasions, Bioterrorism, and Biosecurity," 307–14.

105. Masco, *The Theater of Operations*, 41. See also Hinchliffe, "The Insecurity of Biosecurity," 199–213.

106. Clive Potter, "A Neoliberal Biosecurity? The WTO, Free Trade and the Governance of Plant Health," in *Biosecurity: The Socio-politics of Invasive Species and Infectious Diseases*, ed. Andrew Dobson, Kezia Barker, and Sarah L. Taylor (London: Routledge, 2013), 123–35. See also May, Dibden, Higgins, and Potter, "Governing Biosecurity in a Neoliberal World," 152ff.

107. "Infectious diseases thrive in conditions of poverty, social exclusion and inequality" (Washer, *Emerging Infectious Diseases and Society*, 167).

108. Dobson, Barker, and Taylor, *Biosecurity*, 19–20. Dobson and colleagues also point out that there are discrepancies in how uncertainty is managed. The Convention on Biological Diversity follows a precautionary approach, while the World Trade Organization only accepts strict scientific evidence.

109. The prevailing paradigm is "an image of total danger, one that cannot be deterred but only pre-empted through anticipatory action" (Masco, *The Theater of Operations*, 37).

110. Three people died as a result of the smallpox vaccinations. See Hillel W. Cohen, Robert M. Gould, and Victor W. Sidel, "The Pitfalls of Bioterrorism Preparedness: The Anthrax and Smallpox Experiences," *American Journal of Public Health* 94 (2004): 1668. For a historical perspective, see Elizabeth Fee, "Preemptive Biopreparedness: Can We Learn Anything from History?," *American Journal of Public Health* 91 (2001): 721–26.

111. Tom Jefferson et al., "Oseltamivir for Influenza in Adults and Children: Systematic Review of Clinical Study Reports and Summary of Regulatory Comments," *British Medical Journal* 348 (2014): g2545.

112. Ben Goldacre, "What the Tamiflu Saga Tells Us about Drug Trials and Big Pharma," *Guardian*, April 10, 2014; Deborah Cohen and Philip Carter, "WHO and the Pandemic Flu 'Conspiracies,'" *British Medical Journal* 340 (2010): c2912.

113. Jaro Kotalik, "Preparing for an Influenza Pandemic: Ethical Issues," in *Ethics and Infectious Disease*, ed. Michael J. Selgelid, Margaret P. Battin, and Charles B. Smith (Malden, MA: Blackwell Publishing, 2006), 95–104.

114. Masco, *The Theater of Operations*, 170. For an analysis of the effects of militarization of biosecurity, see Neel Ahuja, *Bioinsecurities: Disease Interventions, Empire, and the Governments of Species* (Durham, NC: Duke University Press, 2016).

115. Rappert and Lentzos, "Biosecurity and Bioterror," 155.

116. DURC is "research intended to benefit human health, but which can also inadvertently or deliberately harm people" (MacIntyre, "Biopreparedness in the Age of Genetically Engineered Pathogens and Open Access Science," 943).

117. Ryan, *Biosecurity and Bioterrorism*, 348–49.

118. Two cases in the early 2000s are discussed by Michael J. Selgelid, "A Tale of Two Studies: Ethics, Bioterrorism, and the Censorship of Science," *Hastings Center Report*

37 (2007): 35–43. He concludes that censorship of publications may sometimes be appropriate.

119. David B. Resnik, "H5N1 Avian Flu Research and the Ethics of Knowledge," *Hastings Center Report* 43 (2013): 22–33; Norman K. Swazo, "Engaging the Normative Question in the H5N1 Avian Influenza Mutation Experiments," *Philosophy, Ethics, and Humanities in Medicine* 8 (2013): https://doi.org/10.1186/1747-5341-8-12.

120. See Resnik, "H5N1 Avian Flu Research and the Ethics of Knowledge," 22–33. Resnick concludes that redacted publication should be a viable option but that it is currently impossible because of legal and practical obstacles.

121. Ruth R. Faden and Ruth A. Karron, "The Obligation to Prevent the Next Dual-Use Controversy," *Science* 335 (2012): 803.

122. MacIntyre argues that "Infectious disease is an industry, with many vested interest groups who could potentially profit from outbreaks." The conclusion is that "the rights of scientists should not outweigh the rights of the public if there is a risk of harm" (MacIntyre, "Biopreparedness in the Age of Genetically Engineered Pathogens and Open Access Science," 947, 948).

123. Michael J. Imperiale and Arturo Casadevall, "A New Synthesis for Dual Use Research of Concern," *PLOS Medicine* 12 (2015): e1001813; Robin Fears and Volker ter Meulen, "European Academies Advise on Gain-of-Function Studies in Influenza Virus Research," *Journal of Virology* 90 (2016): 2162–64.

124. See Faden and Karron, "The Obligation to Prevent the Next Dual-Use Controversy," 802–4; Ori Lev and Limor Samimian-Darash, "Biosecurity Policy in the US: A Critical Assessment," *Frontiers in Public Health* 2 (2014): 110, https://doi.org/10.3389/fpubh.2014.00110; Brett Edwards, "Taking Stock of Security Concerns Related to Synthetic Biology in an Age of Responsible Innovation," *Frontiers in Public Health* 2 (2014): 79, https://doi.org/10.3389/fpubh.2014.00079.

125. Michelle M. Mello, Naria W. Merritt, and Scott D. Halpern, "Supporting Those Who Go to Fight Ebola," *PLOS Medicine* 12 (2015): e1001781, https://doi.org/10.1371/journal.pmed.1001781.

126. Lawrence O. Gostin et al., "Toward a Common Secure Future: Four Global Commissions in the Wake of Ebola," *PLOS Medicine* 13 (2016): e1002042, https://doi.org/10.1371/journal.pmed.1002042.

127. WHO, "The Top 10 Causes of Death," fact sheet, May 2016, http://www.who.int/news-room/fact-sheets/detail/the-top-10-causes-of-death.

128. Sheikh Mohammed Shariful Islam et al., "Non-communicable Diseases (NCDs) in Developing Countries: A Symposium Report," *Globalization and Health* 10 (2014): 81.

129. Rachel Carson, *Silent Spring* (New York: Houghton Mifflin, 1962).

130. Marco Armiero and Lise Sedrez, eds., *A History of Environmentalism: Local Struggles, Global Histories* (London: Bloomsbury Academic, 2014), 129ff. See also Henk ten Have, *Global Bioethics: An Introduction* (London: Routledge, 2016), 228ff.

131. See chapter 1.

132. This number of deaths is higher than the mortality of HIV/AIDS, tuberculosis, and road injuries combined. See "Air Pollution: Crossing Borders" (editorial), *Lancet* 388 (2016): 103; International Energy Agency, *Energy and Air Pollution* (Paris: OECD/IEA, 2016), 32ff.

133. World Wildlife Federation (WWF) European Office, Sandbag, Climate Action Network (CAN) Europe, and Health and Environment Alliance (HEAL), *Europe's Dark Cloud: How Coal-Burning Countries Are Making Their Neighbours Sick* (Brussels, 2016), 12, 15.

134. International Energy Agency, *Energy and Air Pollution*, 125ff.

135. WWF European Office, Sandbag, CAN Europe, and HEAL, *Europe's Dark Cloud*, 47.

136. Stephen Vines, "How Will China Deal with Growing Anger over Pollution?," *Aljazeera*, August 1, 2014; BBC News, "What Is China Doing to Tackle Its Air Pollution?," January 20, 2016; Zachary Davies Boren, "China to Suspend New Coal Power Plant Approvals," Greenpeace Energy Desk, April 13, 2016.

137. Eiichiro Ochiae, *Chemicals for Life and Living* (Heidelberg: Springer, 2011), viii, 271.

138. Ellena Atienza Macias and Emilio José Armaza Armaza, "Rethinking the Bioethical Implications of Chemistry: The Case of DDT; A Paradigmatic Example of the Interconnection between Bioethics and Chemistry," *International Journal of Bioethics and Health Policy* 1 (2016): 47–57. For the toxic effects on human health, see Joseph V. Rodricks, *Calculated Risks: The Toxicity and Human Health Risks of Chemicals in Our Environment*, 2nd ed. (Cambridge: Cambridge University Press, 2007); John F. Sieckhaus, *Chemicals, Human Health, and the Environment: A Guide to the Development and Control of Chemical and Energy Technology* (Bloomington, IN: Xlibris Corp., 2009).

139. Richard Elliot Benedick, "Montreal Protocol on Substances That Deplete the Ozone Layer," *International Negotiation* 1 (1996): 231–46; Chris Peloso, "Crafting an International Climate Change Protocol: Applying the Lessons Learned from the Success of the Montreal Protocol, and the Ozone Depletion Problem," *Journal of Land Use and Environmental Law* 25 (2010): 305–29.

140. Ovnair Sepai, Clare Collier, Birgit van Tongelen, and Ludwine Casteleyn, "Human Biomonitoring Data Interpretation and Ethics: Obstacles or Surmountable Challenges?," *Environmental Health* 7 (suppl. 1, 2008): S13, https://doi.org/10.1186/1476-069X-7-S1-S13.

141. "Approximately 70% of the public tries to avoid chemicals and chemical products in daily life." Nanomaterials are perceived more positively, as less risky. See Adam Capon, Margaret Rolfe, James Gillespie, and Wayne Smith, "Is the Risk from Nanomaterials Perceived as Different from the Risk of 'Chemicals' by the Australian Public?," *Public Health Research and Practice* 26 (2016): 3.

142. By 1990, more than half of all African countries had been approached to accept hazardous waste. See Jennifer Clapp, "The Toxic Waste Trade with Less-Industrialised Countries: Economic Linkages and Political Alliances," *Third World Quarterly* 15 (1994): 507; Clapp, *Toxic Exports: The Transfer of Hazardous Wastes from Rich to Poor Countries* (Ithaca, NY: Cornell University Press, 2001).

143. The official name of this convention is the Basel Convention on the Control of Transboundary Movement of Hazardous Waste and Its Disposal. It came into force in 1992. The United States has not ratified it. It is, together with China and the Russian Federation, the world's biggest producer of toxic waste. See Henrik Selin, *Global Governance of Hazardous Chemicals: Challenges of Multilevel Management* (Cambridge, MA: MIT Press, 2010), 63–84.

144. Basel Convention, Amendment to the Basel Convention on the Control of Trans-boundary Movements of Hazardous Wastes and Their Disposal, 1995. See Ken Conca, *Governing Water: Contentious Transnational Politics and Global Institution Building* (Cambridge, MA: MIT Press, 2006), 34ff.

145. Ian Birrell, "Mafia, Toxic Waste, and a Deadly Cover Up in an Italian Paradise: 'They've Poisoned Our Land and Stolen Our Children,'" *Telegraph*, June 24, 2016. For toxic waste pollution in Côte d'Ivoire, see ten Have, *Global Bioethics*, 141–42.

146. Daniel Hoornweg and Perinaz Bhada-Tata, *What a Waste: A Global Review of Solid Waste Management* (Washington DC: World Bank, 2012), 8.

147. Sean Paquette, *International Toxic Waste Trade: Mobile Phones Growth, Flow and Solutions of International Electronic Waste Trade*, 2014, http://www.ewp.rpi.edu/hartford /~paques/SHWPCE/Research%20Paper%20-%20%231%20/Sean%20Paquette _SolidWaterPaper_DoubleSpaced.pdf.

148. Leyla Acaroglu, "Where Do Old Cell Phones Go to Die?," *New York Times*, May 4, 2013. Damon Beres, "Obama Still Has Time to Curb a Toxic Waste Crisis Caused by Gadgets," *Huffington Post*, July 27, 2016; Basel Action Network, "Federal Agencies Must Stop Exporting Their Toxic E-waste to Developing Countries," 2016, https://www.change.org /p/federal-agencies-must-stop-exporting-their-toxic-e-waste-to-developing-countries.

149. David B. Resnik and Kevin C. Elliott, "Bisphenol A and Risk Management Ethics," *Bioethics* 29 (2015): 182–89.

150. Resnik and Elliott, "Bisphenol A and Risk Management Ethics," 184ff.

151. The Rio Declaration formulates the precautionary principle as follows: "Where there are threats of serious or irreversible damage, lack of full scientific certainty shall not be used as a reason for postponing cost-effective measures to prevent environmental degradation" (Principle 15, UN, Rio Declaration on Environment and Development, 1992), http://www.unep.org/Documents.Multilingual/Default.asp?documentid=78&articleid =1163.

152. See Rosie Cooney and Barney Dickson, eds., *Biodiversity and the Precautionary Principle: Risk and Uncertainty in Conservation and Sustainable Use* (London: Earthscan, 2005); Nicolas de Sadeleer, ed., *Implementing the Precautionary Principle: Approaches from the Nordic Countries, EU and USA* (London: Earthscan, 2007); Christian Munthe, *The Price of Precaution and the Ethics of Risk* (Dordrecht: Springer, 2011); David Vogel, *The Politics of Precaution: Regulating Health, Safety, and Environmental Risks in Europe and the United States* (Princeton, NJ: Princeton University Press, 2012); Daniel Steel, *Philosophy and the Precautionary Principle: Science, Evidence and Environmental Policy* (Cambridge: Cambridge University Press, 2015).

153. *Berlaymont Declaration on Endocrine Disruptors*, 2013. http://www.brunel.ac.uk /_data/assets/pdf_file/0005/300200/The_Berlaymont_Declaration_on_Endocrine _Disrupters.pdf.

154. European Commission, *Report on Public Consultation on Defining Criteria for Identifying Endocrine Disruptors in the Context of the Implementation of the Plant Protection Product Regulation and Biocidal Products Regulation*, Brussels, July 22, 2015, http://ec.europa .eu/health/endocrine_disruptors/docs/2015_public_consultation_report_en.pdf. See also Arthur Neslin, "Two-Thirds of Europeans Support a Ban on Glyphosate, says Yougov Poll," *Guardian*, April 11, 2016. In August 2016, the Endocrine Society called on the European

Commission to adopt policies to regulate endocrine disruptors. More than 1,300 scientific studies have linked these chemicals to diabetes, obesity, infertility, cancers, and neurological disorders. The public and future generations need protection. Inadequate regulation will have disastrous consequences. It is estimated that adverse health effects will cost the European Union 157 billion euro each year in healthcare expenses and lost productivity. The human costs cannot even be calculated. See Jean-Pierre Bourguignon et al., "Science-Based Regulation of Endocrine Disrupting Chemicals in Europe: Which Approach?," *Lancet Diabetes and Endocrinology* 4 (2016): 643–46.

155. Testing of members of the European Parliament in May 2016 demonstrated that they have levels of glyphosate in their urine that are higher than European norms for drinking water. Arthur Neslin, "Controversial Chemical in Roundup Weedkiller Escapes Immediate Ban," *Guardian*, June 29, 2016. The European Food Safety Authority (EFSA) in the EU concluded in October 2015 that glyphosate was not carcinogenic. The EFSA, however, is only using data provided by the industry; these are confidential and not publicly shared. The "Monsanto Papers" (documents released since March 2017 in response to more than 100 lawsuits currently pending against Monsanto in California) show how authorities in Monsanto were worried about possible toxicity of glyphosate since 1999. Their internal e-mails show how they have systematically tried to influence the scientific debate, to spin research, and pay scientists to cover up information about cancers (https://usrtk.org/pesticides/mdl-monsanto-glyphosate-cancer-case-key-documents-analysis).

156. Robert D. Bullard, ed., *The Quest for Environmental Justice: Human Rights and the Politics of Pollution* (Berkeley, CA: Counterpoint, 2005), 17–42, 207–21, 279–97.

157. Each year more and more waste is produced. The United States produces 19% of the world's waste, while it has 5% of the global population. See David Naguib Pellow, *Resisting Global Toxics: Transnational Movements for Environmental Justice* (Cambridge, MA: MIT Press, 2007), 100.

158. Rob Nixon, *Slow Violence and the Environmentalism of the Poor* (Cambridge, MA: Harvard University Press, 2011), 2–3.

159. Pellow, *Transnational Movements for Environmental Justice*, 107–23.

160. Kevin J. O'Brien, *An Ethics of Biodiversity: Christianity, Ecology, and the Variety of Life* (Washington, DC: Georgetown University Press, 2010), 45.

161. This argument is elaborated in two previous books: ten Have, *Global Bioethics*; ten Have, *Vulnerability*.

CHAPTER FIVE: **Drugs**

1. For opposing views on the uses of silphion, see John M. Riddle and J. Worth Estes, "Oral Contraceptives in Ancient and Medieval Times," *American Scientist* 80 (1992): 226–33; Monika Kiehn, "Silphion Revisited," *IUCN Medicinal Plant Conservation* 13 (2007): 3–4.

2. Kara Rogers, *Out of Nature: Why Drugs from Plants Matter to the Future of Humanity* (Tucson: University of Arizona Press, 2012); David J. Newman, John Kilama, Aaron Bernstein, and Eric Chivian, "Medicines from Nature," in *Sustaining Life: How Human Health Depends on Biodiversity*, ed. Eric Chivian and Aaron Bernstein (Oxford: Oxford University Press, 2008), 119–21. See also Michael J. Balick and Paul Alan Cox, *Plants, People, and Culture: The Science of Ethnobotany* (New York: Scientific American Library, 1996).

3. Rogers, *Out of Nature*, 131–35.

4. I. S. Whitaker, J. Rao, D. Izadi, and P. E. Butler, "*Hirudo medicinalis*: Ancient Origins of, and Trends in the Use of Medicinal Leeches throughout History," *British Journal of Oral and Maxillofacial Surgery* 42 (2004): 133–37. See also Newman, Kilama, Bernstein, and Chivian, "Medicines from Nature," 133–35.

5. Newman, Kilama, Bernstein, and Chivian give many examples of medicinal plants that have been used since ancient times, as well as of compounds from plants that have been chemically extracted since the nineteenth century—among others, morphine (1804), atropine (1831), and cocaine (1860). See "Medicines from Nature," 117–61.

6. Akhtar Husain, "Exploitation of Medicinal Plants," in *Medical Plants: Their Role in Health and Biodiversity*, ed. Timothy R. Tomlinson and Olayiwola Akerele (Philadelphia: University of Pennsylvania Press, 1998), 82ff.

7. William McKinley Klein, "The Role of Botanical Gardens and Arboreta in Traditional Medicine: A Personal Reflection and Case Study," in *Medical Plants: Their Role in Health and Biodiversity*, ed. Timothy R. Tomlinson and Olayiwola Akerele (Philadelphia: University of Pennsylvania Press, 1998), 120–33.

8. Rogers, *Out of Nature*, 57–72.

9. Jeremy N. Norman, "William Withering and the Purple Foxglove: A Bicentennial Tribute," *Journal of Clinical Pharmacology* 25 (1985): 479–83. See also Edward Shorter, "Primary Care," in *The Cambridge History of Medicine*, ed. Roy Porter (Cambridge: Cambridge University Press, 2006), 116–17.

10. Newman, Kilama, Bernstein, and Chivian, "Medicines from Nature," 117.

11. Alan Harvey, "Strategies for Discovering Drugs from Previously Unexplored Natural Products," *Drug Discovery Today* 5 (2000): 294.

12. David J. Newman and Gordon M. Cragg, "Natural Products as Sources of New Drugs from 1981 to 2014," *Journal of Natural Products* 79 (2016): 632–33.

13. Newman and Cragg, "Natural Products as Sources of New Drugs from 1981 to 2014," 645, 653.

14. This is also the general argument of Victor Kuete and Thomas Efferth, eds., *Biodiversity, Natural Products and Cancer Treatment* (Singapore: World Scientific Publishing, 2014).

15. Eric Chivian, Aaron Bernstein, and Joshua P. Rosenthal, "Biodiversity and Biomedical Research," in *Sustaining Life: How Human Health Depends on Biodiversity*, ed. Eric Chivian and Aaron Bernstein (Oxford: Oxford University Press, 2008), 179–80.

16. Eric S. Lander, "The Heroes of CRISPR," *Cell* 164 (2016): 18–28; Carl Zimmer, "Breakthrough DNA Editor Born of Bacteria," *Quanta Magazine*, February 6, 2015.

17. Harvey, "Strategies for Discovering Drugs from Previously Unexplored Natural Products," 296; Newman, Kilama, Bernstein, and Chivian, "Medicines from Nature," 117–18.

18. Rogers, *Out of Nature*, 102.

19. Christopher Rojas, "The Southern Gastric-Brooding Frog," *Embryo Project Encyclopedia* (January 26, 2015): http://embryo.asu.edu/handle/10776/8282; Ed Yong, "Resurrecting the Extinct Frog with a Stomach for a Womb," *National Geographic, Not Exactly Rocket Science* (blog), March 15, 2013, https://www.nationalgeographic.com/science/phenomena/2013/03/15/resurrecting-the-extinct-frog-with-a-stomach-for-a-womb. See

also Chivian and Bernstein, eds., *Sustaining Life*, xiii. For "de-extinction," see chapter 2 ("Values and Ethics" section).

20. Gordon M. Cragg and David J. Newman, "Plants as a Source of Anti-cancer Agents," *Journal of Ethno-Pharmacology* 100 (2005a): 72–79.

21. Alan L. Harvey, "Natural Products in Drug Discovery," *Drug Discovery Today* 13 (2008): 895. See also Rogers, *Out of Nature*, 123–24.

22. Geoffrey A. Cordell, "Biodiversity and Drug Discovery: A Symbiotic Relationship," *Phytochemistry* 55 (2000): 463–80.

23. Harvey, "Natural Products in Drug Discovery," 895.

24. Newman, Kilama, Bernstein, and Chivian, "Medicines from Nature," 130.

25. Peter H. Canter, Howard Thomas, and Edzard Ernst, "Bringing Medicinal Plants into Cultivation: Opportunities and Challenges for Biotechnology," *Trends in Biotechnology* 23 (2005): 181.

26. Cragg and Newman, "Plants as a Source of Anti-cancer Agents," 78. See also Alan L. Harvey, Ruangelie Edrada-Ebel, and Ronald J. Quinn, "The Re-emergence of Natural Products for Drug Discovery in the Genomics Era," *Nature Reviews Drug Discovery* 14 (2015): 111–29; Yongyuth Yuthavong, *Tapping Molecular Wilderness: Drugs from Chemistry-Biology-Biodiversity Interface* (Singapore: Pan Stanford Publishing, 2016).

27. Harvey, "Strategies for Discovering Drugs from Previously Unexplored Natural Products," 294. See also Paul Alan Cox, "Biodiversity and the Search for New Medicines," in *Biodiversity Change and Human Health: From Ecosystem Services to Spread of Disease*, ed. Osvaldo E. Sala, Laura A. Meyerson, and Camille Parmesan (Washington, DC: Island Press, 2009), 269–80; R. David Simpson, Roger A. Sedjo, and John W. Reid, "Valuing Biodiversity for Use in Pharmaceutical Research," *Journal of Political Economy* 104 (1996): 166ff.

28. Joshua J. Gagne and Niteesh K. Ghoudry, "How Many 'Me-Too' Drugs Is Too Many?," *Journal of the American Medical Association* 305 (2011): 711–12.

29. Robert Kneller, "The Importance of New Companies for Drug Discovery: Origins of a Decade of New Drugs," *Nature Reviews Drug Discovery* 9 (2010): 869–70.

30. Ashley J. Stevens, Jonathan J. Jensen, Katrine Wyller, Patrick C. Kilgore, Sabarni Chatterjee, and Mark L. Rohrbaugh, "The Role of Public-Sector Research in the Discovery of Drugs and Vaccines," *New England Journal of Medicine* 364 (2011): 535–41.

31. Rita R. Colwell, "Fulfilling the Promise of Biotechnology," *Biotechnology Advances* 20 (2002): 220–21. Harvey, "Natural Products in Drug Discovery," 895.

32. Gordon M. Cragg and David J. Newman, "Biodiversity: A Continuing Source of Novel Drug Leads," *Pure and Applied Chemistry* 77 (2005): 15ff.; Tobias A. M. Gulder and René Richarz, "Nature's Chemical Treasure Trove: Biodiversity and Pharmaceuticals," in *Concepts and Values in Biodiversity*, ed. Dirk Lanzerath and Minou Friele (London: Routledge, 2014), 319.

33. Youyou Tu, "Artemisinin: A Gift from Traditional Chinese Medicine to the World," Nobel lecture, December 7, 2015, 311, https://www.nobelprize.org/nobel_prizes/medicine/laureates/2015/tu-lecture.pdf.

34. Emilia Janska, Mihaela Serbulea, and Brendan Tobin, "The Importance of Traditional Knowledge for Meeting Public Health Needs in Developing Countries," *Work in*

Progress 17 (2005): 29. Percentage examples of populations that rely on traditional health-care: Africa, 80%; India, 67%; China, 40%; Chile, 71%.

35. Samuel S. Antwi-Baffour, Ajediran I. Bello, David N. Adjei, Seidu A. Mahmood, and Patrick F. Ayeh-Kumi, "The Place of Traditional Medicine in the African Society: The Science, Acceptance and Support," *American Journal of Health Research* 2 (2014): 50.

36. WHO, "Traditional Medicine," fact sheet no. 134 (Geneva: WHO, 2003), http://www.who.int/mediacentre/factsheets/2003/fs134/en.

37. Kanakulya Dickson and Paola de Cuzzani, *Making Bioethics Global: Ethical Challenges in Indigenous Medicine* (Saarbrücken, DE: VDM Verlag Dr Müller, 2010), 46–47.

38. See Workineh Kelbessa, "Indigenous Knowledge and Its Contribution to Biodiversity Conservation," *International Social Science Journal* 64 (2013): 143–52; Marcia Langton and Zane Ma Rhea, "Traditional Indigenous Biodiversity-Related Knowledge,"*Australian Academic and Research Libraries* 36 (2005): 45–69. The connection between traditional knowledge and protection of the environment and biodiversity is also assumed in the UNESCO Universal Declaration on Bioethics and Human Rights, which includes "respect for traditional knowledge" in article 17.

39. Luc Hens, "Indigenous Knowledge and Biodiversity Conservation and Management in Ghana," *Journal of Human Ecology* 20 (2006): 23–25.

40. United Nations, Convention on Biological Diversity, 1992, article 8j, https://www.cbd.int/doc/legal/cbd-en.pdf. See also article 10c: "Protect and encourage customary use of biological resources in accordance with traditional cultural practices that are compatible with conservation or sustainable use requirements."

41. Beijing Declaration, adopted by the WHO Congress on Traditional Medicine, Beijing, China, 8 November 2008.

42. World Health Assembly, "Traditional Medicine," agenda item 12.4, May 22, 2009, http://www.who.int/mediacentre/factsheets/2003/fs134/en.

43. Manuel Pardo-de-Santayana and Manual J. Macia, "The Benefits of Traditional Knowledge," *Nature* 518 (2015): 487–88.

44. Donald Worster, *Nature's Economy: A History of Ecological Ideas*, 2nd ed. (Cambridge: Cambridge University Press, 1994), 2. See also chapter 2 of this book.

45. Worster, *Nature's Economy*, 304.

46. Often no distinction is made between local knowledge, traditional knowledge, and indigenous knowledge. However, indigenous knowledge is characterized by "ecological embeddedness," or a sense of place, while traditional knowledge transcends the idea of place; see David Orozco and Latha Poonamallee, "The Role of Ethics in the Commercialization of Indigenous Knowledge," *Journal of Business Ethics* 119 (2014): 276. See also United Nations Permanent Forum on Indigenous Issues, "Indigenous Peoples, Indigenous Voices," fact sheet, http://www.un.org/esa/socdev/unpfii/documents/5session_factsheet1.pdf; World Intellectual Property Organization (WIPO), "Roundtable on Intellectual Property and Traditional Knowledge," Geneva, November 1 and 2, 1999. WIPO/IPTK/RT/99/3, http://www.wipo.int/edocs/mdocs/tk/en/wipo_iptk_rt_99/wipo_iptk_rt_99_3.pdf; WHO, "Traditional Medicine" fact sheet.

47. This is 6% of the world's population. See Richard Horton, "Indigenous Peoples: Time to Act Now for Equity and Health," *Lancet* 367 (2006): 1705. Especially Latin Amer-

ica and the Caribbean have significant indigenous populations. In Bolivia, 71% of the population is indigenous; in Guatemala, 66%; Peru, 47%; Belize, 43%; Honduras, 15%; and Mexico, 14% (see Raul A. Montenegro and Carolyn Stephens, "Indigenous Health in Latin America and the Caribbean," *Lancet* 376 [2006]: 1860).

48. Janska, Serbulea, and Tobin, "The Importance of Traditional Knowledge for Meeting Public Health Needs in Developing Countries," 27–28. See the example of the Maori worldview: Mere Roberts, Waerete Norman, Nganeko Minhinnick, Del Wihing, and Carmen Kirkwood, "Kaitiakitanga: Maori Perspectives on Conservation," *Pacific Conservation Biology* 2 (1995): 7–20. A very interesting explanation of indigenous wisdom is provided by Robin Wall Kimmerer, *Braiding Sweetgrass: Indigenous Wisdom, Scientific Knowledge, and the Teaching of Plants* (Minneapolis, MN: Milkweed Editions, 2013). See also Lawrence Kalinoe, "Ascertaining the Nature of Indigenous Intellectual and Cultural Property and Traditional Knowledge and the Search for Legal Options in Regulating Access in Papua New Guinea," *Melanesian Law Journal* 27 (2000): 1–22. See also Marlene Brant Castellano, "Ethics of Aboriginal Research," *Journal of Aboriginal Health* (January 2004): 98–114; Francesca Merlan, "Indigeneity Global and Local," *Current Anthropology* 50 (2009): 303–33; Darrell A. Posey, "Commodification of the Sacred through Intellectual Property Rights," *Journal of Ethnopharmacology* 83 (2002): 3–12. For the differences between indigenous and scientific knowledge, see Anne Ross et al., *Indigenous Peoples and the Collaborative Stewardship of Nature: Knowledge Binds and Institutional Conflicts.* (Walnut Creek, CA: Left Coast Press, 2011), 31–58.

49. Geoffrey A. Cordell and Michael D. Colvard, "Some Thoughts on the Future of Ethnopharmacology," *Journal of Ethnopharmacology* 100 (2005): 5–14; Aceme Nyika, "Ethical and Regulatory Issues Surrounding African Traditional Medicine in the Context of HIV/AIDS," *Developing World Bioethics* 7 (2007): 25–34.

50. Ana K. Jäger suggests that we should "sideline the chase of new drug leads, and rather concentrate on comprehensive toxicity studies of traditional medicinal plants and on clinical trials of single plants or remedies containing mixtures of plants." This would provide more benefits to a larger population than continuous production of "new" medication. See Jäger, "Is Traditional Medicine Better Off 25 Years Later?," *Journal of Ethnopharmacology* 100 (2005): 3–4.

51. International Bioethics Committee, *Report of the IBC on Traditional Medicine Systems and Their Ethical Implications* (Paris: UNESCO, 2013), 11. For Chile as an example, see Maria Costanza Torri, "Intercultural Health Practices: Toward an Equal Recognition between Indigenous Medicine and Biomedicine? A Case Study from Chile," *Health Care Analysis* 20 (2012): 31–49.

52. David Orozco and Latha Poonamallee, "The Role of Ethics in the Commercialization of Indigenous Knowledge," 277.

53. Shane Mulligan and Peter Stoett, "A Global Bio-prospecting Regime: Partnership or Piracy?," *International Journal* 55 (2000): 227.

54. See, among many other publications, Shivendu K. Srivastava, *Commercial Use of Biodiversity: Resolving the Access and Benefit Sharing Issues* (New Delhi: Sage Publications India, 2016); Andrew Mushita and Carol B. Thompson, *Biopiracy of Biodiversity: Global Exchange as Enclosure* (Trenton, NJ: Eritrea, 2007); Padmashree Gehl Sampath, *Regulating Bioprospecting: Institutions for Drug Research, Access, and Benefit-Sharing* (Tokyo: United

Nations University Press, 2005); Sarah A. Laird, *Biodiversity and Traditional Knowledge: Equitable Partnerships in Practice* (London: Earthscan, 2002).

55. Nicolas Mateo, Werner Nader, and Giselle Tamayo, "Bioprospecting," in *Encyclopedia of Biodiversity*, ed. Simon A. Levin (New York: Academic Press, 2001), 1:475.

56. Simpson, Sedjo, and Reid, "Valuing Biodiversity for Use in Pharmaceutical Research," 164–65.

57. Daniel F. Robinson, *Confronting Biopiracy: Challenges, Cases, and International Debates* (London: Earthscan, 2010), 11.

58. Vandana Shiva, *Biopiracy: The Plunder of Nature and Knowledge* (Boston: South End Press, 1997); Shiva, *Protect or Plunder? Understanding Intellectual Property Rights* (London: Zed Books, 2001).

59. Tade M. Spranger, "Access and Benefit Sharing as a Challenge for International Law," in *Concepts and Values in Biodiversity*, ed. Dirk Lanzerath and Minou Friele (London: Routledge, 2014), 163–80.

60. Tim K. Mackey and Bryan A. Liang, "Integrating Biodiversity Management and Indigenous Biopiracy Protection to Promote Environmental Justice and Global Health," *American Journal of Public Health* 102 (2012): 1091.

61. Hanne Svarstad, Hans Chr. Bugge, and Shivcharn S. Dhillion, "From Norway to Novartis: Cyclosporine from *Tolypocladium inflatum* in an Open Access Bioprospecting Regime," *Biodiversity and Conservation* 9 (2000): 1521–41.

62. For example, over 40 countries, including India, did not patent drugs before the enforcement of the TRIPS agreement. See Lisa Forman, "'Rights' and Wrongs: What Utility for the Right to Health in Reforming Trade Rules on Medicine?," *Health and Human Rights* 10 (2008): 38.

63. Robinson, *Confronting Biopiracy*, 47–49.

64. David Conforto, "Traditional and Modern-Day Biopiracy: Redefining the Biopiracy Debate," *Journal of Environmental Law and Litigation* 19 (2004): 361.

65. Conforto, "Traditional and Modern-Day Biopiracy," 369.

66. Susan K. Sell, *Private Power, Public Law: The Globalization of Intellectual Property Rights* (Cambridge: Cambridge University Press, 2003), 75ff.

67. Sell, *Private Power, Public Law*, 104ff.

68. Sell, *Private Power, Public Law*, 118–19. See also Susan K. Sell and Aseem Prakash, "Using Ideas Strategically: The Contest between Business and NGO Networks in Intellectual Property Rights," *International Studies Quarterly* 48 (2004): 143–75.

69. Ellen 't Hoen, Jonathan Berger, Alexandra Calmy, and Suerie Moon, "Driving a Decade of Change: HIV, AIDS, Patents and Access to Medicine for All," *Journal of the International AIDS Society* 14 (2011): 15, https://doi.org/10.1186/1758-2652-14-15.

70. Henk ten Have, *Global Bioethics: An Introduction* (London: Routledge, 2016), 120ff. See also Fred Powledge, "Patenting, Piracy, and the Global Commons," *BioScience* 51 (2001): 273–77.

71. Graham Dutfield, "The Public and Private Domains: Intellectual Property Rights in Traditional Knowledge," *Science Communication* 21 (2000): 287.

72. Dutfield, "The Public and Private Domains."

73. Claudia Finetti, "Traditional Knowledge and the Patent System: Two Worlds Apart?," *World Patent Information* 33 (2011): 58.

74. Dutfield, "The Public and Private Domains," 281; Ross et al., *Indigenous Peoples and the Collaborative Stewardship of Nature*.

75. Dutfield, "The Public and Private Domains," 276.

76. United Nations, Declaration of the Rights of Indigenous Peoples, 2007, article 24.1 states: "Indigenous peoples have the right to their traditional medicines and to maintain their health practices, including the conservation of their vital medicinal plants, animals and minerals" (http://www.un.org/esa/socdev/unpfii/documents/DRIPS_en.pdf).

77. United Nations, Declaration of the Rights of Indigenous Peoples, 2007, article 31.1: "Indigenous peoples have the right to maintain, control, protect and develop their cultural heritage, traditional knowledge and traditional cultural expressions, as well as the manifestations of their sciences, technologies and cultures, including human and genetic resources, seeds, medicines, knowledge of the properties of fauna and flora, oral traditions, literatures, designs, sports and traditional games and visual and performing arts. They also have the right to maintain, control, protect and develop their intellectual property over such cultural heritage, traditional knowledge, and traditional cultural expressions" (http://www.un.org/esa/socdev/unpfii/documents/DRIPS_en.pdf).

78. See Paul Gepts, "Who Owns Biodiversity, and How Should the Owners Be Compensated?," *Plant Physiology* 134 (2004): 1295–307.

79. ten Have, *Global Bioethics*, 132ff.

80. Zinatul A. Zainol, Latifah Amin, Frank Akpoviri, and Rosli Ramli, "Biopiracy and States' Sovereignty over Their Biological Resources," *African Journal of Biotechnology* 10 (2011): 12395–408.

81. United Nations, Convention on Biological Diversity, preamble.

82. Ian James Kidd, "Biopiracy and the Ethics of Medical Heritage: The Case of India's Traditional Knowledge Digital Library," *Journal of Medical Humanities* 33 (2012): 175–83. See also Dan L. Perlman and Glenn Adelson, *Biodiversity: Exploring Values and Priorities in Conservation* (Malden, MA: Blackwell Science, 1997), 1–17; Fred Powledge, "Patenting, Piracy, and the Global Commons," *BioScience* 51 (2001): 273–77; Alex Riordan and John Schofield, "Beyond Biomedicine: Traditional Medicine as Cultural Heritage," *International Journal of Heritage Studies* 21 (2015): 280–99.

83. Graham Dutfield, "TRIPS-Related Aspects of Traditional Knowledge," *Case Western Reserve Journal of International Law* 233 (2001): 237. See also, particularly, Shiva, *Biopiracy* and Shiva, *Protect or Plunder?*

84. Joseph Millum, "How Should the Benefits of Bioprospecting Be Shared?," *Hastings Center Report* 40 (2010): 24–33; Bram De Jonge, "What Is Fair and Equitable Benefit-Sharing?," *Journal of Agricultural and Environmental Ethics* 24 (2011): 127–46; Reji K. Jospeh, "International Regime on Access and Benefit Sharing: Where Are We Now?," *Asian Biotechnology and Development Review* 12 (2010): 77–94.

85. This agreement became a signature case in benefit sharing in the early 1990s. See David Takacs, *The Idea of Biodiversity: Philosophies of Paradise* (Baltimore, MD: Johns Hopkins University Press, 1996), 288–308.

86. Human Genome Organization Ethics Committee, "Statement on Benefit Sharing," *Eubios Journal of Asian and International Bioethics* 10 (2000): 70–72.

87. Shane P. Mulligan, "For Whose Benefit? Limits to Sharing in the Bioprospecting 'Regime,'" *Environmental Politics* 8 (1999): 35–65.

88. Jospeh, "International Regime on Access and Benefit Sharing," 77–94.

89. Adam Withnall, "Antibiotics: World Leaders Sign Groundbreaking UN Declaration to Tackle 'Biggest Global Health Threat,'" *Independent*, September 20, 2016.

90. Thomas Efferth, Mita Banerjee, Norbert W. Paul, et al., "Biopiracy of Natural Products and Good Bioprospecting Practice," *Phytomedicine* 23 (2016): 166–73. See also Mateo, Nader, and Tamayo, "Bioprospecting," 1:471–88.

91. D. D. Soejarto, H.H.S. Fong, G. T. Tan, et al., "Ethnobotany/Ethnopharmacology and Mass Bioprospecting: Issues on Intellectual Property and Benefit-Sharing," *Journal of Ethnopharmacology* 100 (2005): 15–22. See also a critical self-evaluation of the European Patent Office: *Scenarios for the Future: How Might IP Regimes Evolve by 2025? What Global Legitimacy Might Such Regimes Have?* (Munich: EPO, 2007).

92. Bryan A. Liang, "Global Governance: Promoting Biodiversity and Protecting Indigenous Communities against Biopiracy," *Journal of Commercial Biotechnology* 17 (2011): 248–53; Rex Dalton, "Bioprospects Less Than Golden," *Nature* 429 (2004): 598–600; Richard D. Firn, "Bioprospecting: Why Is It So Unrewarding?," *Biodiversity and Conservation* 12 (2003): 207–16.

93. Richard Conniff, "A Bitter Pill," *Conservation* 13 (Spring 2012).

94. It takes on average 15 years to bring a sample to market, while only one sample in 240,000 will deliver a commercial drug. See Colin Macilwain, "When Rhetoric Hits Reality in Debate on Bioprospecting," *Nature* 392 (1998): 535–40.

95. Arturo Gomez-Pompa, "The Role of Biodiversity: Scientists in a Troubled World," *BioScience* 54 (2004): 224.

96. Christopher B. Barrett and Travis J. Lybbert, "Is Bioprospecting a Viable Strategy for Conserving Tropical Ecosystems?," *Ecological Economics* 34 (2000): 293–300.

97. Tony Simpson and Vanessa Jackson, "Effective Protection for Indigenous Cultural Knowledge: A Challenge for the Next Millennium," *Indigenous Affairs* 3 (1998): 44–56.

98. See Liang, "Global Governance"; Dalton, "Bioprospects Less Than Golden"; Firn, "Bioprospecting."

99. Mohd Shoaib Ansari, "Evaluation of Role of Traditional Knowledge Digital Library and Traditional Chinese Medicine Database on Preservation of Tradition Medicinal Knowledge," *DESIDOC Journal of Library and Information Technology* 36 (2016): 75; Charles Akwe Masango and Victor W. A. Mbarika, "Documenting Indigenous Knowledge about Africa's Complementary and Alternative Medicine: A Cause for Concern?," *Mousaion* 33 (2015): 43.

100. Darrell A. Posey, "Protecting Indigenous Peoples' Rights to Biodiversity," *Environment: Science and Policy for Sustainable Development* 38 (1996): 9.

101. Jospeh, "International Regime on Access and Benefit Sharing," 77–94.

102. Mackey and Liang, "Integrating Biodiversity Management and Indigenous Biopiracy Protection to Promote Environmental Justice and Global Health," 1091.

103. Daniel F. Robinson, *Biodiversity, Access and Benefit-Sharing: Global Case Studies* (Abingdon, UK: Routledge, 2015), 74.

104. Henk ten Have and Bert Gordijn, "Rotten Context: The Unaffordability of Technological Advances," *Medicine, Health Care and Philosophy* 18 (2015): 459–61.

105. See, specifically, Shiva, *Biopiracy*.

106. The three approaches are clarified by Chris Hamilton, "Biodiversity, Biopiracy and Benefits: What Allegations of Biopiracy Tell Us about Intellectual Property," *Developing World Bioethics* 6 (2006): 158–73.

107. Dutfield, "The Public and Private Domains," 258.

108. Dutfield, "The Public and Private Domains," 274.

109. Ameera Haider, "Reconciling Patent Law and Traditional Knowledge: Strategies for Countries with Traditional Knowledge to Successfully Protect Their Knowledge from Abuse," *Case Western Reserve Journal of International Law* 48 (2016): 347–70.

110. Gustavo Ghidini, Rudolph J. R. Peritz, and Marco Ricolfi, eds. *TRIPS and Developing Countries: Towards a New IP World Order?* (Cheltenham, UK: Edward Elgar, 2014).

111. For Salk, see Richard Carter, *Breakthrough: The Saga of Jonas Salk* (New York: Trident Press, 1966), 2ff. For Musk, see "All Our Patent Are Belong [sic] to You," Tesla, June 12, 2014, https://www.tesla.com/blog/all-our-patent-are-belong-you.

112. Robinson, *Biodiversity, Access and Benefit-Sharing*, 48–88.

113. See Hamilton, "Biodiversity, Biopiracy and Benefits."

114. Robinson, *Biodiversity, Access and Benefit-Sharing*, 41.

115. Hamilton, "Biodiversity, Biopiracy and Benefits," 173.

116. Shiva, *Protect or Plunder?*, 96ff.; see also Sell, *Private Power, Public Law*.

117. Shiva, *Biopiracy*, 5.

118. Conforto, "Traditional and Modern-Day Biopiracy."

119. Gerard Bodeker, "Traditional Medical Knowledge, Intellectual Property Rights and Benefit Sharing," *Cardozo Journal of International and Comparative Law* 11 (2003): 785.

120. Conforto, "Traditional and Modern-Day Biopiracy," 388.

121. Barbara Tedlock, "Indigenous Heritage and Biopiracy in the Age of Intellectual Property Rights," *Explore: The Journal of Science and Healing* 2 (2006): 258.

122. ten Have, *Global Bioethics*, 191ff.

123. Klara H. Stumpf, "Reconstructing the 'Biopiracy' Debate from a Justice Perspective," in *Concepts and Values in Biodiversity*, ed. Dirk Lanzerath and Minou Friele (London: Routledge, 2014), 225–42.

124. United Nations, Convention on Biological Diversity, preamble.

125. Doris Schroeder, "Justice and Benefit Sharing," in *Indigenous Peoples, Consent and Benefit Sharing: Lessons from the San-Hoodia Case*, ed. Rachel Wynberg, Doris Schroeder, Roger Chennels (Dordrecht: Springer, 2009), 18ff. See also Henk ten Have, *Bioethiek zonder grenzen: Mondialisering van gezondheid, ethiek en wetenschap* (Nijmegen, NL: Valkhof Pers, 2011).

126. United Nations, Declaration of the Rights of Indigenous Peoples, 2007, preamble.

127. Sell, *Private Power, Public Law*, 173. See also Sell and Prakash, "Using Ideas Strategically"; Nitsan Chorev, "The Institutional Project of Neo-liberal Globalism: The Case of the WTO," *Theory and Society* 34 (2005): 317–55.

128. See Sell, *Private Power, Public Law*.

129. Chumundeeswari Kuppuswamy, "The Ethics of Intellectual Property Rights: The Impact of Traditional Knowledge and Health on International Intellectual Property Law," *Asian Medicine* 5 (2009): 346.

130. Sharon Friel, Deborah Gleeson, Anne Marie Thow, Ronald Labonté, David Stuckler, Adrian Kay, and Wendy Snowdon, "A New Generation of Trade Policy: Potential Risks to Diet-Related Health from the Trans-Pacific Partnership Agreement," *Globalization and Health* 9 (2013): 46, https://doi.org/10.1186/1744-8603-9-46.

131. Forman, "'Rights' and Wrongs," 38.

132. 't Hoen, Berger, Calmy, and Moon, "Driving a Decade of Change," 15. See also Jonathan D. Alpern, John Song, and William M. Stauffer, "Essential Medicine in the United States: Why Access Is Diminishing," *New England Journal of Medicine* 374 (2016): 1904–7.

133. Lisa Forman, "Trading Health for Profit: The Impact of Bilateral and Regional Free Trade Agreements on Domestic Intellectual Property Rules on Pharmaceuticals," in *The Power of Pills: Social, Ethical and Legal Issues in Drug Development, Marketing and Pricing*, ed. Jillian Clare Cohen, Patricia Illingworth, and Udo Schüklenk (London: Pluto Press, 2006), 129. See also Nitsan Chorev, "The Institutional Project of Neo-liberal Globalism." For a critical analysis, see Ghidini, Peritz, and Ricolfi, *TRIPS and Developing Countries*.

134. Jeremy A. Greene, *Generic: The Unbranding of Modern Medicine* (Baltimore, MD: Johns Hopkins University Press, 2014), 243–60.

135. Shiva, *Protect or Plunder?*,114.

136. Marcus Colchester, "Conservation Policy and Indigenous Peoples," in *The Earthscan Reader in Poverty and Biodiversity Conservation*, ed. Dilys Roe and Joanna Elliott (London: Earthscan, 2010), 154.

137. Manuel Ruiz and Ronnie Vernooy, eds., *The Custodians of Biodiversity: Sharing Access to and Benefits of Genetic Resources* (Abingdon, UK: Earthscan, 2012).

138. Leonardo Boff, *Ecology and Liberation: A New Paradigm* (Maryknoll, NY: Orbis Books, 1996), 46.

139. "There will always be some aspect of the nonhuman world that will not entirely mold to human and social intentions" (Zsuzsa Gille, "Materiality: Transnational Materiality," in *Framing the Global: Entry Points for Research*, ed. Hilary E. Kahn [Bloomington: Indiana University Press, 2014], 158).

140. Universal Declaration on the Common Good of the Earth and Humanity, November 2011, http://www.humiliationstudies.org/documents/HoutartUniversalDeclarationo ftheCommonGoodofHumanity.pdf. The declaration has been adopted and promoted by the Rosa Luxemburg Foundation, established in Berlin in 1990. Its purpose is to support international dialogue and cooperation.

141. Kimmerer, *Braiding Sweetgrass*, 376.

142. David Schlosberg, *Defining Environmental Justice: Theories, Movements, and Nature* (Oxford: Oxford University Press, 2007).

CHAPTER SIX: **Food**

1. See, for these examples, Dan Koeppel, *Banana: The Fate of the Fruit That Changed the World* (London: Penguin Books, 2008); and John Reader, *Potato: A History of the Propitious Esculent* (New Haven, CT: Yale University Press, 2009).

2. Jules Pretty, *Agri-Culture: Reconnecting People, Land and Nature* (London: Earthscan, 2002).

3. The association between food production and civilizations is elaborated in Evan D. G. Fraser and Andrew Rimas, *Empires of Food: Feast, Famine, and the Rise and Fall of Civilizations* (New York: Free Press, 2010).

4. Daniel Hillel and Cynthia Rozenzweig, "Biodiversity and Food Production," in *Sustaining Life: How Human Health Depends on Biodiversity*, ed. Eric Chivian and Aaron Bernstein (Oxford: Oxford University Press, 2008), 327. See also Paul Gepts, Thomas R. Famula, Robert L. Bettinger, Stephen B. Brush, Ardeshir B. Damania, Patrick E. McGuire, and Calvin O. Qualset, eds., *Biodiversity in Agriculture: Domestication, Evolution, and Sustainability* (Cambridge: Cambridge University Press, 2012).

5. T.C.H. Sunderland, "Food Security: Why Is Biodiversity Important?," *International Forestry Review* 13 (2011): 267–68. See also Frederic Baudron and Ken E. Giller, "Agriculture and Nature: Trouble and Strife?," *Biological Conservation* 170 (2014): 232–45; Emile A. Frison, Jeremy Cherfas, and Toby Hodgkin, "Agricultural Biodiversity Is Essential for a Sustainable Improvement in Food and Nutrition Security," *Sustainability* 3 (2011): 238–53; Alice M. L. Li, "Ecological Determinants of Health: Food and Environment on Human Health," *Environmental Science and Pollution Research* 24 (2017): 9002–15.

6. Michael Jahi Chappell and Liliana A. LaValle, "Food Security and Biodiversity: Can We Have Both? An Agroecological Analysis," *Agriculture and Human Values* 28 (2011): 4.

7. Food and Agriculture Organization (FAO), International Fund for Agricultural Development (IFAD), and World Food Programme (WFP), *The State of Food Insecurity in the World 2015: Meeting the 2015 International Hunger Targets; Taking Stock of Uneven Progress* (Rome: FAO, 2015), 1.

8. Tim Lang and Michael Heasman, *Food Wars: The Global Battle for Mouths, Minds and Markets*, 2nd ed. (London: Routledge, 2015), 1. See also Lester R. Brown, *Full Planet, Empty Plates: The New Geopolitics of Food Scarcity* (New York: W. W. Norton, 2012), who concludes that "food is the weak link in our modern civilization" (122).

9. FAO, IFAD, and WFP, *The State of Food Insecurity in the World 2015*, 17.

10. See Global Hunger Index 2016, http://ghi.ifpri.org.

11. For developed countries this will be 3,490 calories; for developing countries, 3,000 calories. Nikos Alexandratos and Jelle Bruinsma, *World Agriculture towards 2030/2050: The 2012 Revision*, ESA Working Paper no. 12-03 (Rome: FAO, 2012), 23.

12. Alexandratos and Bruinsma, *World Agriculture towards 2030/2050*, 19.

13. United Nations, *Millennium Development Goals Report 2015*, 4, 20, http://www.un.org/sustainabledevelopment/sustainable-development-goals.

14. United Nations, *Millennium Development Goals Report 2015*, 22.

15. United Nations, *Millennium Development Goals Report 2015*, 8–9.

16. The notion of food security was introduced at the World Food Summit in 1996. FAO uses the following definition: "Food security exists when all people, at all times, have physical and economic access to sufficient, safe and nutritious food that meets their dietary needs and food preferences for an active and healthy life." See FAO, "Rome Declaration on World Food Security," http://www.fao.org/docrep/003/w3613e/w3613e00.HTM.

17. Quentin Farmar-Bowers, "Food Security: One of a Number of 'Securities' We Need for a Full Life; An Australian Perspective," *Journal of Agricultural and Environmental Ethics* 27 (2014): 813.

18. Chappell and LaValle, "Food Security and Biodiversity," 6.

19. Alexandratos and Bruinsma, *World Agriculture towards 2030/2050*, 19.

20. Timothy Johns and Bhuwon R. Sthapit, "Biocultural Diversity in the Sustainability of Developing Country Food Systems," *Food and Nutrition Bulletin* 25 (2004): 146.

21. Hillel and Rozenzweig, "Biodiversity and Food Production," 330; Sunderland, "Food Security," 267.

22. Steve Marquardt, "Green Havoc: Panama Disease, Environmental Change, and Labor Process in the Central American Banana Industry," *American Historical Review* 106 (2001): 49–80; Koeppel, *Banana*; Ioannis Stergiopoulos, André Drenth, and Gert Kema, "Can Science Stop the Looming Banana Extinction?," CNN, *The Conversation*, October 25, 2016, https://www.cnn.com/2016/10/25/health/banana-extinction/index.html.

23. Charles F. Hutchinson and Eric Weiss, "Remote Sensing Contribution to an Early Warning System for Genetic Erosion of Agricultural Crops," in *Proceedings of the Technical Meeting on the Methodology of the FAO World Information and Early Warning System on Plant Genetic Resources*, ed. J. Serwinski and J. Faberová, conference held at the Research Institute of Crop Production, Prague, June 21–23, 1999 (Rome: FAO, 1999). For more data, see also Lori Ann Thrupp, "Linking Agricultural Biodiversity and Food Security: The Valuable Role of Sustainable Agriculture," *International Affairs* 76 (2000): 265–81. For the ethical implications, see José Esquinas-Alcázar, "Protecting Crop Genetic Diversity for Food Security: Political, Ethical and Technical Challenges," *Nature Reviews/Genetics* 6 (2005): 946–53.

24. Sunderland, "Food Security," 268. See also Paul Gepts, "Plant Genetic Resources Conservation and Utilization: The Accomplishments and Future of a Societal Insurance Policy," *Crop Science* 46 (2006): 2278–92.

25. Hutchinson and Weiss, "Remote Sensing Contribution to an Early Warning System for Genetic Erosion of Agricultural Crops."

26. Carolyn Fry, *Seeds: A Natural History* (Chicago: University of Chicago Press, 2016), 162–64. See also Marte Qvenild, "Svalbard Global Seed Vault: A 'Noah's Ark' for the World's Seeds," *Development in Practice* 18 (2008): 110–16. For Syria, see Nick Robins-Early, "Syrian War Causes the Global Doomsday Seed Vault's First Withdrawal," *Huffington Post*, September 22, 2015.

27. Katherine M. Shea, American Academy of Pediatrics Committee on Environmental Health, and American Academy of Pediatrics Committee on Infectious Diseases, "Nontherapeutic Use of Antimicrobial Agents in Animal Agriculture: Implications for Pediatrics," *Pediatrics* 114 (2004): 862–68. See also Ellen K. Silbergeld, *Chickenizing Farms and Food: How Industrial Meat Production Endangers Workers, Animals, and Consumers* (Baltimore, MD: Johns Hopkins University Press, 2016), 89ff.

28. Baudron and Giller, "Agriculture and Nature," 233.

29. Sonja J. Vermeulen, Bruce M. Campbell, and John S. I. Ingram, "Climate Change and Food Systems," *Annual Review of Environment and Resources* 37 (2012): 200.

30. Lang and Heasman, *Food Wars*, 96–130; Jan Willem Erisman, Nick van Eeckeren, Jan de Wit, Chris Koopman, Willemijn Cuijpers, Natasja Oerlemans, and Ben J. Koks, "Agriculture and Biodiversity: A Better Balance Benefits Both," *AIMS Agriculture and Food* 1 (2016): 157–74; Lijbert Brussard, Patrick Caron, Bruce Campbell, Leslie Lipper, Susan Mainka, Rudy Rabbinge, Didier Babin, and Mirjam Pulleman, "Reconciling Biodiversity

Conservation and Food Security: Scientific Challenges for a New Agriculture," *Current Opinion in Environmental Sustainability* 2 (2010): 34–42.

31. Lang and Heasman, *Food Wars*, 58–95. See: Timothy Jones and Pablo B. Eyzaguirre, "Linking Biodiversity, Diet and Health in Policy and Practice," *Proceedings of the Nutrition Society* 65 (2006): 182–89. See also Andrew Wilby, Charles Mitchell, Dana Blumenthal, Peter Daszak, et al., "Biodiversity, Food Provision, and Human Health," in *Biodiversity Change and Human Health: From Ecosystem Services to Spread of Disease*, ed. Osvaldo Sala, Laura A. Meyerson, and Camille Parmesan (Washington, DC: Island Press, 2009), 13–39; David Wallinga, "Today's Food System: How Healthy Is It?," *Journal of Hunger and Environmental Nutrition* 4 (2009): 252–81.

32. David Tilman and Michael Clark, "Global Diets Link Environmental Sustainability and Human Health," *Nature* 515 (2014): 518–22.

33. John Zarocostas, "Q&A: WHO's Keiji Fukuda Discusses Global Plans for Antibiotic Resistance," *Pharmaceutical Journal*, November 25, 2015, http://www.pharmaceutical-journal.com/opinion/qa/qa-whos-keiji-fukuda-discusses-global-plans-for-antibiotic-resistance/20200140.article; Eleanor Goldberg, "Drug Companies Are Slowly Starting to Care More about Poor People," *Huffington Post*, September 26, 2016, http://www.huffingtonpost.com/entry/pharmaceutical-companies-need-to-start-caring-about-illnesses-that-affect-poor-people_us_57dc1c00e4b0071a6e06f6f7.

34. Laura H. Kahn, *One Health and the Politics of Antimicrobial Resistance* (Baltimore, MD: Johns Hopkins University Press, 2016), 97. See also Silbergeld, *Chickenizing Farms and Food*, 89.

35. FAO, IFAD, and WFP, *The State of Food Insecurity in the World 2015*, 31.

36. Raj Patel, *Stuffed and Starved: The Hidden Battle for the World Food System*, 2nd ed. (Brooklyn, NY: Melville House, 2012), 107–8. Another estimate is that 60% is controlled by transnational companies; see Marie-Josée Massicotte, "La Via Campesina, Brazilian Peasants, and the Agribusiness Model of Agriculture: Towards an Alternative Model of Agrarian Democratic Governance," *Studies in Political Economy* 85 (2010): 79.

37. Pretty, *Agri-Culture*, 106–7.

38. Tilman and Clark, "Global Diets Link Environmental Sustainability and Human Health," 520.

39. Vernon H. Heywood, "Ethnopharmacology, Food Production, Nutrition and Biodiversity Conservation: Towards a Sustainable Future for Indigenous Peoples," *Journal of Ethnopharmacology* 137 (2011): 1–15.

40. Vermeulen, Campbell, and Ingram, "Climate Change and Food Systems," 204.

41. Egbert Hardeman and Henk Jochemsen, "Are There Ideological Aspects to the Modernization of Agriculture?," *Journal of Agricultural and Environmental Ethics* 25 (2012): 657. See also Pretty, *Agri-Culture*; Erisman, van Eeckeren, de Wit, et al., "Agriculture and Biodiversity," 157–74.

42. Chappell and LaValle, "Food Security and Biodiversity," 4.

43. Tim Lang and David Barling, "Food Security and Food Sustainability: Reformulating the Debate," *Geographical Journal* 178 (2012): 313.

44. Robert Paarlberg, "The Ethics of Modern Agriculture," *Society* 46 (2009): 6. For a contrary view, see Paul B. Thompson, ed., *The Ethics of Intensification: Agricultural Development and Cultural Change* (Dordrecht: Springer, 2010).

45. Brussard, Caron, Campbell, et al., "Reconciling Biodiversity Conservation and Food Security," 34–42.

46. See, for example, Thrupp, "Linking Agricultural Biodiversity and Food Security"; Sara J. Scherr and Jeffrey A. McNeely, "Biodiversity Conservation and Agricultural Sustainability: Towards a New Paradigm of 'Ecoagriculture' Landscapes," *Philosophical Transactions of the Royal Society B* 363 (2008): 477–94; Tony Weis, "The Accelerating Biophysical Contradictions of Industrial Capitalist Agriculture," *Journal of Agrarian Change* 10 (2010): 315–41; report submitted by the Special Rapporteur on the right to food, Olivier De Schutter, United Nations General Assembly, 16th session, A/HRC/16/49, December 2010; Egbert Hardeman and Henk Jochemsen, "Are There Ideological Aspects to the Modernization of Agriculture?," *Journal of Agricultural and Environmental Ethics* 25 (2012): 657–74; Chappell and LaValle, "Food Security and Biodiversity," 3–26.

47. Hardeman and Jochemsen, "Are There Ideological Aspects to the Modernization of Agriculture?," 659–61.

48. Ben Phalan, Andrew Balmford, Rhys E. Green, and Jörn P. W. Scharlemann, "Minimising the Harm to Biodiversity of Producing More Food Globally," *Food Policy* 36 (2011): 562–71; Brussard, Caron, Campbell, et al., "Reconciling Biodiversity Conservation and Food Security," 34–42.

49. Baudron and Giller, "Agriculture and Nature," 237.

50. Hardeman and Jochemsen, "Are There Ideological Aspects to the Modernization of Agriculture?," 670.

51. Jules Pretty, "Agroecological Approaches to Agricultural Development," background paper for the World Development Report 2008, November 2006, https://open knowledge.worldbank.org/bitstream/handle/10986/9044/WDR2008_0031.pdf; sequence=1; Jeffrey A. McNeely and Sara J. Scherr, *Ecoagriculture: Strategies to Feed the World and Save Wild Biodiversity* (Washington, DC: Island Press, 2003).

52. Lang and Heasman, *Food Wars*, 35ff.

53. Hayo Apotheker, "Is Agriculture in Need of Ethics?," *Journal of Agricultural and Environmental Ethics* 12 (2000): 10.

54. Pretty, *Agri-Culture*, 114–15. See also René Shaw Hughner, Pierre McDonagh, Andrea Prothero, Clifford J. Schultz II, and Julie Stanton, "Who Are Organic Food Consumers? A Compilation and Review of Why People Purchase Organic Food," *Journal of Consumer Behavior* 6 (2007): 94–110; Eric Chivian and Aaron Bernstein, "Genetically Modified Foods and Organic Farming," in *Sustaining Life: How Human Health Depends on Biodiversity*, ed. Eric Chivian and Aaron Bernstein (Oxford: Oxford University Press, 2008), 395–405; Harvey Levenstein, *Fear of Food: A History of Why We Worry about What We Eat* (Chicago: University of Chicago Press, 2012), 107–24; William Lockeretz, ed. *Organic Farming: An International History* (Oxfordshire, UK: CAB International, 2007).

55. Chappell and LaValle, "Food Security and Biodiversity," 7, 11.

56. A. Whitney Sanford, "Ethics, Narrative, and Agriculture: Transforming Agricultural Practice through Ecological Imagination," *Journal of Agricultural and Environmental Ethics* 24 (2011): 284–85.

57. Erisman, van Eeckeren, de Wit, et al., "Agriculture and Biodiversity," 158.

58. Pretty, *Agri-Culture*, 29, 157ff.

59. Chappell and LaValle, "Food Security and Biodiversity," 8.

60. Baudron and Giller, "Agriculture and Nature," 242.

61. Hugh Lehman, *Rationality and Ethics in Agriculture* (Moscow: University of Idaho Press, 1995), 153.

62. Lehman, *Rationality and Ethics in Agriculture*, 153, 160.

63. Genetic modification is "the deliberate and specific manipulation of an organism's genome in order to modify aspects of its biology" (Chivian and Bernstein, "Genetically Modified Foods and Organic Farming," 385). This is the life sciences integrated paradigm in the framework of Lang and Heasman (*Food Wars*, 31–35).

64. Per Sandin and Payam Moula, "Modern Biotechnology, Agriculture, and Ethics," *Journal of Agricultural and Environmental Ethics* 28 (2015): 804. Sven-Erik Jacobsen, Marten Sørensen, Søre Marcus Pedersen, and Jocon Weiner, "Feeding the World: Genetically Modified Crops versus Agricultural Biodiversity," *Agronomy for Sustainable Development* 33 (2013): 652–53.

65. European Commission, Directorate-General for Research, *Europeans and Biotechnology in 2010: Winds of Change?* (Brussels : European Commission, 2010), 36–39.

66. Emily Waltz, "Vitamin A Super Banana in Human Trials," *Nature Biotechnology* 32 (2014): 857.

67. Norman E. Borlaug, "Ending World Hunger: The Promise of Biotechnology and the Threat of Antiscience Zealotry," *Plant Physiology* 124 (2000): 490.

68. Paul B. Thompson and William Hannah, "Food and Agricultural Biotechnology: A Summary and Analysis of Ethical Concerns," *Advances in Biochemical Engineering/Biotechnology* 111 (2008): 229–64. McKay Jenkins, *Food Fight: GMOs and the Future of the American Diet* (New York: Avery, 2017).

69. Jacobsen, Sørensen, Pedersen, and Weiner, "Feeding the World," 654. See also Doug Gurian-Sherman, *Failure to Yield: Evaluating the Performance of Genetically Engineered Crops* (Cambridge, MA: Union of Concerned Scientists, 2009).

70. Ksenia Gerasimova, "Debates on Genetically Modified Crops in the Context of Sustainable Development," *Science and Engineering Ethics* 22 (2016): 539.

71. Bernice Bovenkerk, *The Biotechnology Debate: Democracy in the Face of Intractable Disagreement* (Dordrecht: Springer, 2012), 302–5.

72. Michelle Marvier and Rene C. Van Acker, "Can Crop Transgenes Be Kept on a Leash?," *Frontiers in Ecology and the Environment* 3 (2005): 102.

73. Thompson and Hannah, "Food and Agricultural Biotechnology," 249–50.

74. Marcello Buiatti, "Biologies, Agricultures, Biotechnologies," *Tailoring Biotechnologies* 1 (2005): 16.

75. See chapter 5.

76. Bovenkerk, *The Biotechnology Debate*, 262ff. She names the five major players: Monsanto (United States), DuPont (United States), Novartis (Switzerland), AstraZeneca (United Kingdom / Sweden), and Aventis (Germany/France) (267).

77. Jeffrey Burkhardt, "Biotechnology, Ethics, and the Structure of Agriculture," *Journal of Agricultural Economics Research* 41 (1988): 53–60. See also Michael Blakeley, "Recent Developments in Intellectual Property and Power in the Private Sector Related to Food and Agriculture," *Food Policy* 36 (2011): S109–S113.

78. Jacobsen, Sørensen, Pedersen, and Weiner, "Feeding the World," 653.

79. Jacobsen, Sørensen, Pedersen, and Weiner, "Feeding the World," 659.

80. Bovenkerk, *The Biotechnology Debate*, 268.

81. Buiatti, "Biologies, Agricultures, Biotechnologies," 20–21.

82. Ronald L. Sandler defines food systems as "the complex network of processes, infrastructures and actors that produce the food we eat and deliver it to where we eat it" (Sandler, *Food Ethics: The Basics* [London: Routledge, 2015], 4). See also Harriet Friedmann, "Discussion: Moving Food Regimes Forward: Reflections on Symposium Essays," *Agriculture and Human Values* 26 (2009): 335–44; Philip McMichael, "A Food Regime Analysis of the 'World Food Crisis,'" *Agriculture and Human Values* 26 (2009): 281–95.

83. Colin Khoury quoted in Chelsea Harvey, "How the World's Most Popular Foods Have Traveled All over the Planet," *Independent*, June 10, 2016. Khoury concludes from his study: "The main finding really is that everybody's connected to everywhere else to some degree or another." For the original publication, see Colin K. Khoury, Harold A. Achicanoy, Anne D. Bjorkman, Carlos Navarro-Racines, et al., "Origins of Food Crops Connect Countries Worldwide," *Proceedings of the Royal Society B* 283 (2016): 20160792.

84. Sarah J. Martin and Peter Andrée, "From Food Security to Food Sovereignty in Canada: Resistance and Authority in the Context of Neoliberalism," in *Globalization and Food Sovereignty: Global and Local Change in the New Politics of Food*, ed. Peter Andrée, Jeffrey Ayres, Michael J. Bosia, and Marie-Josée Massicotte (Toronto: University of Toronto Press, 2014), 187.

85. See the various chapters in Andrée, Ayres, Bosia, and Massicotte, *Globalization and Food Sovereignty*.

86. This is also the conclusion of an editorial in the *Lancet*: "The global architecture of the international nutrition system needs to be radically reformed and become more accountable and inclusive" ("Tackling Global Food Insecurity," *Lancet* 371 [2008]: 532).

87. Jessica Franzo, "Ethical Issues for Human Nutrition in the Context of Global Food Security and Sustainable Development," *Global Food Security* 7 (2015): 17.

88. Franzo, "Ethical Issues for Human Nutrition in the Context of Global Food Security and Sustainable Development," 18.

89. See Sandler, *Food Ethics*; and Paul B. Thompson, *From Field to Fork: Food Ethics for Everyone* (Oxford: Oxford University Press, 2015).

90. Corinne Pelluchon has elaborated these broader meanings of food as nourishment; see Pelluchon, *Les nourritures: Philosophie du corps politique* (Paris: Editions du Seuil, 2015).

91. Marion Nestle, *Food Politics: How the Food Industry Influences Nutrition and Health* (Berkeley: University of California Press, 2007), 28.

92. Elke Stehfest, "Food Choices for Health and Planet," *Nature* 515 (2014): 502.

93. Julie Beaulac, Elizabeth Kristjansson, and Steven Cummins, "A Systematic Review of Food Deserts, 1966–2007," *Preventing Chronic Diseases* 6 (2009): 1–8. The US Department of Agriculture estimates that 23.5 million US residents live in food deserts; see J. M. Dieterle, "Food Deserts and Lockean Property," in *Just Food: Philosophy, Justice and Food*, ed. J. M. Dieterle (London: Rowman and Littlefield International, 2015), 39–55.

94. Amy E. Guptill, Denise A. Copelton, and Betsy Lucal, *Food and Society: Principles and Paradoxes*, 2nd ed. (Cambridge: Polity Press, 2017), 85.

95. Nestle, *Food Politics*.

96. Michele Simon, *Appetite for Profit: How the Food Industry Undermines Our Health and How to Fight Back* (New York: Nation Books, 2006), 10–13; Jennifer L. Harris, Jennifer L. Pomeranz, Tim Lobstein, and Kelly D. Brownell, "A Crisis in the Marketplace: How Food Marketing Contributes to Childhood Obesity and What Can Be Done," *Annual Review of Public Health* 30 (2009): 211–25.

97. Nestle, *Food Politics*, 28.

98. J. Michael McGinnis, Jennifer Appleton Gootman, and Vivica I. Kraak, eds., *Food Marketing to Children and Youth: Threat or Opportunity?* (Washington, DC: National Academies Press, 2006); Nicholas Nisbett, Stuart Gillespie, Lawrence Haddad, and Jody Harris, "Why Worry about the Politics of Childhood Undernutrition?," *World Development* 64 (2014): 420–33.

99. See Harris, Pomeranz, Lobstein, and Brownell, "A Crisis in the Marketplace," 218, 220.

100. "Majority of Americans Do Not Trust Food Industry, Survey Finds," *Processing*, March 29, 2013, https://www.processingmagazine.com/majority-of-americans-do-not-trust-food-industry-survey-finds.

101. Frans W. A. Brom has distinguished the three types of consumer concerns. See Brom, "Food, Consumer Concerns, and Trust: Food Ethics for a Globalizing Market," *Journal of Agricultural and Environmental Ethics* 12 (2000): 127–39. See also Lucia Reisch, Ulrike Eberle, and Sylvia Lorek, "Sustainable Food Consumption: An Overview of Contemporary Issues and Policies," *Sustainability: Science, Practice and Policy* 9 (2013): 11.

102. Levenstein, *Fear of Food*. See also Andrew F. Smith, *Fast Food: The Good, the Bad and the Hungry* (London: Reaktion Books, 2016); Ching-Fu Lin, "Global Food Safety: Exploring Key Elements for an International Regulatory Strategy," *Virginia Journal of International Law* 51 (2011): 637–96.

103. "Horsemeat Labelled as Beef Found in Romania," *Huffington Post*, February 20, 2013; Hunter Stuart, "Frozen Burger Sales Plummet in UK after Horse Meat Controversy, Report Says," *Huffington Post*, February 27, 2013; Paul Mooney, "The Story behind China's Tainted Milk Scandal," *US News*, October 9, 2008.

104. Yarlini Balarajan and Michael R. Reich, "Political Economy Challenges in Nutrition," *Globalization and Health* 12 (November 2016): https://doi.org/10.1186/s12992-016-0204-6.

105. Thompson, *From Field to Fork*, 24ff. See also Hub Zwart, "A Short History of Food Ethics," *Journal of Agricultural and Environmental Ethics* 12 (2000): 113–26.

106. Nestle, *Food Politics*, 360.

107. Sandler, *Food Ethics*, 134ff.

108. Elizabeth Smythe, "Food Sovereignty, Trade Rules, and the Struggle to Know the Origins of Food," in *Globalization and Food Sovereignty: Global and Local Change in the New Politics of Food*, ed. Peter Andrée, Jeffrey Ayres, Michael J. Bosia, and Marie-Josée Massicotte (Toronto: University of Toronto Press, 2014), 298.

109. Mellman Group, "Voters Want GMO Food Labels Printed on Packaging," November 23, 2015, http://4bgr3aepis44c9bxt1ulxsyq.wpengine.netdna-cdn.com/wp-content/uploads/2015/12/15memn20-JLI-d6.pdf.

110. Sandler, *Food Ethics*, 168ff.

111. See Zwart, "A Short History of Food Ethics," 113–26. See also Rob Irvine, "Food Ethics: Issues of Consumption and Production," *Bioethical Inquiry* 10 (2013): 145–48.

112. Every day hunger kills more people than HIV/AIDS, tuberculosis, and malaria together. Martín Caparrós, *Hunger: The Mortal Crisis of Our Time* (New York: Other Press, 2017); Patel, *Stuffed and Starved*.

113. "The contemporary age is not short of terrible and nasty happenings, but the persistence of extensive hunger in a world of unprecedented prosperity is surely one of the worst." Amartya Sen, *Development as Freedom* (New York: Anchor Books, 1999), 204.

114. James Vernon, *Hunger: A Modern History* (Cambridge, MA: Belknap Press, 2007), 273.

115. Vernon, *Hunger*, 18. See also Robert I. Rotberg and Theodore K. Rabb, *Hunger and History: The Impact of Changing Food Production and Consumption Patterns on Society* (Cambridge: Cambridge University Press, 1983).

116. Peter Singer, "Famine, Affluence and Morality," *Philosophy and Public Affairs* 1 (1972): 229–43.

117. Amartya Sen, *Poverty and Famine: An Essay on Entitlement and Deprivation* (New York: Oxford University Press, 1981).

118. John Shaw, *Food Security: A History since 1945* (Houndmills, UK: Palgrave Macmillan, 2007).

119. George Kent, *Freedom from Want: The Human Right to Adequate Food* (Washington, DC: Georgetown University Press, 2005).

120. Lucy Jarosz, "Defining World Hunger: Scale and Neoliberal Ideology in International Food Security Policy Discourse," *Food, Culture and Society* 14 (2011): 134.

121. See Onora O'Neill, *Faces of Hunger: An Essay on Poverty, Justice and Development* (London: Allen and Unwin, 1986), 113ff.

122. This argument is developed in George Kent, ed., *Global Obligations for the Right to Food* (Lanham, MD: Rowman and Littlefield, 2008).

123. This is in fact the McKeown thesis discussed in chapter 3: that health follows wealth. Alleviating hunger and malnourishment is the effect of rising income rather than better health. See Lang and Heasman, *Food Wars*; and Lang and Barling, "Food Security and Food Sustainability," 313–26.

124. Jarosz, "Defining World Hunger," 123.

125. Jarosz, "Defining World Hunger," 126. See also Philip McMichael, "Food Security and Social Reproduction: Issues and Contradictions," in *Power, Production and Social Reproduction: Human In/security in the Global Political Economy*, ed. Isabella Bakker and Stephen Gill (Houndmills, UK: Palgrave Macmillan, 2003), 169–89.

126. See note 16. The definition of food security adopted at this summit is usually quoted as the authoritative one.

127. Jarosz, "Defining World Hunger," 130. See, for different framings of food security, Madeleine Fairbairn, "Framing Resistance: International Food Regimes and the Roots of Food Sovereignty," in *Food Sovereignty in Canada: Creating Just and Sustainable Food Systems*, ed. Hannah Wittman, Annette Aurélie Desmarais, and Nettie Wiebe (Halifax: Fernwood Publishing, 2011), 15–32.

128. John Wilkinson, "Food Security and the Global Agrifood System: Ethical Issues in Historical and Sociological Perspective," *Global Food Security* 7 (2015): 9.

129. Fred Pearce, *The Land Grabbers: The New Fight over Who Owns the Earth* (Boston: Beacon Press, 2012).

130. See Brown, *Full Planet, Empty Plates*. He concludes: "Food is the weak link in our modern civilization" (122).

131. See Christopher R. Mayes and Donald B. Thompson, "What Should We Eat? Biopolitics, Ethics and Nutritional Scientism," *Bioethical Inquiry* 12 (2015): 587–99. See also Thompson, *From Field to Fork*, 43–46.

132. Guptill, Copelton, and Lucal, *Food and Society*, 19, 61ff.

133. Martha McMahon, "Local Food: Food Sovereignty or Myth of Alternative Consumer Sovereignty?," in *Globalization and Food Sovereignty: Global and Local Change in the New Politics of Food*, ed. Peter Andrée, Jeffrey Ayres, Michael J. Bosia, and Marie-Josée Massicotte (Toronto: University of Toronto Press, 2014), 125. See also Wittman, Desmarais, and Wiebe, *Food Sovereignty in Canada*.

134. Michael Menser defines food sovereignty as "the right of all peoples, their nations, or unions of states to define their respective agricultural and food policies." See Menser, "The Territory of Self-Determination: Social Reproduction, Agro-ecology, and the Role of the State," in *Globalization and Food Sovereignty: Global and Local Change in the New Politics of Food*, ed. Peter Andrée, Jeffrey Ayres, Michael J. Bosia, and Marie-Josée Massicotte (Toronto: University of Toronto Press, 2014), 53.

135. Kees Blokland, "The Peasant Road towards an Alternative Development Model: The Managua Declaration," *Development in Practice* 3 (1993): 61–63; Massicotte, "La Via Campesina, Brazilian Peasants, and the Agribusiness Model of Agriculture," 69–98; Philip McMichael, "Historicizing Food Sovereignty," *Journal of Peasant Studies* 41 (2014): 933–57.

136. Paul Nicholson explains this agenda: "We propose local food markets, the right of any country to protect its borders from imported food, sustainable agriculture and the defence of biodiversity, health food, jobs and strong livelihood in rural areas." See Nicholson, "Via Campesina: Responding to Global Systemic Crisis," *Development* 51 (2008): 457. See also Mark Navin, "Food Sovereignty and Gender Justice: The Case of La Via Campesina," in *Just Food: Philosophy, Justice and Food*, ed. J. M. Dieterle (London: Rowman and Littlefield International, 2015), 87–100; Annette Aurélie Desmarais, *La Via Campesina: Globalization and the Power of Peasants* (Halifax: Fernwood Publishing, 2007).

137. Managua Declaration, https://viacampesina.org/en/managua-declaration.

138. Jennifer Clapp and Doris Fuchs, *Corporate Power in Global Agrifood Governance* (Cambridge, MA: MIT Press, 2009), 7–12.

139. Kent, *Freedom from Want*, 47. See also Ian Werkheiser, Shakara Tyler, and Paul B. Thompson, "Food Sovereignty: Two Conceptions of Food Justice," in *Just Food: Philosophy, Justice and Food*, ed. J. M. Dieterle (London: Rowman and Littlefield International, 2015), 77.

140. McMichael, "Historicizing Food Sovereignty," 942ff.

141. Four companies (Nestlé, Sara Lee, Kraft, and Procter & Gamble) combined with Tchibo buy half of the world's harvest. See Mara Fridell, Ian Hudson, and Mark Hudson, "With Friends like These: The Corporate Response to Fair Trade Coffee," *Review of Radical Political Economics* 40 (2008): 10.

142. Between 2000 and 2004, sales have tripled in the United States and doubled in the United Kingdom. See Fridell, Hudson, and Hudson, "With Friends like These," 14.

Since 2008, fair trade certified sales continued to grow 22% per year; see Noah Zerbe, "Exploring the Limits of Fair Trade: The Local Food Movement in the Context of Late Capitalism," in *Globalization and Food Sovereignty: Global and Local Change in the New Politics of Food*, ed. Peter Andrée, Jeffrey Ayres, Michael J. Bosia, and Marie-Josée Massicotte (Toronto: University of Toronto Press, 2014), 95.

143. Werkheiser, Tyler, and Thompson, "Food Sovereignty," 71–86.

144. Martín Caparrós, *Hunger*, 533.

145. Grace Sung, "Cows Can Fly While World's Poor Starve," YaleGlobal Online, October 3, 2002, http://yaleglobal.yale.edu/content/cows-can-fly-while-worlds-poor-starve.

146. For the United States, see Chris Edwards, "Agricultural Subsidies," Downsizing the Federal Government, October 7, 2016, https://www.downsizinggovernment.org/agriculture/subsidies; for the European Union, see FarmSubsidy.org, http://farmsubsidy.openspending.org.

147. Lou Keune, "Development Cooperation: A Hindrance for Self-Sustainable Development," *Tailoring Biotechnology* 1 (2005): 48. See for the case of Mexico after the North American Free Trade Agreement, see Steve Tammelleo, "Food Policy, Mexican Migration and Collective Responsibility," in *Just Food: Philosophy, Justice and Food*, ed. J. M. Dieterle (London: Rowman and Littlefield International, 2015), 101–18.

148. Keune, "Development Cooperation," 54.

149. "The market, through competition, is supposed to bring increased efficiency and lower prices. But the effect of turning food production over to the market has been to produce less competition, and offer more structural power to the largest companies" (Patel, *Stuffed and Starved*, 112).

150. Dieterle, *Just Food*, xvii.

151. These and many other examples are discussed and elaborated in Robert Gottlieb and Anupama Joshi, *Food Justice* (Cambridge, MA: MIT Press, 2013).

152. Werkheiser, Tyler, and Thompson, "Food Sovereignty," 75. See also Navin, "Food Sovereignty and Gender Justice," 87–100.

153. Overall, world food prices in 2008 were 83% higher compared with 2005. See Ray Bush and Giulliano Martiniello, "Food Riots and Protest: Agrarian Modernizations and Structural Crisis," *World Development* 91 (2017): 193.

154. Clyde Lee Miller, "World Hunger, Poverty and Ethics," *Cross Currents* (Fall 1977): 308.

155. Krishna Mani Pathak denounces a lack of collective ethics. The global food regime is an "ethical failure"; it should focus on the "forms of inequality that exist in the world today" (Pathak, "Poverty and Hunger in the Developing World: Ethics, the Global Economy, and Human Survival," *Asia Journal of Global Studies* 3 [2010]: 89, 90, 99). Martin McLaughlin has pointed out that this is exactly the view of the Catholic Church. The problems with the food system are the result of "outmoded strictures that perpetuate unbearable injustices" (McLaughlin, *World Food Security: A Catholic View of Food Policy in the New Millennium* [Washington, DC: Center of Concern, 2002], 117). See also James McCarthy and Scott Prudham, "Neoliberal Nature and the Nature of Neoliberalism," *GeoForum* 35 (2004): 275–83.

156. Apotheker, "Is Agriculture in Need of Ethics?," 13.

157. Wilkinson, "Food Security and the Global Agrifood System," 9.

158. Pretty, *Agri-Culture*, 52–53.

159. Valeria Negri, "Agro-biodiversity Conservation in Europe: Ethical Issues," *Journal of Agricultural and Environmental Ethics* 18 (2005): 10–11.

160. Lang and Heasman, *Food Wars*, 20.

161. This is Noah Zerbe's argument, using the notion of embeddedness (see Zerbe, "Exploring the Limits of Fair Trade," 92). The argument is based on the well-known book by Karl Polanyi, originally published in 1944, *The Great Transformation: The Political and Economic Origins of Our Time* (Boston: Beacon Press, 1971). Harvey James argues that even Adam Smith has pointed out that economic considerations (self-interest and the pursuit of profit) need to be balanced with noneconomic considerations. There are limits to the pursuit of self-interest. In Smith's system, competition should be combined with sympathy and justice. Harvey S. James, "Sustainable Agriculture and Free Market Economics: Finding Common Ground in Adam Smith," *Agriculture and Human Values* 23 (2006): 427–38.

162. OWINFS, "About Our World Is Not for Sale," http://notforsale.mayfirst.org/en /about.

163. Henk ten Have, *Global Bioethics: An Introduction* (Abingdon, UK: Routledge, 2016), 211ff.

164. This argument is advanced by Onora O'Neill: the right to food implies that a hungry person has a legitimate claim against specifiable others. Human rights empower the powerless. But without counterpart obligations from specific agents and agencies, not much will happen (O'Neill, *Faces of Hunger*, 101, 119).

165. For example, Ari Paloviita, Teea Kortetmäki, Antti Puupponen, and Tiina Silvasti, "Vulnerability Matrix of the Food System: Operationalizing Vulnerability and Addressing Food Security," *Journal of Cleaner Production* 135 (2016): 1242–55; Glenn Fieldman, "Neoliberalism, the Production of Vulnerability and the Hobbled State: Systemic Barriers to Climate Adaptation," *Climate and Development* 3 (2011): 159–74; Polly J. Ericksen, "What Is the Vulnerability of a Food System to Global Environmental Change?," *Ecology and Society* 13 (2008): 14; Michael J. Watts and Hans G. Bohle, "The Space of Vulnerability: The Causal Structure of Hunger and Famine," *Progress in Human Geography* 17 (1993): 43–67.

166. Henk ten Have, *Vulnerability: Challenging Bioethics* (Abingdon, UK: Routledge, 2016).

167. ten Have, *Vulnerability*, 149–66.

168. See Jampel Dell'Angelo, Paolo d'Odorico, Maria Cristina Rulli, and Philippe Marchand, "The Tragedy of the Grabbed Commons: Coercion and Dispossession in the Global Land Rush," *World Development* 92 (2017): 1–12.

169. Stephan Barthel, Carole L. Crumley, and Uni Svedin, "Biocultural Refugia: Combating the Erosion of Diversity in Landscapes and Food Production," *Ecology and Society* 18 (2013): 71, http://dx.doi.org/10.5751/ES-06207-180471.

170. Peter Andrée, "Citizen-Farmers: The Possibilities and the Limits of Australia's Emerging Alternative Food Networks," in *Globalization and Food Sovereignty: Global and Local Change in the New Politics of Food*, ed. Peter Andrée, Jeffrey Ayres, Michael J. Bosia, and Marie-Josée Massicotte (Toronto: University of Toronto Press, 2014), 141ff., 153.

171. Ronald Sandler discusses the relationship between food and cultural diversity (*Food Ethics*, 168–75).

172. Three crops—rice, wheat, and maize—account for more than 55% of human energy intake. Thomas Allen, Paolo Prosperi, Bruce Cogill, and Guillerma Flichman, "Agricultural Biodiversity, Social-Ecological Systems and Sustainable Diets," *Proceedings of the Nutrition Society* 73 (2014): 501.

173. Khoury et al., "Origins of Food Crops Connect Countries Worldwide," 7.

174. Sarah Wright describes a positive example of solidarity. A network of small farmers in the Philippines achieved better control over production, better access to food, better quality and diversity of food, less vulnerability, increasing biodiversity, empowerment, and self-reliance. See Wright, "Food Sovereignty in Practice: A Study of Farmer-Led Sustainable Agriculture in the Philippines," in *Globalization and Food Sovereignty: Global and Local Change in the New Politics of Food*, ed. Peter Andrée, Jeffrey Ayres, Michael J. Bosia, and Marie-Josée Massicotte (Toronto: University of Toronto Press, 2014), 209ff. La Via Campesina used a smart strategy to broaden its bases by engaging consumer organizations. It also succeeded in engaging FAO and in influencing governmental meetings and international platforms (see Desmarais, *La Via Campesina*).

175. Daniel J. Hurst, "Restoring a Reputation: Invoking the UNESCO Universal Declaration on Bioethics and Human Rights to Bear on Pharmaceutical Pricing," *Medicine, Health Care and Philosophy* 20 (2016): 105–17. See also ten Have, *Global Bioethics*, 222–24.

176. Unilever (2000) and Nestlé (2001) joined early; Coca Cola (2006), Pepsico (2008), General Mills (2008), Campbell Soup (2009), and Monsanto (2009) later; and Bunge (2015), Archer Daniels Midland (2016), and Kellogg (2016) only recently. However, many large food industries (such as Kraft, Cargill, Tyson Food, ConAgra and McDonald's) have not joined. See https://www.unglobalcompact.org/what-is-gc.

177. Thompson, *From Field to Fork*, 174–92. The argument is elaborated in Paul B. Thompson, *The Agrarian Vision: Sustainability and Environmental Ethics* (Lexington: University Press of Kentucky, 2010).

178. FAO, *Ethical Issues in Food and Agriculture*, FAO Ethics Series 1 (Rome: FAO, 2001), 13. See also Johannes M. M. Engels, Hannes Dempewold, and Victoria Henson-Apollonia, "Ethical Considerations in Agro-biodiversity Research, Collecting, and Use," *Journal of Agricultural and Environmental Ethics* 24 (2011): 107–26.

179. Fairbairn, "Framing Resistance," 24ff.

180. This is what Sanford calls "ecological imagination," driven by interdependence, reciprocity, balance, and mutual obligation (A. Whitney Sanford, "Ethics, Narrative, and Agriculture: Transforming Agricultural Practice through Ecological Imagination," *Journal of Agricultural and Environmental Ethics* 24 [2011]: 283–303).

181. Simon, *Appetite for Profit*, 85, 299ff.

182. McLaughlin, *World Food Security*, 117.

183. For this food moralization framework, see Michael Heasman, "Toward an Ethical System," *Harvard International Review* (September 2006). See also Terry Newholm, Sandra Newholm, and Deirdre Shaw, "A History for Consumption Ethics," *Business History* 57 (2015): 290–310.

184. "Ecological citizenship" is on the rise. See Gill Seyfang, "Ecological Citizenship and Sustainable Consumption: Examining Local Organic Food Networks," *Journal of Rural Studies* 22 (2006): 383–95.

185. J. I. R. Castanon, "History of the Use of Antibiotic [*sic*] as Growth Promoters in European Poultry Feeds," *Poultry Science* 86 (2007): 2466–71. See also Silberstein, *Chickenizing Farms and Food*.

186. Janet M. Conway, *Edges of Global Justice: The World Social Forum and Its "Others"* (London: Routledge, 2013), 67–74.

187. Adrian Smith, Alison Stenning, and Katie Willis, eds., *Social Justice and Neoliberalism: Global Perspectives* (London: Zed Books, 2008) give many examples of how neoliberalism can be contested and domesticated.

188. Beth Kowitt, "The War on Big Food," *Fortune* 171 (June 2015): 61–70.

CHAPTER SEVEN: **Water**

1. Cynthia Barnett, *Rain: A Natural and Cultural History* (New York: Crown Publishers, 2015).

2. James Salzman, *Drinking Water: A History* (New York: Overlook Duckworth, 2012); Lida Schelwald-van der Kley and Linda Reijerkerk, *Water: A Way of Life; Sustainable Water Management in a Cultural Context* (Leiden: CRC Press/Balkema, 2009); Veronica Strang, *The Meaning of Water* (Oxford: Berg, 2004).

3. Eran Feitelson, "What Is Water? A Normative Perspective," *Water Policy* 14 (2012): 52–64; Christopher Hamlin, "'Waters' or 'Water'? Master Narratives in Water History and Their Implications for Contemporary Water Policy," *Water Policy* 2 (2000): 313–25.

4. Steven Solomon, *Water: The Epic Struggle for Wealth, Power, and Civilization* (New York: HarperCollins, 2010).

5. David Lewis Feldman, *Water* (Cambridge: Polity, 2012), 1.

6. Mike Acreman, "Ethical Aspects of Water and Ecosystems," *Water Policy* 3 (2001): 257–65.

7. WHO, "Drinking-Water," fact sheet, November 2016, http://www.who.int/mediacentre/factsheets/fs391/en.

8. WHO, "Mortality and Burden of Disease from Water and Sanitation," http://www.who.int/gho/phe/water_sanitation/burden/en.

9. WHO, "Mortality and Burden of Disease from Water and Sanitation," http://www.who.int/gho/phe/water_sanitation/burden_text/en.

10. A. Prüss-Ustün, J. Wolf, C. Corvalán, R. Bos, and M. Neira, "Preventing Diseases through Health Environments: A Global Assessment of the Burden of Disease from Environmental Risks" (Geneva: WHO, 2016), x, 1.

11. UNICEF, *Thirsting for a Future: Water and Children in a Changing Climate* (New York: UNICEF, 2017), 14.

12. United Nations (UN), *Report of the United Nations Water Conference*, Mar del Plata, Argentina, March 14–25, 1977 (New York: United Nations, 1977):11, 107.

13. UN, *Report of the United Nations Water Conference*, 97.

14. UN, *Report of the United Nations Water Conference*, 66.

15. UN, *Report of the United Nations Water Conference*, 67.

16. Asit K. Biswas, "From Mar del Plata to Kyoto: An Analysis of Global Water Policy Dialogues," http://www.doccentre.net/docsweb/water1/water-biswas.htm.

17. World Commission on Water, *World Water Vision: A Water Secure World; Vision for Water, Life, and the Environment* (Marseille: World Water Council 2000). The Commission

argues that "there is growing recognition that the world is now beginning to feel the first pangs of a more chronic and systemic water crisis" (12).

18. Philip Micklin, "The Aral Sea Disaster," *Annual Review of Earth and Planetary Sciences* 35 (2007): 47–72. See also Michael Gross, "The World's Vanishing Lakes," *Current Biology* 27 (2017): R43–R56.

19. Mesfin M. Mekonnen and Arjen Y. Hoekstra, "Four Billion People Facing Severe Water Scarcity," *Science Advances* 2 (2016): https://doi.org/10.1126/sciadv.1500323.

20. UNICEF, *Thirsting for a Future*, 20–21.

21. Munir A. Hanjra and M. Ejaz Qureshi, "Global Water Crisis and Future Food Security in an Era of Climate Change," *Food Policy* 35 (2010): 365–77. See also Charles J. Vörösmarty, P. Green, J. Salisbury, and R. B. Lammers, "Global Water Resources: Vulnerability from Climate Change and Population Growth," *Science* 289 (2000): 284–88.

22. World Commission on Water, *World Water Vision*, 34ff.

23. World Commission on Water, *World Water Vision*, 53.

24. For the crisis language regarding water, see Robin Clarke, *Water: The International Crisis* (Cambridge, MA: MIT Press, 1993); Rob Bowden, *What if We Do Nothing? Earth's Water Crisis* (Milwaukee, WI: World Almanac Library, 2007).

25. Frank R. Rijsberman, "Water Scarcity: Fact or Fiction?," *Agricultural Water Management* 80 (2006): 5–22. Barbara Rose Johnston, "The Commodification of Water and the Human Dimensions of Manufactured Scarcity," in *Globalization, Water and Health: Resource Management in Times of Scarcity*, ed. Linda Whiteford and Scott Whiteford (Santa Fe, NM: School of American Research Press, 2005), 133–52.

26. "For the water crisis is mainly one of distribution of water, knowledge and resources and not one of absolute scarcity" (Lord Selborne, *The Ethics of Freshwater Use: A Survey* [Paris: UNESCO, 2000], 5).

27. Mesfin M. Mekonnen and A. Y. Hoekstra, *National Water Footprint Accounts: The Green, Blue, and Grey Water Footprint of Production and Consumption*, vol. 1, main report (Delft, NL: UNESCO-IHE Institute for Water Education, 2011).

28. Mike Acreman, *Water and Ecology* (Paris, UNESCO, 2004), 6.

29. Mekonnen and Hoekstra, *National Water Footprint Accounts*, 7, 10–20.

30. Rijsberman, "Water Scarcity," 9–10.

31. Peter H. Gleick, "Water and Conflict: Fresh Water Resources and International Security," *International Security* 18 (1993): 79–112. See also David L. Feldman, *Water Politics* (Cambridge: Polity Press, 2017), 164–83.

32. John Fleck, *Water Is for Fighting over and Other Myths about Water in the West* (Washington, DC: Island Press, 2016). See also Jim Lochhead and Pat Mulroy, "The Colorado River Story," in *The Water Problem: Climate Change and Water Policy in the United States*, ed. Pat Mulroy (Washington, DC: Brookings Institution Press, 2017), 69–86.

33. Vandana Shiva, *Water Wars: Privatization, Pollution, and Profit* (2002; Berkeley, CA: North Atlantic Books, 2016), viii, 27.

34. Shiva, *Water Wars*, 99.

35. Conflicts are the result of "politics of inequality." See Jeremy Allouche, "The Sustainability and Resilience of Global Water and Food Systems: Political Analysis of the Interplay between Security, Resource Scarcity, Political Systems and Global Trade," *Food Policy* 36 (2011): S4.

36. "The human right to water is indispensable for leading a life in human dignity." UN Committee on Economic, Social and Cultural Rights, *General Comment No. 15: The Right to Water*, January 20, 2003, UN Doc. E/C.12/2002/11, http://www.refworld.org/docid /4538838d11.html.

37. Donald Worster, *Nature's Economy: A History of Ecological Ideas*, 2nd ed. (Cambridge: Cambridge University Press, 1994). See also chapters 2 and 5 of this book.

38. Jeremy J. Schmidt and Dan Shrubsole, "Modern Water Ethics: Implications for Shared Governance," *Environmental Values* 22 (2013): 359–79.

39. Janos J. Bogardi, David Dudgeon, Richard Lawford, et al., "Water Security for a Planet under Pressure: Interconnected Challenges of a Changing World Call for Sustainable Solutions," *Current Opinion in Environmental Sustainability* 4 (2012): 35–43. David Groenfeldt clearly identifies the water crisis as a moral crisis (Groenfeldt, *Water Ethics: A Values Approach to Solving the Water Crisis* [New York: Routledge, 2013], x).

40. Ravichandran Moorthy and Ganesan Jeyabalan, "Ethics and Sustainability: A Review of Water Policy and Management," *American Journal of Applied Sciences* 9 (2012): 24–31; Francisco Garcia Novo, "Moral Drought: The Ethics of Water Use," *Water Policy* 14 (2012): 65–72; Acreman, "Ethical Aspects of Water and Ecosystems," 257–65.

41. See Groenfeldt, *Water Ethics*; Peter G. Brown and Jeremy J. Schmidt, eds., *Water Ethics: Foundational Readings for Students and Professionals* (Washington, DC: Island Press, 2010); Adrian C. Armstrong, "Further Ideas towards a Water Ethic," *Water Alternatives* 2 (2009): 138–47; Armstrong, "Ethical Issues in Water Use and Sustainability," *Area* 38 (2006): 9–15; Mohammed Shail and Sue Cavill, "Ethics: Making It the Heart of Water Supply," *Civil Engineering* 159 (2006): 11–15; Metta Winter, "The Science and Ethics of Clean Water," *Human Ecology* (2003): 16–17; Robert Sandford and Merrell-Ann Phare, "A New Water Ethic," *Alternatives Journal* 37 (2011): 12–14.

42. The commission is known as COMEST, the abbreviation of its French name, Commission mondiale d'éthique des connaissances scientifiques et des technologies.

43. "The Commission . . . was to serve as a forum of reflection and was mandated to formulate principles that could provide decision-makers in sensitive areas with criteria that went beyond the purely economic or scientific" (Selborne, *The Ethics of Freshwater Use*, 2).

44. Jeremy J. Schmidt, "Water Ethics and Water Management," in *Water Ethics*, ed. Peter G. Brown and Jeremy J. Schmidt (Washington, DC: Island Press, 2010), 5; Groenfeldt, *Water Ethics*, 7.

45. Jerome Delli Priscoli, James Dooge, and Ramón Llamas, *Water and Ethics: Overview* (Paris: UNESCO, 2004), 8. From the Islamic perspective, water is a common good that should not be for sale. See Faraj Al-Awar, Mohammad J. Abdulrazzak, and Radwan Al-Weshah, "Water Ethics Perspectives in the Arab Region," in *Water Ethics*, ed. Peter G. Brown and Jeremy J. Schmidt (Washington, DC: Island Press, 2010), 33; Magdy A. Hefni, "Water Commoditization: An Ethical Perspective for a Sustainable Use and Management of Water Resources, with Special Reference to the Arab Region," in *Re-thinking Water and Food Security: Fourth Botin Foundation Water Workshop*, ed. Luis Martinez-Cortina, Alberto Garrido, and Elena Lopez-Gunn (London: Taylor and Francis, 2010), 159–80.

46. Selborne, *The Ethics of Freshwater Use*, 7.

47. "[A]mong the participants influencing regional management are powerful international corporations whose agendas must be adjusted to serve rather than dominate regional needs" (Selborne, *The Ethics of Freshwater Use*, 38).

48. World Commission on the Ethics of Scientific Knowledge and Technology (COMEST), *Best Ethical Practice in Water Use* (Paris: UNESCO, 2004), 15.

49. Selborne, *The Ethics of Freshwater Use*, 30.

50. Selborne, *The Ethics of Freshwater Use*, 6–7. See, for a somewhat different range of ethical principles for water management, Darryl R. J. Macer, *Water Ethics and Water Resource Management* (Bangkok: UNESCO, 2011).

51. Dan Tarlock and Patricia Wouters, "Reframing the Water Security Dialogue," *Journal of Water Law* 20 (2010): 53–60; Christina Cook and Karen Bakker, "Water Security: Debating an Emerging Paradigm," *Global Environmental Change* 22 (2012): 94–102.

52. Tim Ford, "Water and Health," in *Environmental Health: From Global to Local*, ed. Howard Frumkin, 2nd ed. (San Francisco: Jossey-Bass, 2010), 487–555; David Resnik, *Environmental Health Ethics* (Cambridge: Cambridge University Press, 2012), 141–48.

53. Salzman, *Drinking Water*, 131.

54. Salzman, *Drinking Water*, 108.

55. Robert Langkjaer-Bain, "The Murky Tale of Flint's Deceptive Water Data," *Significance* 14 (2017): 16–21; Victoria Morckel, "Why the Flint, Michigan, USA Water Crisis Is an Urban Planning Failure," *Cities* 62 (2017): 23–27; Henry A. Giroux, "Poisoned City in the Age of Casino Capitalism," *Theory in Action* 10 (2017): 7–31; Lawrence O. Gostin, "Politics and Public Health: The Flint Drinking Water Crisis," *Hastings Center Report* 46 (2016): 5–6.

56. Solomon, *Water: The Epic Struggle for Wealth, Power, and Civilization*.

57. Fleck, *Water Is for Fighting over and Other Myths about Water in the West*.

58. Elena Lopez-Gunn, Lucia De Stefano, and M. Ramón Llamas, "The Role of Ethics in Water and Food Security: Balancing Utilitarian and Intangible Values," *Water Policy* 14 (2012): 89–105.

59. Allouche, "The Sustainability and Resilience of Global Water and Food Systems," S4; Aaron T. Wolf, "Conflict and Cooperation along International Waterways," *Water Policy* 1 (1998): 251–65.

60. Allouche, "The Sustainability and Resilience of Global Water and Food Systems," S5. See also Wendy Barnaby, "Do Nations Go to War over Water?," *Nature* 458 (2009): 282–83.

61. Shiva, *Water Wars*, xv.

62. It is surprising that the UNESCO report on freshwater ethics briefly mentions water as "common property" but does not refer at all to the notion of common heritage that plays such an important role in the UNESCO tradition. See Selborne, *The Ethics of Freshwater Use*, 30.

63. *Directive 2000/60/EC of the European Parliament and of the Council Establishing a Framework for Community Action in the Field of Water Policy* October 23, 2000, 1, https://www.bmu.de/fileadmin/bmu-import/files/pdfs/allgemein/application/pdf/water_framework_directive.pdf.

64. Henk ten Have, *Global Bioethics: An Introduction* (Abingdon, UK: Routledge, 2016), 117ff.

65. See Kemal Baslar, *The Concept of the Common Heritage of Mankind in International Law* (The Hague: Martinus Nijhoff, 1998); Alexandre Kiss, "The Common Heritage of Mankind: Utopia or Reality?," *International Journal* 40 (1985): 423–41.

66. Derek Wall, *The Commons in History: Culture, Conflict, and Ecology* (Cambridge, MA: MIT Press, 2014).

67. Joseph L. Sax, "Understanding Transfers: Community Rights and the Privatization of Water," in *Water Ethics: Foundational Readings for Students and Professionals*, ed. Peter G. Brown and Jeremy J. Schmidt (Washington, DC: Island Press, 2010), 117–24.

68. Garrett Hardin, "The Tragedy of the Commons," *Science* 162 (1968): 1244.

69. Elinor Ostrom, *Governing the Commons: The Evolution of Institutions for Collective Action* (New York: Cambridge University Press, 1990).

70. W. M. Hanemann, "The Economic Conception of Water," in *Water Crisis: Myth or Reality?*, ed. Peter P. Rogers, M. Ramón Llamas, and Luis Martinez-Cortina (London: Taylor and Francis, 2006), 61–91; Hubert H. G. Savenije, "Why Water Is Not an Ordinary Economic Good, or Why the Girl Is Special," *Physics and Chemistry of the Earth* 27 (2002): 741–44; P. van der Zaag and H.H.G. Savenije, *Water as an Economic Good: The Value of Pricing and the Failure of Markets* (Delft, NL: UNESCO-IHE, 2006).

71. Marcela Olivera and Jorge Viaña, "Winning the Water War," *Human Rights* (Spring 2003): 10–11. See also ten Have, *Global Bioethics*, 123; Madeline Baer, "The Global Water Crisis, Privatization, and the Bolivian Water War," in *Water, Place, and Equity*, ed. John M. Whiteley, Helen Ingram, and Richard Perry (Cambridge, MA: MIT Press, 2008), 195–224; Andrew Nickson and Claudia Vargas, "The Limitations of Water Regulations: The Failure of the Cochabamba Concession in Bolivia," *Bulletin of Latin American Research* 21 (2002): 99–120.

72. The Convention on the Right of the Child, adopted in 1989, and the Convention on the Rights of Persons with Disabilities, adopted in 2006.

73. WHO, *The Right to Water* (Geneva: World Health Organization, 2003).

74. UN Committee on Economic, Social and Cultural Rights, *General Comment No. 15: The Right to Water*, article 2.

75. UN General Assembly, 64th session, resolution adopted on July 28, 2010: The Human Right to Water and Sanitation, UN A/RES/64/292, 2, http://www.un.org/en/ga/search/view_doc.asp?symbol=A/RES/64/292.

76. UN, General Assembly, Human Rights Council, 15th session, resolution adopted on October 6, 2010: Human Rights and Access to Safe Drinking Water and Sanitation, UN A/HRC/RES/15/9, http://www.right2water.eu/sites/water/files/UNHRC%20Resolution%202015-9.pdf.

77. Erik B. Bluemel, "The Implications of Formulating a Right to Water," *Ecology Law Quarterly* 31 (2004): 957–1006; Salman M. A. Salman and Siobhan McInerney-Lankford, *The Human Right to Water: Legal and Policy Dimensions* (Washington, DC: World Bank, 2004).

78. Peter H. Gleick, "The Human Right to Water," *Water Policy* 1 (1998): 496; Gleick, "Water in Crisis: Path to Sustainable Water Use," *Ecological Applications* 8 (1998): 575; Jonathan Chenoweth, "Minimum Water Requirement for Social and Economic Development," *Desalination* 229 (2008): 245–56. The basic question is what constitutes a dignified

level of living; there are different estimates of a basic water requirement. See Eran Feitelson, "What Is Water? A Normative Perspective," *Water Policy* 14 (2012): 54.

79. Gleick, "The Human Right to Water," 489.

80. Wouter Vandenhole and Tamara Wielders, "Water as a Human Right—Water as an Essential Service: Does It Matter?," *Netherlands Quarterly of Human Rights* 26 (2008): 424.

81. Chad Staddon, Thomas Appleby, and Evadne Grant, "A Right to Water? Geographico-legal Perspectives," in *The Right to Water: Politics, Governance and Social Struggles*, ed. Farhana Sultana and Alex Loftus (London: Earthscan, 2012), 61–77.

82. Tim Hayward, "A Global Right to Water," *Midwest Studies in Philosophy* 40 (2016): 217–33; Mathias Risse, "The Human Right to Water and Common Ownership of the Earth," *Journal of Political Philosophy* 22 (2014): 178–203.

83. Jamie Linton, "The Human Right to What? Water, Rights, Humans and the Relation of Things," in *The Right to Water: Politics, Governance and Social Struggles*, ed. Farhana Sultana and Alex Loftus (London: Earthscan, 2012), 45–60.

84. Karen Bakker, "The 'Commons' versus 'Commodity': Alter-globalization, Anti-privatization and the Human Right to Water in the Global South," *Antipode* 39 (2007): 439.

85. Jessica Budds and Gordon McGranahan, "Are the Debates on Water Privatization Missing the Point? Experiences from Africa, Asia and Latin America," *Environment and Urbanization* 15 (2003): 87–114.

86. Maude Barlow, *Blue Covenant: The Global Water Crisis and the Coming Battle for the Right to Water* (New York: New Press, 2007), 5.

87. Christiana Z. Peppard, *Just Water: Theology, Ethics, and the Global Water Crisis* (Maryknoll, NY: Orbis Books, 2014); Neelke Doorn, "Water and Justice: Towards an Ethics of Water Governance," *Public Reason* 5 (2013): 97–114; Friends of the Earth International, *Water Justice for All: Global and Local Resistance to the Control and Commodification of Water* (Amsterdam: Friends of the Earth International, 2003); Helen Ingram, John M. Whiteley, and Richard Perry, "The Importance of Equity and the Limits of Efficiency in Water Resources," in *Water, Place, and Equity*, ed. John M. Whiteley, Helen Ingram, and Richard Perry (Cambridge, MA: MIT Press, 2008), 1–32; Adam Davidson-Harden, Anil Naidoo, and Andi Harden, "The Geopolitics of the Water Justice Movement," *Peace Conflict and Development* 11 (2007): 1–34.

88. Kemal Derviş, foreword to *Beyond Scarcity: Power, Poverty and the Global Water Crisis*, United Nations Development Programme, Human Development Report 2006 (New York: UNDP, 2006), v.

89. Farhana Sultana and Alex Loftus, "The Human Right to Water: Critiques and Condition of Possibility," *WIREs Water* 2 (2015): 97–105.

90. In the United Kingdom, disconnection is legally prohibited (Staddon, Appleby, and Grant, "A Right to Water," 70). In general, see Adrian Walsh, "The Commodification of the Public Service of Water," *Public Reason* 3 (2011): 90–106.

91. Davidson-Harden, Naidoo, and Harden, "The Geopolitics of the Water Justice Movement," 1–34.

92. Anand Kumar, "Coco-Cola Is in Troubled Waters (Again) for a Factory It Was Forced to Shut Down 12 Years Ago," *Yahoo News*, July 21, 2016, https://www.yahoo.com/news/coca-cola-troubled-waters-again-120000412.html; Environmental Justice Atlas,

Coca Cola Plant in Plachimada, Kerala, India, June 201, https://ejatlas.org/conflict/coca-cola
-kerala-india.

93. Charles Fishman, "If Bottled Water Is So Bad, Why Are Sales Hitting Records?,"
National Geographic, April 20, 2016.

94. David L. Feldman, *Water Politics: Governing Our Most Precious Resource* (Cambridge: Polity Press, 2017), 111.

95. Peter H. Gleick, *Bottled and Sold: The Story behind Our Obsession with Bottled Water*
(Washington, DC: Island Press, 2010).

96. UN, *Sustainable Development Goals*, August 2015, 16, http://www.un.org/sustain
abledevelopment/sustainable-development-goals.

97. WHO, *Financing Universal Water, Sanitation and Hygiene under the Sustainable Development Goals*, UN-Water Global Analysis and Assessment of Sanitation and Drinking-
Water (GLAAS) 2017 report (Geneva: WHO, 2017).

98. Feldman, *Water Politics*, 38.

99. Feldman, *Water Politics*, 91; Budds and McGranahan, "Are the Debates on Water
Privatization Missing the Point?"; Erik Swyngedouw, "Dispossessing H_2O: The Contested
Terrain of Water Privatization," *Capitalism Nature Socialism* 16 (2005): 81–98.

100. Budds and McGranahan, "Are the Debates on Water Privatization Missing the
Point?," 102. See also Karen Bakker, *Privatizing Water: Governance Failure and the World's
Urban Water Crisis* (Ithaca, NY: Cornell University Press, 2010), 92–97.

101. Patrick Bond, "Water Commodification and Decommodification Narratives: Pric-
ing and Policy Debates from Johannesburg to Kyoto to Cancun and Back," *Capitalism
Nature Socialism* 15 (2004): 17; Johnston, "The Commodification of Water and the Human
Dimensions of Manufactured Scarcity," 133–52.

102. It is interesting that in the United States the public utility model is the most common one. In France, water is mostly provided by private companies. The major transnational water companies are French (e.g., Suez and Veolia). See Karen J. Bakker, "A Political
Ecology of Water Privatization," *Studies in Political Economy* 70 (2003): 35–58.

103. Alexander J. Loftus and David A. McDonald, "Of Liquid Dreams: A Political
Ecology of Water Privatization in Buenos Aires," *Environment and Urbanization* 13 (2001):
179–99.

104. Karen J. Bakker, *An Uncooperative Commodity: Privatizing Water in England and
Wales* (Oxford: Oxford University Press, 2003).

105. Barbara Rose Johnston, "The Political Ecology of Water: An Introduction," *Capitalism Nature Socialism* 14 (2003): 84–85.

106. Belen Balanya, Brid Brennan, Olivier Hoedeman, Satoko Kishimoto, and Philipp
Terhorst, eds., *Reclaiming Public Water: Achievements, Struggles and Visions from around the
World* (Amsterdam: Transnational Institute and Corporate Europe Observatory, 2005).

107. Karen Bakker, *Privatizing Water*.

108. David Groenfeldt and Jeremy J. Schmidt, "Ethics and Water Governance," *Ecology
and Society* 18 (2013): 14. See also Muhammad Mizanur Rahaman and Olli Varis, "Integrated Water Resources Management: Evolution, Prospects and Future Challenges,"
Sustainability: Science, Practice, and Policy 1 (2005): 15–21.

109. Flora Ijjas, "Social Indicators and Ethics in Sustainable Water Management," *Pe-
riodica Polytechnica Social and Management Sciences* 23 (2015): 113–20.

110. Andrea K. Gerlak, Robert G. Varady, Olivier Petit, and Arin C. Haverland, "Hydrosolidarity and Beyond: Can Ethics and Equity Find a Place in Today's Water Resource Management?," *Water International* 36 (2011): 252. See also Andrea K. Gerlak, Robert G. Varady, and Arin C. Haverland, "Hydrosolidarity and International Water Governance," *International Negotiation* 14 (2009): 311–28; Malin Falkenmark and Carl Folke, "Ecohydrosolidarity: A New Ethics for Stewardship of Value-Adding Rainfall," in *Water Ethics: Foundational Readings for Students and Professionals*, ed. Peter G. Brown and Jeremy J. Schmidt (Washington, DC: Island Press, 2010), 247–64; William J. Cosgrove, "Fulfilling the World Water Vision: Hydrosolidarity," *Water International* 28 (2003): 527–32.

111. Andres Cózar, Elisa Marti, Carlos M. Duarte, et al., "The Arctic Ocean as a Dead End for Floating Plastics in the North Atlantic Branch of the Thermophaline Circulation," *Science Advances* 3 (2017): https://doi.org/10.1126/sciadv.1600582; Dominique Mosbergen, "The Oceans Are Drowning in Plastic—and No One's Paying Attention," *Huffington Post*, April 27, 2017, https://www.huffingtonpost.com/entry/plastic-waste-oceans_us _58fcd37be4b0c46f0781d426; Ian Johnston, "Once-Pristine Arctic Ocean Contains 300 Billion Pieces of Plastic, Study Suggests," *Independent*, April 19, 2017; World Economic Forum, *The New Plastics Economy: Rethinking the Future of Plastics* (Cowes, UK: Ellen MacArthur Foundation, 2016). A recent study in Malaysia found microplastics in fishes sold to human beings; see Ali Karami, Abolfazi Golieskardi, Yu Bin Ho, Vincent Larat, and Babak Salamatinia, "Microplastics in Eviscerated Flesh and Excised Organs of Dried Fish," *Scientific Reports* 7 (2017): https://doi.org/10.1038/s41598-017-05828-6.

112. World Economic Forum, *The New Plastics Economy*, 6.

113. See Kate Kershner, "Americans Use 500 Million Straws Every Day: Would You Pledge to Go Strawless?," How Stuff Works, January 6, 2017, http://now.howstuffworks .com/2017/01/06/americans-use-500-million-straws-every-day-would-you-pledge-go -strawless.

114. As argued by Evo Morales, president of Bolivia, in his opening speech to the Third World Water Forum. See UNESCO, *Water and Indigenous Peoples*, ed. Rutgerd Boelens, Moe Chiba, and Douglas Nakashima (Paris: UNESCO, 2006), 22–23; Kenichi Matsui, "Water Ethics for First Nations and Biodiversity in Western Canada," *International Indigenous Policy Journal* 3 (2012): https://doi.org/10.18584/iipj.2012.3.3.4.

115. Victoria Tauli Corpuz, "Indigenous Peoples and International Debates on Water: Reflections and Challenges," in *Water and Indigenous Peoples*, ed. Rutgerd Boelens, Moe Chiba, and Douglas Nakashima (Paris: UNESCO, 2006), 24–35.

116. Robert William Sandford and Merrell-Ann S. Phare, *Ethical Water: Learning to Value What Matters Most* (Victoria, BC: Rocky Mountain Books, 2011).

117. See Joyeeta Gupta, Aziza Akhmouch, William Cosgrove, Zachary Hurwitz, Josephina Maestu, and Olcay Ünver, "Policymakers' Reflections on Water Governance Issues," *Ecology and Society* 18 (2013): http://dx.doi.org/10.5751/ES-05086-180135.

118. The World Water Council is, in the words of Maude Barlow, "a major vehicle for the corporate takeover of the world's water." See Barlow, *Blue Covenant*, 50; Maude Barlow and Tony Clarke, *Blue Gold: The Fight to Stop the Corporate Theft of the World's Water* (New York: New Press, 2002), 157–60; Bond, "Water Commodification and Decommodification Narratives," 20.

119. Selborne, *The Ethics of Freshwater Use*, 6–7. See also Jerome Delli Priscoli, "What Is Public Participation in Water Resources Management and Why Is It Important?," *Water International* 29 (2004): 1–7.

120. Jeremy J. Schmidt and Dan Shrubsole, "Modern Water Ethics: Implications for Shared Governance," *Environmental Values* 22 (2013): 359–79.

121. Peppard, *Just Water*, 35–185.

122. Feldman, *Water Politics*.

123. Sandra Postel, *Last Oasis: Facing Water Scarcity* (1992; New York: W. W. Norton, 1997), 187.

124. Feldman, *Water Politics*, 126ff.

125. Shiva, *Water Wars*, 67.

126. Patrick Bond, "Privatisation, Participation and Protest in the Restructuring of Municipal Services: Grounds for Opposing World Bank Promotion of 'Public-Private Partnerships,'" *Urban Forum* 9 (1998): 37–75.

127. David Dudgeon, Angela H. Arthington, Mark O. Gessner, et al., "Freshwater Biodiversity: Importance, Threats, Status and Conservation Challenges," *Biological Review* 81 (2006): 164.

128. R. Rheingans, R. Dreibelbis, and M. C. Freeman, "Beyond the Millennium Development Goals: Public Health Challenges in Water and Sanitation," *Global Public Health* 1 (2006): 31–48.

129. Malin Falkenmark and Johan Rockström, *Balancing Water for Humans and Nature: The New Approach in Ecohydrology* (London: Earthscan, 2004).

130. Jeremy J. Schmidt, *Water: Abundance, Scarcity, and Security in the Age of Humanity* (New York: New York University Press, 2017), 41.

131. Jamie Linton, *What Is Water? The History of a Modern Abstraction* (Vancouver: UBC Press, 2010), 70.

132. Adrian Walsh, "The Commodification of the Public Service of Water," *Public Reason* 3 (2011): 90–106.

133. Linton, *What Is Water?*, 49.

134. Walsh, "The Commodification of the Public Service of Water," 101.

135. Daniel Finn, "Water Wars in Ireland," *New Left Review* 95 (2015): 49–63.

136. Hefni, "Water Commoditization," 162.

CHAPTER EIGHT: **Global Bioethics in Practice**

1. See Robert C. Cassidy and Alan R. Fleischman, eds., *Pediatric Ethics: From Principles to Practice* (Amsterdam: Harwood Academic Publishers, 1996); Richard B. Miller, *Children, Ethics and Modern Medicine* (Bloomington: Indiana University Press, 2003); Geoffrey Miller, ed., *Pediatric Bioethics* (Cambridge: Cambridge University Press, 2010); Alan R. Fleischman, *Pediatric Ethics: Protecting the Interests of Children* (Oxford: Oxford University Press, 2016).

2. American Academy of Pediatrics Council on Community Pediatrics, "Poverty and Child Health in the United States," *Pediatrics* 137 (2016): e20260339.

3. Gary Rischitelli, Peggy Nygren, Christina Bougatsos, Michelle Freeman, and Mark Helfand, *Screening for Elevated Lead Levels in Childhood and Pregnancy: Update of a 1996 U.S.*

Preventive Services Task Force Review, Evidence Syntheses, no. 44 (Rockville, MD: Agency for Healthcare Research and Quality, 2006).

4. Mattie Kahn, "Meet the Woman Building the 'Model Public Health Program' in Flint, Michigan," *Elle*, February 16, 2016.

5. David B. Resnik and Gerard Roman, "Health, Justice, and the Environment," *Bioethics* 21 (2007): 10.

6. Van Rensselaer Potter, *Global Bioethics: Building on the Leopold Legacy* (East Lansing: Michigan State University Press, 1988), 78.

7. Van Rensselaer Potter, "Moving the Culture toward More Vivid Utopias with Survival as Goal," *Global Bioethics* 14 (2001): 19–30.

8. Potter, *Global Bioethics*, 60.

9. Pankaj Mishra, *Age of Anger: A History of the Present* (Milton Keynes, UK: Allen Lane, 2017), 11.

10. Ulrike Schuerkens, *Social Changes in a Global World* (London: Sage, 2017), 21ff.

11. For this argument, see Henk ten Have, *Global Bioethics: An Introduction* (London: Routledge, 2016), 102–3.

12. Naomi Grimley, "Identity 2016: 'Global Citizenship' Rising, Poll Suggests," *BBC News*, April 28, 2016.

13. ten Have, *Global Bioethics*, 47–49 107–9.

14. Max Bearak, "Theresa May Criticized the Term 'Citizen of the World': But Half the World Identifies That Way," *Washington Post*, October 5, 2016.

15. See Wolfgang Sachs, *Planet Dialectics: Explorations in Environment and Development* (London: Zed Books, 2015).

16. Doreen Massey, "Geographies of Responsibility," *Geografiska Annaler: Series B; Human Geography* 86 (2004): 5–18.

17. Tim Ingold, "Globes and Spheres: The Topology of Environmentalism," in *Environmentalism: The View from Anthropology*, ed. Kay Milton (London: Routledge, 1993), 31–42.

18. Philip McMichael, *Development and Social Change: A Global Perspective* (Los Angeles: Sage, 2017), 4.

19. For the transition from development project into globalization project, see McMichael, *Development and Social Change*.

20. ten Have, *Global Bioethics*, 66–71.

21. Mishra, *Age of Anger*, 324.

22. For recent publications, see Wolfgang Streeck, *How Will Capitalism End? Essays on a Failing System* (London: Verso, 2016); Sanford F. Schram, *The Return of Ordinary Capitalism: Neoliberalism, Precarity, Occupy* (Oxford: Oxford University Press, 2015); Simon Reid-Henry, *The Political Origins of Inequality: Why a More Equal World Is Better for Us All* (Chicago: University of Chicago Press, 2015); Brad Evans and Henry A. Giroux, *Disposable Futures: The Seduction of Violence in the Age of Spectacle* (San Francisco: Open Media Series/City Lights Books, 2015). For a Christian critique, see Ryan LaMothe, "Neoliberal Capitalism and the Corruption of Society: A Pastoral Political Analysis," *Pastoral Psychology* 65 (2016): 5–21. See also Pope Francis, *Laudato si: On Care for Our Common Home*, encyclical letter (Vatican City, May 2015).

23. ten Have, *Global Bioethics*, 61ff.

24. United Nations High Commissioner for Refugees, *Global Trends: Forced Displacement in 2015* (Geneva: UNHCR, 2016).

25. McMichael, *Development and Social Change*, 148.

26. GBD Mortality and Causes of Death Collaborators, "Global, Regional, and National Life Expectancy, All-Cause Mortality, and Cause-Specific Mortality for 249 Causes of Death, 1980–2015: A Systematic Analysis for the Global Burden of Disease Study 2015," *Lancet* 388 (2016): 1459–544.

27. Amanda Glassman and Miriam Temin, eds., *Millions Saved: New Cases of Proven Success in Global Health* (Washington, DC: Center for Global Development, 2016), 1.

28. WHO, *World Health Statistics* (Geneva: WHO, 2006).

29. Samuel L. Dickman et al., "Health Spending for Low-, Middle-, and High-Income Americans, 1963–2012," *Health Affairs* 35 (2016): 1189–96.

30. ten Have, *Vulnerability*, 149ff.

31. McMichael, *Development and Social Change*, 158. Zygmunt Bauman describes how human beings have become the "collateral casualties" of progress. See Bauman, *Wasted Lives: Modernity and Its Outcasts* (Cambridge: Polity Press, 2004), 15, 132. Brad Evans and Henry Giroux examine the politics of disposability in *Disposable Futures.*

32. McMichael, *Development and Social Change*, 218.

33. Will Steffen, Asa Persson, Lisa Deutsch, et al., "The Anthropocene: From Global Change to Planetary Stewardship," *Ambio* 40 (2011): 749.

34. Ramachandra Guha, *Environmentalism: A Global History* (New York: Longman, 2000).

35. Reuters, "Five Pacific Islands Lost to Rising Seas as Climate Change Hits," *Guardian*, May 10, 2016.

36. James Garvey, *The Ethics of Climate Change: Right and Wrong in a Warming World* (London: Continuum, 2008).

37. Lee Hannah, Thomas E. Lovejoy, and Stephen H. Schneider, "Biodiversity and Climate Change in Context," in *Climate Change and Biodiversity*, ed. Thomas E. Lovejoy and Lee Hannah (New Haven, CT: Yale University Press, 2005), 12.

38. Garvey, *The Ethics of Climate Change*, 27.

39. United Nations High Commissioner for Refugees, "Frequently Asked Questions on Climate Change and Disaster Displacement," November 6, 2016, http://www.unhcr.org /news/latest/2016/11/581f52dc4/frequently-asked-questions-climate-change-disaster -displacement.html.

40. Ariel Scotti, "Two Billion People May Become Refugees from Climate Change by the End of the Century," *New York Daily News*, June 27, 2017.

41. Alastair Greig, David Hulme, and Mark Turner, *Challenging Global Inequality: Development Theory and Practice in the 21st Century* (Houndmills, UK: Palgrave Macmillan, 2007). See also Vicente Navarro, "Neoliberalism as a Class Ideology; or, the Political Causes of the Growth of Inequalities," *International Journal of Health Services* 37 (2007): 47–62.

42. Jonathan D. Ostry, Prakash Loungani, and Davide Furceri, "Neoliberalism: Oversold?," *Finance and Development* 53 (2016): 38–41.

43. Rebecca Riffkin, "In US, 67% Dissatisfied with Income, Wealth Distribution," Gallup, January 20, 2014, http://www.gallup.com/poll/166904/dissatisfied-income-wealth -distribution.aspx.

44. The IMF, for example, in 2017 optimistically announced that global economic growth is increasing by 3.5%. But a diminishing part goes to employees, further increasing inequality in income, so that fewer people are benefiting. See International Monetary Fund, *World Economic Outlook April 2017: Gaining Momentum?* (Washington, DC: IMF, 2017).

45. Joseph E. Stiglitz, *The Price of Inequality. How Today's Divided Society Endangers Our Future* (New York: W. W. Norton, 2012), 74.

46. ten Have, *Global Bioethics*, 67ff. See also Noel Castree, "Neoliberalising Nature: The Logics of Deregulation and Reregulation," *Environment and Planning A*, 40 (2008): 131–52.

47. Schuerkens, *Social Changes in a Global World*.

48. See, for example, David Parkin, John Appleby, and Alan Maynard, "Economics: The Biggest Fraud Ever Perpetrated on the World?," *Lancet* 382 (2013): e11–e15; Thomas Piketty, *Capital in the Twenty-First Century* (Cambridge, MA: Belknap Press, 2014). Even notorious protagonists of neoliberalism are today arguing that the market should be embedded in other values so that economic policies will benefit and protect all people rather than just the rich and corporations. See the recent interview with Larry Summers, former US secretary of the treasury and president of Harvard University (Nathan Gardels, "Globalization Will Work if We Stop Catering to the Elite, Says Larry Summers," *Huffington Post*, June 22, 2017). As chief economist of the World Bank, Summers in 1991 circulated an internal memo that export of toxic waste and pollution to developing countries (or the Third World) was economically sound, "world-welfare enhancing trade" to be encouraged by the World Bank. Costs of pollution are less in countries with the lowest wages, because in such countries, people usually die before its health impacts, such as cancer, arise. Summers rejected objections against this economic and technical reasoning because such objections are moral and social statements that could be used against all decisions of the Bank. See James Ferguson, *Global Shadows: Africa in the Neoliberal World Order* (Durham, NC: Duke University Press, 2006), 70–71.

49. Anthony B. Atkinson, *Inequality: What Can Be Done?* (Cambridge, MA: Harvard University Press, 2015). Other scholars are more critical and pessimistic about the possibilities of changing and reducing inequalities. In his recent book, Walter Scheidel argues that in the past, inequality has been decreased primarily by violent ruptures. Large-scale violence such as mass mobilization and (total) warfare, transformative revolutions, state failures, and lethal pandemics have been the key mechanisms of violent leveling. See Scheidel, *The Great Leveler: Violence and the History of Inequality from the Stone Age to the Twenty-First Century* (Princeton, NJ: Princeton University Press, 2017).

50. Peter Singer, *One World: The Ethics of Globalization*, 2nd ed. (New Haven, CT: Yale University Press, 2004); see also Garvey, *The Ethics of Climate Change*, 125ff.

51. Singer, *One World*, 35.

52. Ramachandra Guha, *How Much Should a Person Consume? Environmentalism in India and the United States* (Berkeley: University of California Press, 2006).

53. Joan Martinez-Alier, *The Environmentalism of the Poor: A Study of Ecological Conflicts and Valuation* (Cheltenham, UK: Edward Elgar, 2002).

54. Garvey, *The Ethics of Climate Change*, 148.

55. J. Timmons Roberts and Bradley C. Parks, *A Climate of Injustice: Global Inequality, North-South Politics, and Climate Policy* (Cambridge, MA: MIT Press, 2007).

56. Ved P. Nanda, "Climate Change, Developing Countries, and Human Rights: An International Law Perspective," in *Climate Change and Environmental Ethics*, ed. Ved P. Nanda (New Brunswick, NJ: Transaction Publishers, 2013), 165.

57. Singer, *One World*, 56.

58. Freya Mathews, "Moral Ambiguities in the Politics of Climate Change," in *Climate Change and Environmental Ethics*, ed. Ved P. Nanda (New Brunswick, NJ: Transaction Publishers, 2013), 43–76.

59. See Roberts and Parks, *A Climate of Injustice*.

60. Garvey, *The Ethics of Climate Change*, 114ff.

61. World Commission on Environment and Development, *Our Common Future* (Oxford: Oxford University Press, 1987), 45.

62. World Commission on Environment and Development, *Our Common Future*, 216ff.

63. Commission on Global Governance, *Our Global Neighborhood*, 1995, http://www.gdrc.org/u-gov/global-neighbourhood.

64. United Nations, Convention on Biological Diversity, 3, https://www.cbd.int/doc/legal/cbd-en.pdf.

65. Arild Vatn, "The Environment as a Commodity," *Environmental Values* 9 (2000): 493–509.

66. Janna Thompson argues that the environment should also be regarded as cultural heritage. This view implies that we have moral obligations to preserve it for future generations. See Thompson, "Environment as Cultural Heritage," *Environmental Ethics* 22 (2000): 241–58.

67. This is formulated in article 3: "States have, in accordance with the Charter of the United Nations and the principles of international law, the sovereign right to exploit their own resources pursuant to their own environmental policies, and the responsibility to ensure that activities within their jurisdiction or control do not cause damage to the environment of other States or of areas beyond the limits of national jurisdiction." UN, Convention on Biological Diversity, 4, https://www.cbd.int/doc/legal/cbd-en.pdf.

68. Charles Perring, *Our Uncommon Heritage: Biodiversity Change, Ecosystem Services, and Human Wellbeing* (Cambridge: Cambridge University Press, 2014), 124.

69. Gerardo Ceballos, Paul R. Ehrlich, and Rodolfo Dirzo, "Biological Annihilation via the Ongoing Sixth Mass Extinction Signaled by Vertebrate Population Losses and Declines," *Proceedings of the National Academy of Sciences* 114 (July 25, 2017): https://doi.org/10.1073/pnas.1704949114.

70. See, for example, Benjamin Coriat, ed., *Le retour des communs: La crise de l'idéologie propriétaire* (Paris: Editions les liens qui libèrent, 2015); Peter J. Stoett, *Global Ecopolitics: Crisis, Governance, and Justice* (Toronto: University of Toronto Press, 2012); James McCarthy, "Commons as Counterhegemonic Projects," *Capitalism, Nature, Socialism* 16 (2005): 9–24; Maude Barlow and Tony Clarke, *Blue Gold: The Fight to Stop the Corporate Theft of the World's Water* (New York: New Press, 2002); Naomi Klein, "Reclaiming the Commons," *New Left Review* 9 (2001): 81–89; Brian Donahue, *Reclaiming the Commons* (New Haven, CT: Yale University Press, 1999).

71. Geraldine van Bueren, "Take Back Control: A New Commons Charter for the Twenty-First Century Is Overdue, 800 Years after the First," *Times Literary Supplement*, March 10, 2017, 23.

72. Yochai Benkler mentions carpooling. It is the second-largest commuter transportation system in the United States. See Benkler, "Sharing Nicely: On Shareable Goods and the Emergence of Sharing as a Modality of Economic Production," *Yale Law Journal* 114 (2004): 281.

73. ten Have, *Global Bioethics*, 231–32.

74. ten Have, *Global Bioethics*, 132–34.

75. Andrew Dobson, *Citizenship and the Environment* (Oxford: Oxford University Press, 2003), 22.

76. NGO documents for the Earth Summit, 1992, http://www.earthsummit2002.org /toolkits/women/ngo-doku/ngo-conf/ngoearth17-2.html. See also Rikard Warlenius, Gregory Pierce, and Vasna Ramasar, "Reversing the Arrow of Arrears: The Concept of 'Ecological Debt' and Its Value for Environmental Justice," *Global Environmental Change* 30 (2015): 21–30.

77. Ramachandra Guha, "Radical American Environmentalism and Wilderness Preservation: A Third World Critique," *Environmental Ethics* 11 (1989): 71–83. See also chapter 2.

78. Paul W. Wood, *Biodiversity and Democracy: Rethinking Society and Nature* (Vancouver: UBC Press, 2000). See also the ethical debate on the value of biodiversity in chapter 2.

79. Matthew Sparke and Dimitar Anguelov, "H1N1, Globalization and the Epidemiology of Inequality," *Health and Place* 18 (2012): 726–36.

80. Stephen B. Scharper and Hilary Cunningham, "The Genetic Commons: Resisting the Neoliberal Enclosure of Life," in *The Global Idea of "the Commons,"* ed. Donald M. Nonini (New York: Berghahn Books, 2008), 56. Essential medicine patents between 1995 and 2005 were typically filed in countries with higher incomes, higher healthcare expenditures, and with larger populations. See Reed F. Beall, Rosanne Blanchet, and Amir Attaran, "In Which Developing Countries Are Patents on Essential Medicines Being Filed?," *Globalization and Health* (2017): 7.

81. Thomas Pogge, *World Poverty and Human Rights: Cosmopolitan Responsibilities and Reforms* (Cambridge: Polity Press, 2013).

82. David Schlosberg, *Defining Environmental Justice: Theories, Movements, and Nature* (Oxford: Oxford University Press, 2007), 45ff.

83. J. Timmons Roberts, "Globalizing Environmental Justice," in *Environmental Justice and Environmentalism: The Social Justice Challenge to the Environmental Movement*, ed. Ronald Sandler and Phaedra C. Pezzullo (Cambridge, MA: MIT Press, 2007), 285–307. Movements can also lose momentum. See Talia Buford, "Has the Moment for Environmental Justice Been Lost?," *PBS NewsHour*, July 24, 2017.

84. Alison Hope Alkon and Julian Agyeman, eds., *Cultivating Food Justice: Race, Class, and Sustainability* (Cambridge, MA: MIT Press, 2011).

85. Christiana Z. Peppard, *Just Water: Theology, Ethics, and the Global Water Crisis* (Maryknoll, NY: Orbis Books, 2014), 35, 185.

86. Judith N. Shklar, *The Faces of Injustice* (New Haven, CT: Yale University Press, 1990).

87. Shklar, *The Faces of Injustice*, 40ff.

88. Schlosberg, *Defining Environmental Justice*, 165ff.

89. ten Have, *Global Bioethics*, 67–71.

90. Ferguson, *Global Shadows*; James Ferguson, "The Uses of Neoliberalism," *Antipode* 41 (2009): 166–84; Nik Heynen, James McCarthy, Scott Prudham, and Paul Robbins, eds., *Neoliberal Environments: False Promises and Unnatural Consequences* (London: Routledge, 2007).

91. Navarro, "Neoliberalism as a Class Ideology," 47–62.

92. Lisa Forman, "'Rights' and Wrongs: What Utility for the Right to Health in Reforming Trade Rules on Medicine?," *Health and Human Rights* 10 (2008): 40. See also Frances Fox Piven, "Can Power from below Change the World?," *American Sociological Review* 73 (2008): 1–14.

93. United Nations, *Millennium Development Goals Report 2010* (New York: United Nations, 2010), 54. Marco Gonzalez, Kristen N. Taddonio, and Nancy J. Sherman, "The Montreal Protocol: How Today's Successes Offer a Pathway to the Future," *Journal of Environmental Studies and Sciences* 5 (2015): 122–29.

94. Ken Conca, *Governing Water: Contentious Transnational Politics and Global Institution Building* (Cambridge, MA: MIT Press, 2006), 33ff.

95. The G-20 in July 2017 finally made a statement that tuberculosis is a top priority, especially because of growing antimicrobial resistance (Lauren Weber, "The G-20 Declaration Makes a Major Mention of the World's Top Infectious Killer," *Huffington Post* July 10, 2017).

96. Hazel Sheffield, "Sweden's Recycling Is So Revolutionary, the Country Has Run out of Rubbish," *Independent*, December 8, 2016.

97. Jeff D. Colgan and Robert O. Keohane, "The Liberal Order Is Rigged: Fix It Now or Watch It Wither," *Foreign Affairs* (May/June 2017): 36–44.

98. Argued by Dobson, *Citizenship and the Environment*, 135.

99. Sarah Whitmee, Andy Haines, Chris Beyer, Frederick Boltz, et al., "Safeguarding Human Health in the Anthropocene Epoch: Report of the Rockefeller Foundation–*Lancet* Commission on Planetary Health," *Lancet* 386 (2015): 1973–2028.

100. Ronald Labonté and Renee Torgerson, "Interrogating Globalization, Health and Development: Towards a Comprehensive Framework for Research, Policy and Political Action," *Critical Public Health* 15 (2005): 157–79. The Rockefeller Foundation–*Lancet* Commission on planetary health states that "Governance should help . . . to reduce the risks to health and natural systems in a world where vested interests undermine the political will to act and where inequalities have marginalized the voices of many disadvantaged groups." However, it hardly addresses the power systems that support the current governance regimes (Whitmee et al., "Safeguarding Human Health in the Anthropocene Epoch," 2018).

101. Steffen, Persson, Deutsch, et al., "The Anthropocene," 739–61; Evans and Giroux, *Disposable Futures*.

102. Shawn H. E. Harmon argues that WHO *should* be the most authoritative leader. See Harmon, "In Search of Global Health Justice: A Need to Reinvigorate Institutions and Make International Law," *Health Care Analysis* 23 (2015): 358.

103. Julio Frenk and Suerie Moon, "Governance Challenges in Global Health," *New England Journal of Medicine* 368 (2013): 936–942. See also Chelsea Clinton and Devi

Sridhar, *Governing Global Health: Who Runs the World and Why?* (Oxford: Oxford University Press, 2017).

104. Alexander Kaufman, "This Pharma Company Won't Commit to Fairly Pricing a Zika Vaccine You Helped to Pay For," *Huffington Post*, June 9, 2017.

105. ten Have, *Global Bioethics*, 201–5.

106. Jeremy Shiffman, "A Social Explanation for the Rise and Fall of Global Health Issues," *Bulletin of the World Health Organization* 87 (2009): 608–13.

107. See Andrew Dobson, *Citizenship and the Environment*, 83ff. Ecological citizens are people who are "very willing to make sacrifices for the benefit of the environment and poverty reduction"; they exist, at least in Sweden, according to Sverker C. Jagers, "In Search of the Ecological Citizen," *Environmental Politics* 18 (2009): 18.

108. Nirija Gopal Jayal, "Ethics, Politics, Biodiversity: A View from the South," in *Moral and Political Reasoning in Environmental Practice*, ed. Andrew Light and Avner de-Shalit (Cambridge, MA: MIT Press, 2003), 295–316.

109. Whitmee et al., "Safeguarding Human Health in the Anthropocene Epoch."

110. For example, World Social Forum (https://fsm2016.org/en), Third World Network (http://www.twn.my), World Watch Institute (http://www.worldwatch.org), Corporate Watch (https://corporatewatch.org), International Work Group for Indigenous Affairs (http://www.iwgia.org), and Food Policy Council (https://dcfoodpolicy.org).

111. See BioEdge, https://www.bioedge.org; and the Global Health Network's (https://tghn.org), *Global Health Bioethics, Research Ethics and Review*, https://bioethicsresearchreview.tghn.org. See also, for an overview of bioethics centers, councils, and organizations, Bioethics.com, http://www.bioethics.com/bioethics-websites.

112. See chapter 6.

113. See chapter 4. David P. Fidler, *SARS, Governance and the Globalization of Disease* (Houndmills, UK: Palgrave Macmillan, 2004), 107–8.

114. Frenk and Moon, "Governance Challenges in Global Health," 938.

115. See chapter 5. See also M. Gregg Bloche and Elizabeth R. Jungman, "Health Policy and the World Trade Organization," in *Globalization and Health*, ed. Ichiro Kawachi and Sarah Wamala (Oxford: Oxford University Press, 2007), 250–67, especially "WTO's tacit endorsement of protection for health as an interpretive principle . . ." (261); Nitsan Chorev, "The Institutional Project of Neo-liberal Globalism: The Case of the WTO," *Theory and Society* 34 (2005): 317–55.

116. Peppard, *Just Water*, 52ff.

117. UN Human Rights Council, 34th session, *Report of the Special Rapporteur on the Issue of Human Rights Obligations Relating to the Enjoyment of a Safe, Clean, Healthy and Sustainable Environment*, A/HRC/34/49, January 19, 2017, 21, http://www.refworld.org/docid/58ad9dd44.html.

118. Since 2007, 127 environmental activists have been killed in Honduras. See Global Witness, *Defenders of the Earth: Global Killings of Land and Environmental Defenders in 2016* (London: Global Witness, 2017).

119. See chapter 6.

120. Susannah L. Rose, Janelle Highland, Matthew T. Karafa, and Steven Joffe, "Patient Advocacy Organizations, Industry Funding, and Conflicts of Interest," *JAMA Internal Medicine* 117 (2017): 344–50; Matthew S. McCoy, Michael Carmol, Catherine Chockley,

John W. Urwin, Ezekiel J. Emanuel, and Harald Schmidt, "Conflicts of Interest for Patient-Advocacy Organizations," *New England Journal of Medicine* 376 (2017): 880–85.

121. "Neither true democracy nor environmental protection is possible where citizens become mere competitors with no commitment beyond their own narrow self-interests." Walter F. Baber and Robert V. Bartlett, *Deliberative Environmental Politics: Democracy and Ecological Rationality* (Cambridge, MA: MIT Press, 2005), 120.

122. Alex Perullo, "The Rise of Rights and Non-profit Organizations in East African Societies," in *Framing the Global: Entry Points for Research*, ed. Hilary E. Kahn (Bloomington: Indiana University Press, 2014), 206–28; 221.

123. Baber and Bartlett, *Deliberative Environmental Politics*, 133.

124. Jeff Haynes, "Power, Politics and Environmental Movements in the Third World," *Environmental Politics* 8 (1999): 222–42.

125. That this sense of solidarity is important for many people today is exemplified in Paul Ritvo, Kumanan Wilson, J. L. Gibson, et al., "Canadian Survey on Pandemic Flu Preparations," *BMC Public Health* 10 (March 11, 2010): https://doi.org/10.1186/1471-2458-10-125. The majority of the Canadian population supported public health officials in the case of pandemics; they are willing to sacrifice individual's liberties in behalf of the public good.

126. McMichael, *Development and Social Change*, 334.

127. See Streeck, *How Will Capitalism End?*

128. This issue is related to the growing literature on deliberative democracy. See Graham Smith, *Deliberative Democracy and the Environment* (London: Routledge, 2003); Dobson, *Citizenship and the Environment*; Baber and Bartlett, *Deliberative Environmental Politics*.

Index